Accession no.
01070660

LIBRARY
Tel: 01244 375444   Ext: 3301

Chester
A College of the
University of Liverpool

This book is to be returned on or before the last date stamped below. Overdue charges will be incurred by the late return of books.

UCC LIBRARY
- 2 MAR 2001

- 5 JAN 2004

CANCELLED

intellect™

EXETER, ENGLAND
PORTLAND, OR, USA

First Published in Hardback in 1999 by
Intellect Books, FAE, Earl Richards Road North, Exeter EX2 6AS, UK

First Published in USA 1999 by
Intellect Books, ISBS, 5804 N.E. Hassalo St, Portland, Oregon 97213-3644, USA

Copyright ©1999 Intellect Ltd

All rights reserved. No part of this publication may be reproduced, stored in a retrieval system, or transmitted, in any form or by any means, electronic, mechanical, photocopying, recording, or otherwise, without written permission.

Consulting Editor:	Masoud Yazdani
Cover Images:	Peter Anders
Copy Editor:	Wendi Momen

A catalogue record for this book is available from the British Library

ISBN 1-84150-013-5

Printed and bound in Great Britain by Cromwell Press, Wiltshire

Acknowledgements

There are many individuals to thank for their help in bringing this book into being. In addition to the authors themselves, my colleagues and students in CAiiA-STAR, the editorial support team at ACES, and the staff of Intellect, particular thanks are due to Professor Ken Overshott, Principal of the University of Wales College Newport, who has made privileged provision for CAiiA from its inception. His continuing support for the Consciousness Reframed conferences is much appreciated.

Preface iv
Introduction 1

Part I – Mind

1 Models 6
Can there be Non-Embodied Information? – Stephen Jones
Acts Between and Between Acts – Ranulph Glanville
Between Reality and Virtuality: Toward a New
    Consciousness? – Julio Bermudez
The Actualisation of the Virtual – Mark Palmer
We are the Consciousness Musicians – Electronic Art,
    Consciousness and the Western Intellectual Tradition – Mike King
Immersive Computer Art and the Making
    of Consciousness – Laurie McRobert

2 Memory 38
Casablanca and Men in Black: Consciousness, Remembering,
    and Forgetting – Michael Punt
Virtual Space and the Construction of Memory – Andrea Polli
Memory Maps and the Nazca – Bruce Brown

3 Transcendence 52
Parallel Worlds: Representing Consciousness at the Intersection of Art,
    Dissociation, and Multidimensional Awareness – Kristine Stiles
Sacred Art in a Digital Era:
    Or the Internet and the Immanent Place in the Heart –
        Niranjan Rajah and Raman Srinivasan
Negotiating New Systems of Perception: Darshan, Diegesis and Beyond –
    Margot Lovejoy and Preminda Jacob
Seeing Double: Art and the Technology of Transcendence –
    Roy Ascott
Jumping Over the Edge: Consciousness and Culture, the Self
    and Cyberspace – Lily Díaz
Space-time Boundaries in the Xmantic Web – Maria Luiza Fragoso

Part II – Body

4 Post-biological Body 81
Recreating Ourselves: Romeo and Juliet in Hades by Future Movie –
    Naoko Tosa and Ryohei Nakatsu
The Body as Interface – Jill Scott

Art at the Biological Frontier – Eduardo Kac
Images and Imaging – Amy Ione
The Power of Seduction in Biomedicine –
    Nina Czegledy
Stasis: The Creation & Exploration of Subjective Experiential Realities
    through Hypnosis, Psychosynthesis & Digital Technology –
    Richard Brown
Emotion, Interactivity, and Human Measurement –
    Christopher Csikszentmihalyi
Writing the Post-Biological Body – Steve Tomasula

5   Space and Time   119
For a Spatiotemporal Sensitivity – Ginette Daigneault
Virtual Geographies, Borders and Territories:
    GPS Drawings and Visual Spaces – Andrea Wollensak
An Exploration of Spatial/Temporal Slippage –
    Chris Speed
Thinking Through Asynchronous Space – Mike Phillips
The Space Between the Assumed Real and the Digital Virtual –
    Dan Livingstone

Part III – Art

6   Strategies   145
The Work of Art in the Age of Digital Historiography – Brett Terry
Mind Memory Mapping Metaphor: Is Hypermedia Cognitive Art? –
    Dew Harrison
The House that Jack Built: Jack Burnham's Concept of "Software"
    as a Metaphor for Art – Edward Shanken
Streams of Consciousness: Info-Narratives in Networked Art –
    Christiane Paul
Consciousness and Music – Mladen Milicevic
The Irreducibility of Literary Consciousness – Philipp Wolf
Nonsense Logic and Re-embodied Intelligence – Bill Seaman
Conscious Isomorphisms and Young Farmers: Consciousness
    as a Subject of Artistic Research – Stefaan Van Ryssen
Modeling Interpretation – Sharon Daniel
Performing Presence – Barry Edwards

7   Projects   196
Éphémère: Landscape, Earth, Body and Time in Immersive Virtual Space –
    Char Davies
Virtual Environment as Rebus – Margaret Dolinsky

Dynamic Behavioural Spaces with Single and Multi-User Group Interaction – Miroslaw Rogala
Reality, Virtuality and Visuality in the Xmantic Web – Tania Fraga
Xmantic Webdesign – Fatima Bueno
Assigning Handlers to a Shadow – Kieran Lyons

8 Architecture — 223
Human Spatial Orientation in Virtual Worlds – Dimitrios Charitos
An Interactive Architecture – Gillian Hunt
Heterotic Architecture – Ted Krueger
The Cybrid Condition: Implementing Hybrids of Electronic and Physical Space – Peter Anders
Virtual Architecture as Hybrid: Conditions of Virtuality vs. Expectations from Reality – Dace Campbell

9 Creative Process — 248
Acquiring an Artistic Skill: A Multidimensional Network – Christine Hardy
Enlarging the Place for Creative Insight in the Theatre Model of Consciousness – Thomas Draper
An Approach to Creativity as Process – Ernest Edmonds and Michael Quantrill
We are having an idea: Creativity within Distributed Systems – Fred McVittie
To be or not to be.... Conscious – Eva Lindh and John Waterworth

Part IV – Values

10 Values — 273
CyberArt & CyberEthics – Colin Beardon
The Metaphoric Environment of Art and Technology – Carol Gigliotti
The Four Seas: Conquest Colonisation, Consciousness in Cyberspace – Joseph Lewis
Future Present: Reaestheticizing Life through a New Technology of Consciousness – Anna Bonshek and Gurdon Leete
Conspiracies, Computers and Consensus Reality – Paul O'Brien
The Failure and Success of Multimedia – Sean Cubitt
Abstract Virtual Realism – Nik Williams
Beyond Film & Television-whose Consciousness is it Anyway? – Clive Myer
Chimera for the 21st Century: Mis-Construction as Feminist Strategy – Terry Gips

Preface

The articles in this book are representative of over a hundred papers presented at the Second International CAiiA Research Conference, *Consciousness Reframed: art and consciousness in the post-biological era*, which was convened at the University of Wales College, Newport (UWCN) in August 1998. CAiiA, the Centre for Advanced Inquiry in the Interactive Arts, was established at UWCN in 1994 to research and advance new fields of practice emerging from the convergence of art, science and technology; fields in which questions of the mind and consciousness play an important part. In tandem with CAiiA, STAR – the Centre for Science, Technology and Art Research – was created in the School of Computing at the University of Plymouth in 1997. CAiiA-STAR operates as an integrated centre of research, pooling the considerable resources of both universities, and has attracted to its doctoral and postdoctoral research programmes artists who are amongst the principal proponents of their field .

The work of CAiiA-STAR embodies artistic and theoretical research in interactive media and telematics, including aspects of artificial life, telepresence, immersive VR, robotics, non-linear narrative, computer music, intelligent architecture involving a wide range of technological systems, interfaces and material structures. As such it can be seen as a microcosm of a widely based and complex field which is developing internationally. While this means that a great diversity of issues are addressed, it is possible to discern a common thread in the emergent discourse of this field. This commonality of ideas embraces a radical rethinking of the nature of consciousness, awareness, cognition and perception, with mind as both the subject and the object of art.

The CAiiA-STAR conferences have attracted speakers from Australia, Austria, Brazil, Canada, Ecuador, France, Germany, Holland, Hong Kong, Ireland, Israel, Italy, Japan, Malaysia, New Zealand, Norway, Poland, Romania, Spain, Sweden, Switzerland, the USA as well as the United Kingdom. This widespread response supports the contention that issues of mind and consciousness are significant in contemporary thinking about art and technology. In turn, the intensity and fecundity of work produced in this context, and the debates it engenders, insist that the conference be convened on an annual basis.

The approaches represented here are multidisciplinary and multicultural, offering many dynamic, compelling and provocative strategies, projects and lines of inquiry. Their purpose has been to identify key questions rather than to provide definitive answers, to pursue creative implications rather than prescriptive explanations.

Introduction

To most readers it will seem to be no more than a truism to say that issues of consciousness are at the top of the agenda in art today. Surely, it will be argued, art has always been concerned with conscious experience, and with matters of sensation, perception, and cognition; it has always dealt with what is seen, felt and understood, and has always made objects to be seen, felt and understood. Who, if not the artist, explores the mystery of being, and what, if not a work of art, articulates the rich quality of human sensibility? Equally, it would seem to be nothing but perverse to suggest that it is technology, particularly computer technology, that has brought consciousness into particular focus in the art of today. Many will find it difficult to see how technology, apparently cold and alienating, could do anything to advance the subtlety of feeling and vision that art has always demanded.

Historians, however, will know that technology, whether in the form of engineering, chemistry, optics, or pharmacology, has always mediated the vision and aspirations of artists in all parts of the world and at all times. And observers of contemporary culture will confirm that, despite the seeming paradox, artists today are finding in digital technology and telematic media new ways to make consciousness both the subject and object of their work. Their use of interactive media enables the viewer to participate in a shared space of consciousness and to actively participate in the construction and transformation of artistic meaning. This book examines the new artistic sensibility arising from the confluence of art, technology and consciousness research. It is a sensibility that is leading to new visions of what art can become, what the mind can embrace, and how culture might develop.

The word "technoetic" has been coined to describe this sensibility. More than simply marking the meeting of mind and machine (a familiar conjunction in 20th century art), technoetic is intended to signify the symbiosis of technology and consciousness (Gk *noetikos*) which characterises the new cultural paradigm emerging at the turn of the millennium. In this definition, the word symbiosis has been carefully chosen to indicate by analogy "the intimate living together of two dissimilar organisms in a mutually beneficial relationship", as Webster defines it. In the technoetic context, art is seen as consisting in dynamic networks of minds, exploiting the connectivity of telematic media, whose nodal points have both human and artificial attributes; an art set in unfolding fields of consciousness. 'Intimacy', 'life' and 'relationship', are terms which not only fulfil Webster's criteria but constitute the defining qualities of the interactive arts; and, as the articles in this book make clear, either directly or implicitly, art mediated by computer and telecommunications technology is nothing if it is not interactive. Interactive art always involves the user or viewer in an intimate process of transformative action. One can say that all systems of technologically informed art aspire to the condition of transformative process, where, at best, not only the structure of the 'artwork' undergoes change but the consciousness of the viewer is transformed. Such art is technoetic in a profound sense and its value lies in the reframing of the consciousness that it effects and the richness of meaning which it generates.

Central to this process is a concern with relationship, which is to say a concern with identity and being. It has been characteristic of a number of twentieth century artists to pursue in their work what the philosopher David Chalmers calls the hard problem, an understanding of the nature of being. As Nagel has put it, "there is something it is like to be a conscious organism". How can we describe that something? In science, many disciplines are pursuing this intractable problem. For example, one of the principal forums of research, the biannual Tucson conference, "Toward a Science of Consciousness", is notable for the very wide range of disciplines that are brought together in the attempt to unlock the mysteries of consciousness. These are from the philosophy of mind and dream research, to neuropsychology, pharmacology, and molecular dynamics, to neural networks, phenomenological accounts, and even the physics of reality.

To artists, for whom the mystery of consciousness is equally compelling, it is less a matter of seeking to explain consciousness and more a matter of exploring how it might be navigated, altered, or extended; in short, reframed. In parallel to the Tucson meetings, the Consciousness Reframed conferences show that art, technology and science are converging in important ways to produce new strategies, new theories and new forms of creativity, increasingly relying for their advance on a kind of trans-disciplinary consultation and collaboration. It is possible to see this as a paradigmatic shift in creative practice, an emergent aesthetic constituting a truly technoetic paradigm. The provenance for this emergent field is not hard to find. Through the course of this century, there has been a tradition in art of valuing concepts in their own right, even to the exclusion of direct visual reference to the external world at its surface level of appearance. To make the invisible visible is a familiar ambition of artists, an ambition by no means restricted to conceptual art alone. At the same time there is a strand of art practice that eschews representation and self-expression entirely in favour of construction. These conceptual and constructive tendencies exert a huge influence on the artistic strategies of those using new technology today. Similarly, there is a marked provenance in Western art of the spiritual and visionary, of works attempting to transcend their materiality and the materialist view of human nature, to express or evoke other planes of experience and awareness. The effect of Eastern thought in this context is readily apparent, and its place in mediating the new principles of physics, and of scientific ideas more generally, has been particularly beneficial to artists.

The impact of science on art, especially its metaphors and models, and on artists' readings of the world and of human identity and potential, has been no less considerable. Complexity, quantum physics, the cognitive sciences, and new biology, for example, provide fresh perspectives on the eternal questions of being and becoming. Advanced technology has provided opportunities for the exploration of mind and the extension of the body that challenge many preconceptions we have held about our 'innate' nature and the limitations of space and time. We need only look at the effects of connectivity and interaction, in the Net for example, to see how rapidly new technologies are enabling people, places and ideas to come together in entirely new configurations and conjunctions.

At the same time, our processes of perception and cognition are undergoing change and adaptation to the extent that what might be described as the bionic faculty of cyberception is emerging. And, just as consciousness is supervenient upon the material substrate of the living body, so we might expect artificial life to be the substrate for the emergence of artificial consciousness. A sense of bio-telematics, if that is what we might call distributed mind and non-local body, is becoming an important ingredient of the technoetic field. Indeed, there is some evidence to suggest that we may be moving away from our present preoccupation as

Introduction

artists with the immaterial and screen-based digital realities towards a re-materialisation of art invested in artificial life, nanotechnological structures and the artificial consciousness that may thereby emerge.

Artists whose practice is invested in networked hypermedia and virtual reality, in their interactions with artificial agents and avatars, know that personal identity can be endlessly transformed. We see the immutability and unity of the self, so dearly prized in the European tradition, giving way to an understanding of how we each can be involved in our own self-creation. In cyberspace, the self is open to tele-differentiation, distribution and planetary dissemination. In consequence, a kind of non-linear identity is emerging. In searching for new frames of consciousness, some are turning as much to ancient spiritual practices as to the mind altering technologies of digital and post-biological systems. For some, a creative concordance is forming between shamanic culture and telematic society. A further consequence of this is that a number of economically poor or marginal societies – hitherto considered to be no more than mere recipients of technological largess – are now seen quite differently. They are viewed to be in possession, through their tradition and customs, of noetic knowledge and skills that can offer much to the spiritual enlargement of current research and practice.

These many aspects of the technoetic paradigm inform the essays and the structure of this book. They have been assembled in 'Parts' that focus in turn on issues of Mind, Body, Art and Values. These distinctions are not intended to signify boundaries between fields of inquiry. Indeed a seamless flow of ideas characterises the technoetic discourse. But they are indicative of different degrees of emphasis that artists place on issues of mind and matter and the contexts in which they address them.

Under the rubric of Mind in the first Part, three aspects of consciousness are considered. The first comprises those articles dealing with models of cognition and perception; then questions of memory are addressed; finally, various states of transcendence and paranormal perception are discussed through a multi-cultural and multi-temporal filter. In the second Part, notions of the post-biological body are examined, followed by articles dealing with aspects of space and time. The body and the body's representation have both received considerable critical attention in recent decades, due both to the fear of its 'disappearance' into the immaterial, and to the multiple readings of the body that new technologies can provide. The politics and polemics of gender and sexuality inform much of the writing. The third Part deals with Art. Firstly, new strategies in art and in architecture are presented in the context of interactivity, connectivity and transformation, then a number of specific art projects are described; and finally the creative process is addressed. The fourth and final Part is devoted to ethical and aesthetic values. While the question of values challenges all those working in the technoetic domain, and especially artists, whose practice is characteristically unconstrained by convention or orthodoxy, this Part does not constitute a conclusion to the book. No conclusion would be appropriate to a field so much in evolution and in flux. We are in the middle of a process of cultural transformation, the complexity of which is matched only by the transformation that is taking place in our conception of our selves. We are re-thinking what it is to be human, to be conscious, to have identity and meaning in a world that itself is undergoing massive technological re-construction and philosophical and scientific re-definition.

Discussions of technology and consciousness, with or without judgements of value, have not been rare in recent decades, but few have prised open the dialectics of mind and machine to include art. This book takes the view that the discourses developed within and around the triad of art, technology and consciousness are likely to be useful and relevant to our aspirations and anxieties about the future. This is not simply in the domain of artistic culture but in the shape and direction that global society is taking.

40 years ago, a meeting of the New York University Institute of Philosophy was convened under the title *The Dimensions of Mind* to explore consciousness, the brain and the mind in the context of the machine. Norbert Wiener led the discussion, Michael Scriven presented 'The Compleat Robot: A Prolegomena to Androidology' while J. B. Rhine discussed 'Parapsychology and the Nature of Man'. There were papers by Ernest Nagel, B.F. Skinner, Arthur Danto, and others. Forty years on, the questions they raised remain almost completely unanswered. Sidney Hook's observation that "among the problems which philosophy and the various sciences have to deal, the mind-body problem is still the most intriguing" still holds. Michael Scriven correctly surmised that even in the future when intelligent machines are "snapping at our heels", the key question will remain "what is the secret of consciousness?" Art was conspicuously absent from the meeting. No artists were invited to the debate. Nor was it, I believe, thought necessary at that time, by either scientists or artists, that they should be. It took place at the terminal point in art of High Modernism, whose principle proponents' disdain for science and high technology was equalled only by the scientists' contempt for modern art. It was to be a few years before any significant alliance between art, science and technology would be formed, and even more before the conceptual and spiritual provenance of 20th-century art could be understood and articulated.

At *The Dimensions of Mind* meeting, Scriven questioned how a machine could possibly produce "new theories, new conceptual schemes, new works of art". The machine was seen as the technological Other: no sense of bio-technological integration or human-computer symbiosis was envisaged. Only Satosi Watanabe really spoke the language of today, recognising the importance of "emergent" properties and "environment-and-structure-sensitive" properties of matter, and arguing the significance of Niels Bor's principle of complimentarity to the understanding of a new kind of dualism. But the past forty years have seen the gradual convergence of art, science and technology, most notably in the domain of communications and computer technology. The telematisation of society, as Minc and Nora so memorably named it, has profoundly affected not only the distribution and production of art but our very sense of personal identity, our processes of cognition and perception, and the nature of artistic authorship. Thus, the discourse engendered by *Consciousness Reframed: art and consciousness in the post-biological era* seems to be not only timely but necessary if these new developments in art and their theoretical underpinnings are to be more widely understood and disseminated.

Such is the purpose of this book. It draws on the expertise and interests of authors based in seventeen countries around the world. The complexity and creativity of their thinking is embodied in articles that offer a stimulating perspective on current approaches to the new field. While the boundaries of this field are, and must remain, fuzzy and indistinct, there is a well-defined core of research forming which provides an important centre of gravity for future speculation and investigation. It is hoped that publishing this work will not only inform a wider public but that the issues it presents will be extended and enriched as a result of their dissemination. I hope that the reader will share in the excitement and promise of the emerging culture that this book addresses.

Part I – Mind

1   *Models*
2   *Memory*
3   *Transcendence*

1 Models

Can There Be Non-Embodied Information?
Stephen Jones

Introduction
There are, in philosophies of consciousness, two kinds of theory which attempt to explain subjectivity, the phenomenal experience of one's mental representations: Dualist theories in which a mind and a body are independent and combine in making a person and identist theories in which the mind is a function of the living process of the body. Dualism is built from the idea that the mind is primary, existing in some superspace, attached to a body for a lifetime, endowing it with a self. This is the form of most theologies which posit a soul. Identist theories, on the other hand, attempt to show that consciousness is in some way a function of the brain-in-its-context. That is, that the physical system that we are is conscious.

Now, the mind is an information processing system. Subjectivity consists in information gained through perception and reflection on that information. In dualism this information is immaterial and apprehended by an immaterial mind, Cartesian Dualism rests on the idea that the only thing knowable with any certainty is the self, the thinking thing. But since one might be deceived about the actual existence of anything in the material world, this self needs to be completely independent of the physical world. The primary objection to Descartes' view of the mind is how would the perceptual data of the world be transmitted to the immaterial mind? Descartes suggested that it is transmitted through the Pineal body, the only non-twinned part of the brain anatomy.

Identist theories do not pose this problem of immaterial informational stuff. Information is a function of the brain. It is detected by distinguishing differences in things and attaching meaning to them. But of course the problem that identist theories suffer is, just where does the subjective experience come from? Subjectivity appears to have a status quite outside what can be explained by the physical laws of the world. This apparent difference in quality between the phenomenal and the physical is the driving issue of current debate about consciousness., David Chalmers in his book *The Conscious Mind*, uses this apparent difference as the basis for his whole argument.

So subjectivity involves *information* having two kinds of description:
1. First person description, or phenomenology, pertaining to one's experience of things that is private, accessible solely to oneself and having an apparent special *experiential* quality. For dualism this information stands in mysterious relation to the information of the physical world, invoking quantum theories of consciousness.
2. Third person description, or 'physics', pertaining to the reporting, in culturally

consensual ways, of information about one's perception: information which can only ever be physically embodied in the usual materials, brains, television, paper, etc.

What is information?

Information theory as developed by Claude Shannon is about the communication of a signal which should be as noise - and distortion - free as possible. But this is information without meaning, a syntactical information, simply a matter of the accuracy of the transmission through the communication channel. Nevertheless it is embodied information. But we want to know about the *content* of the channel, that aspect of information known as *meaning*.

Our subjectivity is a function of the brain's acquisition of difference.

The only possible ways for minds to have any content is for information to be either innate or to be gained by experience. One has to acknowledge that a complete mind produced *a priori*, even if only revealed over time, looks pretty unlikely given the number of different minds in the world. So information must be *gained* through experience and reflection on that experience.

Gaining information is a matter of detecting differences between things or changes over time. Gregory Bateson defined information as a *difference which makes a difference*. Difference is a product of the *relations between* things, it is not inherent in any particular thing. 'The unit of information is difference . . . [and] the unit of psychological input is difference.' (Bateson 1972). Informationally, what travels along the neuron isn't an impulse but something more like 'news of a difference'. There is a vast range of possible differences, all physically embodied in the object or in the transmission. And they must all impact in some physical way on our brains to be perceived.

In the real world most of what goes on has little importance to a living system, *e.g.*, the random bumping of electrons into atoms in a wire. These things are differences but of no significance and so are background noise. The determination of significance becomes important. For us, there is the biologically significant, the culturally significant and the personally significant. The *significant is the signal*, largely determined by what has meaning, linking the syntactical and semantic aspects of information.

Difference is detected by variations in features in the senses. The various differences arrive by light, by pressure waves, etc. They are the physical embodiments of such information as we read of these transmissions and their variability. Changes are noticed and interpreted by processing nets of the brain and have physical embodiment as transient conditions of these nets as well as producing long-term memory traces. Reflection also produces physical embodiments of information as culture.

At any stage of the perceptual process a cross-section of the neural pathways involved would show that what is present as information about the percept is a transformation of the original physical data presented to the sense organ. Quoting Bateson:

> 'In considering perception, we shall not say, for example, 'I see a tree', because the tree (as such) is not within our explanatory system. At best, it is only possible to see an image which is a complex but systematic transform of the tree. This image, of course, is energised by my metabolism and the nature of the transform is, in part, determined by factors within my neural circuits: 'I' make the image, under various restraints, some of which are imposed by my neural circuits, while others are imposed by the external tree.' (Bateson, 1972).

Korzybsky coined the phrase 'the map is not the territory' to indicate that what we know is not the thing in itself but a representation carried within us, a sequence of transforms of the results of sense organ stimulation. We only ever have our subjective transforms, meaningfully interpreted. But Kant's *ding an sich* is forever unobtainable. There can be no knowledge of ultimate reality; everything we know is mediated, mere inference developed through experience.

Information and the Real World

So the world is informational. What are the implications for theories of consciousness? What do I imply by saying that the phenomenal and the physical are different representations of the same information? As Schopenhauer commented, 'The intellect and matter are correlatives, the one exists only for the other; both stand and fall together; the one is only the other's reflex. They are in fact really one and the same thing, considered from two opposite points of view. . . " (Schopenhauer, 1966).

Kant originally realised that the intrinsic nature of external objects is utterly unknowable, that everything we know is in fact a mediated transform of the physical data originally perceived: the thing in itself, the *ding an sich*. So let's have a closer look at this idea of the unknowability of the 'real' world.

When I *see* or *touch* an object, e.g. a table, what I actually know is the result e.g. of certain neural processes in my brain. I know nothing of the *intrinsic* nature of the table. The only stuff I know *intrinsically* is my neurally produced information about it, my phenomenology. What a physicist *knows* about matter is the informational relations that have been built up over years of consistent inferential processes. For Bertrand Russell, 'the particulars which are [the table's] aspects have to be collected together by their relations to each other, not to it, since it is merely inferred from them' (Russell 1921).

Now, the table *I* know intrinsically will be different from the table anyone else knows. The descriptions individuals provide may well be sufficiently similar as to differentiate this table from that one but this is a function of common language. Ultimately the 'table' will be that collection of brain processes which are one's phenomenology of the table, coupled with inferential knowledge of the relations of those perceived objects.

Michael Lockwood defines 'self-awareness [as]: knowing certain brain events by virtue of their belonging to one's own conscious biography, knowing them . . . as they are in themselves - knowing them "from the inside", by living them or one might almost say, by self-reflectively being them and argues 'that the phenomenal qualities presented in

– Models –

What we know of the world is mediated and can only be inference.

perception . . . are amongst the intrinsic attributes of certain physical states in the brain . . . [they] are realised - that is to say instances of them come into being - by way of being sensed.' (Lockwood 1989).

But information must still be represented in something even if what is represented is intrinsically unknowable. If it is differences that we detect, there must still be something in which those differences are detectable. You can't have a difference between this nothing and that nothing; it is meaningless. A consciousness has to be triggered by some external or internal difference.

Some Physics of Information

Quantum physics acknowledges that it is the formal description and analysis of our *knowledge* of the world, *i.e.* that information we gain from the world by experiment and measurement. Quoting Eugene Wigner, 'Thought processes and consciousness are the primary concepts, . . . our knowledge of the external world is the content of our consciousness . . . we do not know of any phenomenon in which one object is influenced by another without exerting an influence thereupon. . . ' This influence may be infinitesimally small, e.g. the pressure of light on a material object but it nevertheless occurs. 'Light quanta do not influence each other directly but only by influencing material bodies which then influence other light quanta . . . Similarly, consciousnesses never seem to interact with each other directly but only via the physical world. Hence any knowledge about the consciousness of another being must be mediated by the physical world'. (Wigner 1962). However, though light can have a direct effect on a material body, a consciousness is mediated via the physico-chemical system of bodily sensors and effectors to enable any effect on a material body.

Our knowledge of the world, information, is the brain's experience of detectable differences in things, as dimensioned by their ordered relations with each other. We assign qualities to things on the basis of these detections. These brain states also carry logical relations and significance and point to a name. So information enters the physical world as a result of the brain's detection of differences. But can information enter consciousness by means other than as a detected difference? The main argument against the possibility of information not being embodied in some physical system arises from its relation to entropy, as follows.

The information carried in an event is inversely proportional to the probability of that event. When all events in a system are equally probable that system is in equilibrium and its entropy is at a maximum. There will be no apparent order to the system, i.e. there are no differences between one part of the system and another and the system contains no information. If we order the relations of the particles in any way then we reduce the entropy of the system, we expend energy and we can now detect differences in the relations of the particles from which we gather information. The production of information is the production of order in the universe. Information is the converse of entropy and randomness.

The idea that information could exist not physically embodied implies that this is information gained for free, coming out of 'nothing', produced without the aid of energy. Because the level of entropy in the universe has not been changed by the appearance of this information there can have been no changes in the ordered relations of things. For information to appear out of 'nothing', it has to be random, which contradicts the nature of information. Information production increases the order in the universe, subtracting from its entropy. Information cannot come into the universe from nowhere, in all ways it must be a function of the relations between things and so is embodied. If some suprasensual world does exist in the manner usually thought of, *i.e.* as consisting in an immaterial stuff, then the second law of thermodynamics is wrong. As it works so well otherwise, this is unlikely. For information to appear in my brain, without being produced by its activity, appears to be impossible, which forces the psychic realm to partake of the physical stuff of the universe. This is contradictory to what we mean by the immaterial.

On this basis, dualism fails because it requires an immaterial stuff entering my and others brains, violating the second law and generally wreaking havoc with the universe. Identity theories of course do not suffer from this problem.

In Conclusion

There is a lot more that one should say about all this. Although it appears that information must be embodied, the nature of this embodiement has rather shifted from the physical world to brain states within the physical world, brain states which we know intrinsically, unlike our knowledge of anything else in the physical world. So dualistic theories of the mind fail but the more interesting thing is what has happened to identist theories, and that is that they must now rest on the nature of information rather than on the nature of the physical world. Perhaps information is in fact the world in process; and consciousness is our *experience of being that process*.

In fact, some quantum physicists (*e.g.* Charles Bennett) have been suggesting that perhaps information is fundamental to the physical world, not simply a *product* of the world in some way (i.e. in the way that consciousness is said by some, *e.g.* Descartes or Chalmers, to be a product of the brain that is somehow logically separate from the physical sensing systems). We assign information to a particular or we detect a difference in some relation and label that difference as information about that particular. What we are detecting about a particle is information about it. Is it possible that the different relations are more fundamental than the particle which is essentially a system of different relations?

References

Bateson, G. 1972. *Steps to an Ecology of Mind*. St. Albans: Paladin. pp.386 and 457.

Chalmers, D. 1996. *The Conscious Mind*. Oxford: Oxford University Press.

Lockwood, M. 1989. *Mind, Brain and Quantum*. Oxford: Blackwell. pp.159-61.

Russell, B. 1921. *The Analysis of Mind*. London: Allen and Unwin. p 98.

Schopenhauer, A. 1966. *The World as Will and Representation*. New York: Dover. p.16.

Wigner, E. 1962. 'Remarks on the Mind-Body Problem', in Good, I.J. *The Scientist Speculates*. London: Heinemann. p 290-5.

Stephen Jones is an Australian video artist of long standing. For many years he was the videomaker for the electronic music band Severed Heads. He has been involved with the philosophical aspects of the nature of consciousness for almost longer than his involvement in video. He now works as an electronic engineer, on equipment ranging from analogue video synthesisers to motion JPEG compressors. He has been producing The Brain Project web site since August 1996.
email: sjones@merlin.com.au
The Brain Project: URL: http://www.merlin.com.au/brain_proj/

Acts Between and Between Acts[1]

Ranulph Glanville

We observe observing.

Our observing is not of: it is. If we insist it is of, then it is of observing. We do not observe things. We observe observing. If we insist there should be things to be observed, these things come about through our constructing.[2]

When we insist that our observing is of (some thing), we insist there is an object of observing. Call that postulated thing an Object (with initial capital signifying it is an artefact), the Object of our attention, of our observing. Objects are the artefacts of the Theory of Objects.[3]

Because such Objects are fictions that may or may not exist apart from our contrivance, we can pretend/assume/insist that you and I, observing, observe the same Object. Thus, we can pretend/assume/insist our observing is the same.

In this way we can pretend/assume/insist there is common reference, a reality we can know that is independent of our observing.[4]

But, actually, we observe observing.

We contemplate observing in two ways.

When we contemplate observing an Object, we think "as if", for we do not know there is such an Object: we create it, invent it to account for our observing, permitting our observing to be of (some thing). When we think as if, we think not of observing, but of a description of or account for our observing.

In contrast, we may contemplate our observing as our observing. While Objects are mere postulates, we give them the reality of observing: Objects are, we assert. This is the way of Western knowing. It is a useful device. Both ways of thinking are used here.[5]

Accept Objects as if they were (treat them as existing). We will use this device to generate an account of experience.

To enter the universe of observing, each Object must observe. The least it can observe is itself. To observe itself, an Object must be both observer and observed. Yet the Object's self is itself, indivisible.

Still, an Object might be both observer and observed if it treats observer and observed as roles, switching between them, oscillating. Oscillators generate time. Objects switch roles: they have (two) phases: each makes its own time.[6]

Thus, each Object is assumed to generate its own time of and through self-observing.

When not observing themselves (inwardly), this oscillation-mechanism gives Objects a time to observe others - to synchronise, outwardly, with others.

All observing occurs in time. The time of each observing is different. Each observing occurs (to me, observing) in (my) sequence. Each takes time: it starts, it ends.[7]

Since Objects are construct-fictions, all that can distinguish observings, as opposed to explanations of these observings, is the time of observing. The time of observing differentiates each observing.

Observing, beginning and ending, requires and generates time, for it happens in time and time derives from it. That time derives from both the act of observing, and (in the "as if" account) the synchronisation of two Objects, observing and observed, bringing them together by bringing their times and phases together to interact, producing a particular instance, an observing.[8]

We find observings coming together. Much of what we might think were possible observings does not happen. We make connections, observing, only in relatively few instances of an inconceivably vast variety.[9]

The need, in observings, for synchronisation of the self times of distinct Objects accounts for this.

Since, for observing to occur, two Object self times must synchronise, it is not surprising we do not observe many instances of the inconceivably vast potential variety. It is hard to synchronise. Consider the difficulties that face us in life, the devices we have created to help us overcome this.[10]

When we synchronise (observe), we may synchronise one observing within another observing.

We may make several separate observings within the timespan of an overriding (or containing) act of observing. Speaking as if, we say we observe several Objects within the timespan of another (containing) Object.[11]

Thus we can find a relational logic, from synchronised times of observing.

If we consider that we place several Objects into the containing Object, we consider the containing Object already formed: it remains constant through different occasions of observing.[12]

Thus we make different observings into one constant Object.

If we think the containing Object is not formed, we construct a new containing Object to contain the several Objects.

Thus we act creatively: finding the actuality of creativity is nothing to be surprised by.

These acts occur between. Observing (of observing) we take to be by and of Objects: the observing is between them. This is the interface. Observing creates and occupies the place where observing may occur. This place is the space between, cyberspace. To impress the interface onto an observed Object is to deny that Object the space to be in. As we say colloquially, I need my space.

The space between may be taken to be as if it were an Object.

Call how we look at these acts, the collections we make observing (Objects) within the containing observing (Object), events. Events are coherent collections of one or more acts of observing. Events are to acting as Objects are to observing. This usage is accords with everyday usage.

In saying we believe that an Object-constructed as if of an act of observing is the same for two different observers, and thus that an observing is also the same, we postulate and act "as if" there were such a commonality of experience. Similarly, with the collections of acts of observing that constitute events.

Yet, being different, our observings cannot be the same. Nor can the synchronicities that admit of observing, even when we talk as if.

The events, made of my observings, made of your observings, dance between being thought to be the same yet believed to be different, and believed to be the same yet thought to be different.[13]

This difference is the source of the problem of communication-as-coding resolved through conversation a further source of potential novelty.[14]

Sameness reflects our belief that we can share because we make what we treat as if they were the same acts of observing. Thus we come to postulate there is a true, a real world. This is the comforting and convenient delusion we derive when we believe we can see Behind the Curtain, when we forget the "as if" of the Objects we claim to see there.

Events are made by observing. They are the observing that brings together or contains the other observings that we account for through synchronisation.

Events (constituted of synchronised observings) occur in the space between. But they also determine that space. Without observing events, there would be no need for a space between. We would and could not deduce the need for it.

Events make up the interface. The interface is observing. Where observing is, in the space between, is where the interface is. This also holds when we talk of Objects. The space between is the observing.

It is useful to think of the sum of all observing (no matter how defined) by differentiating the space between of observing from the space between of all observing, calling this cyberspace.[15]

The interface lies in the as if universe of Objects, between (the) acts of the Objects. And the acts between are the events made of observings, for there is nothing else, in this universe.

When we assume the interface is "as if" it were on the Object of our observing, we give no space to that Object to help form that interface. Then the interface only admits action and reaction: behaviour determined by the observing Object. We deny exclusive privacy by invading it, forgetting it is a construction and a mystery, the unique selfness distinct from all others.

We forget the how and the why of our invention. We lose the benefits, the magic.

We assume Objects to be around this space between, the interface, the all-pervasive space between. Beyond the interface, the observing, nothing can be observed. Beyond, all is conjecture, designed to allow certain styles of thinking, certain certainties and a certain contentment which is, however, illusory, misleading and dangerous.

We forget this.

We forget that an Object is a surmise deriving from an observing but is not that observing. We observe observing. An act of observing is not an Object, yet we postulate such Objects. We never observe Objects - we cannot. This is the meaning of the metaphor of the Waying Theatre. In this universe, we can never go behind the screen. We cannot know if there is a behind. We never know if there is that puppet and that light source of which we talk so freely—whether there is anything more than the screen. And, if we do not know whether any of what we imagine is there, we no longer know that it is a screen.[16]

We are blessed by ignorance and the space that leaves for the imagination.

You are sitting in a theatre. A performance may be about to begin.

What is Behind the Curtain?

Notes

1. This paper was written while a Visiting Fellow at the School of Design, Hong Kong Polytechnic University. Thank you to them for giving me the time to do this.
2. This is the major thrust of von Foerster's work in epistemology (13, 14, 14).
3. This is a key to the central argument of my doctoral dissertation (1), in which the Theory of Objects is developed. The Theory develops the notions of time and synchronicity in observing that are crucial to this paper. Several of the other papers cited (2, 3, 4) evolve the Theory further. Note, also, that the word object, in English, has changed its meaning. It used to mean what we now call subject. I chose the word because of this ambiguity.
4. I will argue, in (9), that this is the basis of our science, and the concepts that science involves.
5. In a certain respect, we always think "as if". In this paper, we describe observing, and our observing is of observing. This is Wittgenstein's point of the interface, of the Curtain/screen that cuts off that about which we must pass over in silence (16). See also (5) for an extended examination. Both of these raise the problem that we are different: we cannot see as the other.
6. This switching role is discussed in my PhD (1), but also, and more picturesquely, in (2). Every clock we have is based on oscillators. However, the notion of time is not the same to the oscillator itself and to an external observer. What may be a perfectly regular beat (changing from observing to observed etc, in sequence) to an Object might appear far from regular to another Object, inevitably observing externally.
7. Oscillation is essentially circular and hence endless. But it leaves a linear trace, like a cycle wheel on the wet sand.
8. See, especially, (3). And, again, my PhD (1). Observing both generates and requires time—simultaneously!
9. Do not take these numbers literally. Their only value is to make a point, drop a hint.

10. Consider the rituals and devices used in performance—whether in drama, religion, meditation or whatever, to bring everyone together on the one occasion in one event (making synchronicity appear). And think of diaries, timetables etc. It may be much easier and less frustrating if we just fall in with what crops up.
11. Which allows us to create hierarchy from our observing. Such hierarchy is not, of course, "out there", but is a personal construct. In this universe, the world is not hierarchical.
12. This is the crucial point in Piaget's developmental psychology. How do children come to recognise objects as remaining constant in a world of ever changing perceptions (as Piaget imagines them, from the impossible vantage point where he already can conserve such objects)? Although Piaget's objects (12) (and von Foerster's (15)) are not identical to my Objects, there are strong similarities. The question of how to create constancy is, in effect, answered in the statement to which this note is attached. Of course, this account is very over-simplified.
13. See (4). For how this may be used in communicating, see (6).
14. Conversation Theory, perhaps the masterwork of Gordon Pask, is one of very few genuine Theories of interaction. In it, a means of communicating when we do not see the world the same (or the same world) is developed and extended. It is beautiful work, although often difficult to understand. See (11). For an easier, potted version, see (8). For my interpretation in terms of communication and language, see (6).
15. The assertions are based on my arguments at last year's conference (7), and summarise the findings in that paper. This paper is firmly based in that.
16. For an elegant recent exposition see (1), which is heavily based in the work of George Spencer Brown, whose edict "Draw a Distinction" has been so influential in (second-order).

References

1. Glanville, R. 1975. *A Cybernetic Development of Theories of Epistemology and Observation, with reference to Space and Time, as seen in Architecture*. Ph D Thesis, unpublished Brunel University. Also known as The Object of Objects, the Point of Points, - or Something about Things.
2. Glanville, R. 1976. 'What is Memory, that it can remember what it is?' In Trappl, R. et al eds. *Recent Progress in Cybernetics & Systems Research*, vol 7. Washington DC: Hemisphere Press.
3. Glanville, R. 1980. 'Consciousness, and so on'. *Journal of Cybernetics* vol 10.
4. Glanville, R. 1980. 'The Same is Different', In Zeleny, M. ed. *Autopoiesis*. New York: Elsevier.
5. Glanville, R. 1994. '["as if" Radical Objectivism]. in Trappl, R. *Cybernetics and Systems Research '94*. Singapore: World Scientific.
6. Glanville, R. 1996. 'Communication without Coding: Cybernetics, Meaning and Language: How Language, becoming a System, Betrays Itself. *Modern Language Notes, vol. 111 no 3*.
7. Glanville, R. 1997. 'Behind the Curtain,' in Ascott, R. ed. *Consciousness Reframed '97*. Newport: UWCN.
8. Glanville, R. 1998. 'Gordon Pask's Cybernetics'. In Mandel, T. ed. Luminaries Section; *International Society for Systems Science*. Web site http://www.isss.org/lumPask.html
9. Glanville, R. Forthcoming. *Science, Cybernetics and the Wayang Theatre*. Invited paper for IJGS.
10. Kauffman, L. 1998. *Virtual Logic—the Calculus of Indications*. Cybernetics and Human Knowing, vol. 5 no 1.
11. Pask, G. 1975. *Conversation, Cognition and Learning*. Amsterdam: Elsevier.
12. Piaget, J. 1955. *The Child's Conception of Reality*. New York: Basic Books.
13. von Foerster, H. 1972. *Notes on an Epistemology for Living Things*. University of Illinois at Urbana: BCL fiche 104/1.
14. von Foerster, H. 1973. 'On Constructing a Reality.' in Preiser, F. ed. *Environmental Design Research*. Stroudberg: Dowden, Hutchinson and Ross.

15 von Foerster, H. 1976 'Objects: Tokens foreign - Behaviours. *Cybernetics Forum, vol. 8 nos. 3 and 4.*
16 Wittgenstein, L. 1971. *Tractatus Logico-Philosophicus.* 2nd ed. London: Routledge and Kegan Paul.

Ranulph Glanville is an Independent Academic who works on personal projects and visits universities in London, Brussels, Melbourne and Hong Kong helping develop research in design areas. He also teaches cybernetics. Previously, he spent 25 years teaching architecture. He is an associate of CAiiA and the Centre for Interaction and Co-operative Technology in Amsterdam, on the editorial board of several journals and the organising and committees of several conferences. He runs a consultancy and small publishing house and has around 200 academic and other publications to his name. He has the Diploma of the Architectural Association, London, a PhD in Cybernetics and another in Human Learning, and is adjunct professor in the Faculty of the Constructed Environment, Royal Melbourne Institute of Technology. He is a Fellow of the Royal Society for the Arts, Sciences and Commerce, and of the Cybernetics Society, London. He works in a number of distinct and apparently unrelated areas, referring to those he is not currently doing as his hobbies. Currently, he is looking for funding to construct large interactive sound sculptures and develop a large publishing/art work, Secret Pieces. He lectures and has performed around the world since 1964, when he founded and directed the world's first live, interactive electronic music band. He has one son and one wife (not related). He is contactable through CybernEthics Research, 52 Lawrence Road, Southsea, Hants, PO5 1NY, UK: tel +44 1705 737779; fax +44 1705 796617; email ranulph@glanville.co.uk. His url remains a secret.

Between Reality & Virtuality: Toward A New Consciousness?
Julio Bermudez

Art and Consciousness
A close scrutiny of life reveals that living creatures are not material entities separated from their surroundings but rather regulatory interfaces of interactions occurring between their internal and external environments. Life is an emergent condition whenever and wherever certain complex internal and external tensions meet one another and find some dynamic balance. Life is a 'boundary conditions' phenomenon. (Brooks and Wiley 1988).

To maintain continuity under constantly changing circumstances, life must endlessly 1) *monitor* the boundary conditions and 2) *act* towards responding at once to internal and external demands. The resulting *motor-sensory activity* generates a certain *experiential field, life space* (Lewin 1951), or *'ambiance'* (Von Uexkyl 1968) that an organism is aware of. Awareness is thus a motor-sensory function presenting a *synthetic* 'report' of boundary conditions. '*Awareness happens neither "inside" nor "outside" an individual but instead in the threshold between the two, at their interface*' (Bermudez 1994).

Mentations are thus not a particular metaphysical event or mysterious gift but rather a natural phenomenon emerging from relatively complex, sustained yet fluctuating transactions between the inner and outer worlds of a living organism. Brain and mind research seem to support this position (Eccles 1992, 1990; Fischback 1992).

Awareness works very well under close-to-equilibrium conditions, that is when environmental pressures from either side follow patterns that are historically consistent. When boundary conditions go beyond these expected patterns awareness becomes increasingly ineffective and consciousness takes precedence. Consciousness is automatically unleashed whenever we face a surprising outcome that contradicts or happily extends our habitual expectations. In other words, whenever awareness is transcended, consciousness is called in to deal with the situation.

Consciousness may be elicited by natural or artificial circumstances. We can be 'pushed' into a conscious mode by a natural event (e.g. social pressure, personal sickness, a physical threat, etc.) or by artificial means, that is, by intentionally problematising an otherwise not challenging situation. It is the 'artificial' or purposeful creation of conditions leading to the arousal of consciousness where lies an essential secret of all successful art.

Without such arousal, art would go to form part of the background experience accompanying ordinary states of mind. In this phenomenological 'disappearance', there would be no art to be spoken of. As Heidegger (1971) and Dewey (1934) argue, unless the work of art is present in attention, it vanishes, it doesn't 'exist', it is 'anesthetic'.

The paradox behind this attention arousing condition for defining art as art is that the object (the work) must problematise the subject, the ongoing interface . . . Were there no obstruction or difficulty exceeding an individual's existing values, skills and knowledge, there would be no need for adjustment, or change, and thus no aesthetic experience. Emotional, cognitive and/or sensorial dissonance is fundamental to art but it must be within certain limits, otherwise the individual would be overwhelmed and unable to deal with the situation. Hence the artistic craft at designing the 'boundary conditions' necessary for interfacial disturbance.

Good art therefore means to change the existing structures for dealing with the world and the self. It implies confronting frictional boundary conditions that create some degree of anxiety, confusion, insecurity and therefore arouses strong feelings. This is a normal response to the natural resistance of all life to far-from-equilibrium situations.

Of course, we could debate whether a work's 'disappearance' into unconsciousness could be legitimately constructed as defining its anesthetic or artless quality. Such an interpretation has avantgarde and aggressive overtones that may appear quite Western, male and modern from culturally sensitive, feminist and postmodern points of view. For now, I would like to avoid this discussion for the sake of advancing my art work. I believe that the video work to be shown conjures up in itself a possible response to the foreseen criticism.

New Virtuality = New Consciousness?
Although the arts have always used virtual environments (e.g. painting, literature, music, etc.) to conjure up consciousness, recent technological developments have created unprecedented conditions that may open us to new opportunities. Unlike the past, the new *digital* virtuality permits three-dimensionality, interactivity, immersion and multi-media phenomena. This invites us to wonder about the potential new consciousness that may

arise from these circumstances. Does the meeting of a largely reality-trained psychology and a completely enveloping and foreign artificiality lead to new kinds of thought, experience, self? How does the external tension of virtuality affect and demand compensation from internal psychological forces? What is the resulting conscious interface like?

These questions among others launched a design investigation that uses a virtualscape to generate particular conscious states. The design proposal focuses on the aural qualities associated with emotional and situational states and offer one-sided boundary conditions that are as representationally alien (i.e. clear from memory associations) as possible. The virtualscape emphasises the interface between artwork and viewer by creating conditions that fully absorb the observer in the unfolding aesthetic events.

Practically this means to design *space* rather than form and the *environment* and not the 'stuff' within it. (Please refer to the video images). By moving from the solid to the void, by making it impossible to focus on one thing, the virtualscape directs consciousness towards nowhere except the boundary conditions, that is, towards itself. In this way, any separation between subject as ego-observer and object as artwork-observed vanishes. Experience occurs in a non-dualist mode that is closer to meditation and other non-ordinary states of mind than to daily waking experience. It reminds us of Whitehead's claim (1961) that in any moment of experience the subject is that very experience as it is occurring whereas the object for this subject is the multiplicity of the preceding moments of experiences. My work thus attempts to kindly but surely eradicate ego from the artistic equation, hence inviting its most ambitious goal to emerge: the experience of experience as experience. The new virtuality creates a conscious but selfless experience, a 'new' consciousness . . .

Remarkably, the chosen media to explore this matter is video and not electronic. (The video production is the result of a digital manipulation of actual video footage) Arguably, video is the form of art that had the largest influence on human consciousness ever. Surrounding our minds for many hours every day and requiring or stealing our attention, video impacts our lives directly or indirectly, (in)(re)(de)forming our sense of reality to the point of making the virtual natural.

As Steven Johnson (1997) argues, television is, for the great majority of people now living, not an artificial thing anymore but part of the normal, ordinary, hence natural landscape of our civilisation. Video work is therefore guaranteed to receive some degree of attention and response, particularly if one manages to capture the audience. Selecting video as the platform to simulate and investigate the potential of digital virtuality in affecting consciousness appears reasonable, even if odd.

The use of video also provides us with other important advantages. On one hand, video is at a middle point between traditional media (material) and the new computer-based media (digital). This permits the direct investigation of the elusive yet emotionally charged connection of painting, video and computer through the manipulation of analog and electronic media. This flexibility is important at the time of trying to address semiotic and expressive issues associated with finding a novel representation of the virtual. On the other hand, video permits hybrid media iterations that cannot be attempted by traditional or digital means alone, thus allowing the use of a completely 'new' and infinite repertoire of reality to expand the conceptual, design and aesthetic horizons of the virtual.

– Models –

This point brings us to perhaps the major premise of the work: to produce a virtual environment full of embryonic possibilities and un-akin to any that have been seen so that it conjures up an alien consciousness. This meant a world of 'pure' signifiers in which the signifieds are missing and with them all our preconceptions about the virtual. A new mind can only be found at the interface with the yet unencountered, unconceived and therefore truly alien 'otherness' of virtuality.

To accomplish this objective, the work uses a minimalist reinterpretation of tectonics for expressing the qualities of virtuality. This liberates us from having either to copy or to reject reality as a source and, at the same time, allows us to concentrate on "environmental" issues (i.e. space and context) rather than objects. The *'tectonic simulation of the virtual'* has followed the selection and juxtaposition of materials, processes, technologies and states associated with the qualities of information - such as fluidity, lightness, complexity, ambiguity and transparency. By changing the focus of observation, context, scale and speed of real material phenomena, the original is not *re-presented* but *'trans-presented'* in such a way that it keeps only few invisible but still influential qualities of the original. *Trans-presented* tectonics thus generate powerful and completely new visualisations and conceptualisations of the virtual that stubbornly avoid recognition, hence inviting a new consciousness.

A strong tectonic presence, even if virtual, invites the natural projection of bodily sensations that creates a direct kinship between feelings and experience - the sense of touch arouses emotional responses (intimacy, smoothness, etc.). As a result, the design immediately obtains a qualitative character very difficult to elicit by other means. Furthermore, the fact that tectonics are used metaphorically to suggest virtual environments that are by definition immaterial, challenges stereotypes of virtuality creating a very strange and evocative perception.

The strong reactions that this work elicits in its audience is a good indication that technology-based virtuality may bring up higher yet different levels of consciousness than those allowed by art in the past. Surprisingly, however, the fluid interfaces framed by the design work possess a peaceful and surrendering quality. The grainy quality, the blurry definition and the relentlessly changing events create an uncanny reference, even a remembrance of states of mind associated with lucid daydreaming that reveal as much as it conceals aspects of reality hitherto hidden yet somehow deeply intuited . . . Zen concepts such as 'Shunyata' and 'Samadhi' begin to emerge. In contrast to the 'frictional' metaphor utilised to describe attention rising conditions in the earlier arguments), the designed virtualscape manages to achieve the same end using positive (i.e. effortless, non-struggling) means.

The abstract quality of the environment (generated out of reality trans-presented) permits us to intellectually disengage from the known whereas its tectonic nature brings us back to emotional and perceptual engagement at the body level. In this condition of familiar 'otherness' the basic problems of the virtual - placelessness, characterless, etc. (Goldberger 1995) - are addressed while a new mind is aroused.

Although this work may be associated with the visual expressions of 'psychedelic art' of the late 60s and early 70s and more recent media installations (e.g. Bill Viola's), its methodology, media operations and thematics set it apart. The critical iterations between video, computer and imagery create a new hybrid horizon for art.

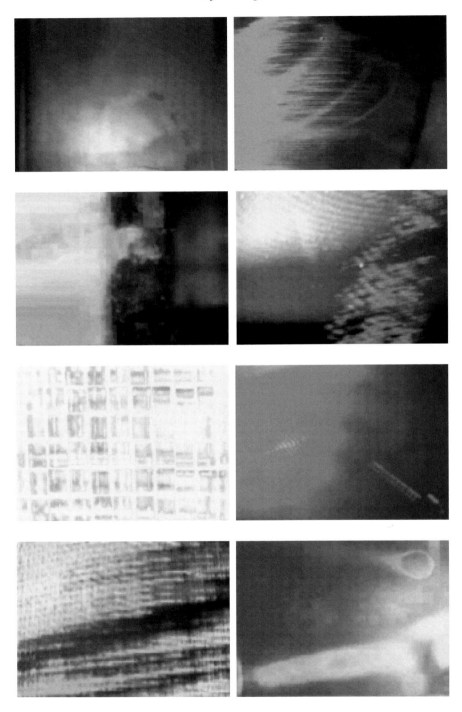

The Artwork (Video)
The eight stills represent different moments of the video narrative.

References
Bermudez, J. 1994. *Interfacial Education*. Ann Arbor, Michigan: University Microfilm Inc.
Brooks, D. and Wiley, E.O. 1988. *Evolution as Entropy: Toward a Unified Theory of Biology*. Chicago, IL: The University of Chicago Press.
Dewey, J. 1934. *Art As Experience*. New York: Wideview/Perigee Book.
Eccles, J. 1992. 'Evolution of Consciousness'.. in Proc. *Natl. Acad. Sci.* USA, vol. 89 (August), pp. 7320.
Eccles, J. 1990. 'The Mind-Brain Problem Revisited: The Microsite Hypothesis'. in *Exp.Brain Res. Ser.*, vol..21, pp.549-68.
Fischbach, G.D. 1992. 'Mind and Brain'. in *Scientific American*, vol. 267, no. 3 (Sep.), pp. 48-57.
Goldberger, P. 1995. 'Cyberspace, Trips to Nowhere Land'. in *The New York Times* (5 October), B-1.
Heidegger, M. 1971. *Poetry, Language, Thought*. New York: Harper & Row. Publishers
Johnson, S. 1997. *Interface Culture*. San Francisco: HarperEdge.
Lewin, K. 1951. *Field Theory in Social Science. Selected Theoretical Papers*. New York: Harper & Row.
Whitehead, A. 1961. *Alfred North Whitehead. His Conceptions on Man and Nature*. (R .N. Anshan.). New York: Harper and Brothers.
Von Uexk½l. Cited in Bertalanffy, L.V. 1968. *General Systems Theory*. New York: Braziller, pp. 228-35.

Julio Bermudez is an Assistant Professor at the University of Utah Graduate School of Architecture (Salt Lake City, USA). He holds a Master's in Architecture and a Ph.D. in Education from the University of Minnesota.

Dr. Bermudez is interested in the role of digital media in architectural design, representation, and spatial visualisation. His research covers virtuality and electronic-based design process from architectural, artistic and interdisciplinary perspectives. Professor Bermudez has been teaching graduate seminars and design studios on digital media since 1991.

Dr Bermudez has presented and/or published work in over 25 national and international conferences in this area. His recent work on visualisation (both pedagogical and research) has been presented and exhibited at national and international (e.g. Canada, Italy, USA and soon in Argentina and Brazil). Professor Bermudez is the chairperson of the digital media topic for the 1998 ACSA National Meeting in Cleveland and will co-direct the ACADIA 1999 Conference in Salt Lake City. He is also the author of Visualize!, a software teaching spatial visualisation and representation. Dr Bermudez has just received the prestigious AIA (American Institute of Architects) Educational Honors Award for 1998.

Dr Julio Bermudez, Assistant Professor, Graduate School of Architecture, University of Utah, Salt Lake City, UT 84112, USA. 801.581.7176 (ph.). 801.581.8217 (fax). bermudez@arch.utah.edu (email) http://www.arch.utah.edu/people/faculty/julio/julio.htm (URL)

The Actualisation of the Virtual
Mark Palmer

What do we mean by Virtual Reality? This term, with its peculiar momentum from the virtual to the real, has become associated with the manifestation of dreams, formed, condensed, not so much like breath against a cold window but the window itself. This concretion of dreams, this transformation from no-*thing* to some-*thing* is the ironic joke, the oxymoron. Matter as the measure of reality, of some-thing real, escapes us, the uncrossable chasm between the virtual and the real. But why let reality stop you when you're dealing in dreams? The move from dream to reality has become so much a part of the hyperbolae of virtual reality that it has consumed its own myth, it has become a product on the way to its own making. Tomorrow will bring us the solutions that we seek today. But how can we form anything that can be called a critical discourse upon such shifting sands? We hit a point of resistance, and the problems of today are swept aside with the assurity of tomorrow. The pace of dramatic revolution is claimed, a revolution that will be discovered upon its arrival rather than in its making. But if revolution were a condition in itself, it would never be known. The revolutionary *transforms* the world we *already* live in, *growing from it*. Thus, intrinsic to the questioning of revolution (in both senses), must be the questioning of the world and our bond with it. It is this paper's aim to investigate how notions of the virtual and matter might be rethought or rediscovered owing to the emergence of virtual reality and, through this, reconsider the claims made for the role of consciousness within discourses concerning digital aesthetics.

But in condemning a discourse based upon the futurity of the virtual, we might ask if there is an underlying sensibility that might be folded back revealing the nature of our experience today. In *idoru* William Gibson begins to elaborate what might be considered such a sensibility, one also paralleled in Neal Stevenson's *The Diamond Age*. Seen cynically, these simply reconcile the virtual with matter, our dreams made real through the science of nanotechnology, compiled from matter rather than code. But, more radically, we can recognise a new sensibility of the virtual, binding matter with a reality that subtends it. In *idoru*, the protagonist Laney has the ability to see the coalescence of flows or nodes of information rather than its units 'an intuitive fisher of patterns of information'[1] akin as the idoru describes to 'seeing faces in the clouds, except the faces are really there?'[2] In a philosophy dealing with chaos, multiplicity and information theory, Michel Serres notes that

> Noise cannot be a phenomenon; every phenomenon is separated from it, a silhouette on a backdrop, like a beacon against the fog, as every message, every cry, every call, every signal must be separated from the hubbub that occupies silence, in order to be, to be perceived, to be known, to be exchanged. As soon *as a phenomenon appears, it leaves the noise*; as soon as a form looms up or pokes through, it reveals itself by veiling noise. *So noise is not a matter of phenomenology*, so it is a matter of being itself.[3]

We can therefore begin to define a 'new' sensibility where the insensible is the basis from which the phenomena of the world emerge. Something which, until the advent of computing, might well have been condemned as a call to metaphysics but through the discovery of emergent systems and the new sciences of chaos and complexity can be thought of differently. If we consider this sensibility which lies behind the sensible we begin to reclaim the original conception of virtuality, which, the Oxford English Dictionary states, is the 'Essential nature or *being, apart from external form or embodiment*' and 'A virtual (*as opposed to actual*) thing . . .'. The virtual is not opposed to the real but to the actual, as the virtual is real in itself, a fundamental facet of being. But how can we begin to understand the grounds upon which we may think of this relationship? It is within the philosophies of Bergson and Deleuze that we encounter a philosophy of the virtual, where the virtual is real.

Through an analysis of the difference between matter and memory, a difference in kind, Bergson introduces us to the notion of the virtual. Our habit is to think of memory as a diminished perception, revealing the assumption of a difference in degree. But for Bergson the crux of this difference is that perception occurs in matter. So when I say that things are real but that I have my view of them, I express two parts of perception. 'My view' is not the separation of perception from the world; it is its inclusion. Perception is not a cerebral modelling of the world; it is embedded in the world. It is in the object itself, and not elsewhere, that its image is formed and perceived. And if perception occurs in matter, we can no longer think of memory in the form of a diminished perception, for how could matter become less, multiplied by its faded images?

So how are we to conceive of memory? For Bergson, memory meets perception in two forms, covering

> ... as it does with a coat of recollections a core of immediate perception, and also contracting a number of external moments into a single internal moment, constitutes the principal share of individual consciousness in perception.[4]

The covering of the core of immediate perception is essential to memory. Memory *borrows from* perception in order to become actualised. Our everyday experience stands in mute testament to this fact, because without the intervention of memory, we would live every experience as if it were our first. But what happens in this borrowing from perception? Bergson states that

> Our distinct perception is really comparable with a closed circle, in which the perception-image, going toward the mind, and the memory-image launched into space, careen the one behind the other.[5]

This is not an act of retrieval that leads away from the perceived object but one where perception and memory are held in mutual tension, always folding back to the object from which it proceeds, a circuit so well closed that any increase in concentration produces new circuits spun around the object, moving from subliminal *re*-cognition to the concomitant details that expand our understanding. Personal recollections are formed at the limits of these circuits, essentially fugitive becoming materialised only by chance. Proust perhaps best describes this capricious nature of memory.

> I feel that there is much to be said for the Celtic belief that the souls of those whom we have lost are held captive in some inferior being, in an animal, in a plant, in some intimate object, and thus effectively lost to us until the day (which to many never comes) when we happen to pass by the tree or to obtain possession of the object which forms their prison. They start and tremble, they call us by our name, and soon as we have recognised them the spell is broken. Delivered by us, they have overcome death and return to share our life.
>
> And so it is with our own past. It is a labour in vain to attempt to recapture it: all the efforts of our intellect must prove futile. The past is hidden outside the realm, beyond the reach of intellect, in some material object (in the sensation which that material object will give us) of which we have no inkling.[6]

Our memories, *real* but *virtual in nature*, only become actualised through perception. But it is memory that borrows from perception. The intellect's attempt to retrieve memory is futile, existing as it does in another realm. But surely my experience tells me that I retrieve memory: I remember how to do, or not do tasks; I read texts and draw on relevant thinkers. But these memories are not sought as such but insert themselves with a view towards action, opening the horizon of the future. But even if memory borrows from the object in this move towards action, don't we find ourselves engaged in an effort of memory? Proust describes this effort as one in which we shut out the extraneous influences of the world and clear a space for the memory to form. One in which we await, rather than seek memory. We detach ourselves from the world, loosening the ties of the body, of perception, so that by distancing its call to action we may approach the state of dreams. Bergson sees this

> Whenever we are trying to recover a recollection, to call up some period of our history, we become conscious of an act *sui generis* by which we detach ourselves from the present in order to replace ourselves, first, in the past in general, then, in a certain region of the past-a work of adjustment, something like the focusing of a camera. But our recollection still remains virtual; we simply prepare ourselves to receive it by adopting the appropriate attitude.[7]

This aspect of memory demonstrates that, before its actualisation, memory cannot be known. We convince ourselves that memories are stored in cerebral matter. But, as Bergson notes

> ... for the sake of action, the real order of things, we are so strongly obsessed by images drawn from space, that we cannot hinder ourselves from asking *where* memories are stored up.[8]

For if memories are not stored in the brain, we immediately ask, where are they are stored? Possessed by the compulsion to identify matter as a necessary condition of reality, we picture our memories stored in some cerebral library. But there can be no question of 'where' in the storage of memory. We must resist the temptation to substantiate that which by its very nature has no substance. Memories do not have, as matter does, the property of containing and of being contained. The very effort of putting oneself into the past- memory- 'in general' is recognition of this. The localisation of matter never has the quality of 'in general'. This virtual memory, from which memory-images are actualised, is a multiplicity. Each part of memory contains the whole.

So even after discovering the reality of the virtual, we have to ask whether this play between virtual and actual memory gives credence to a sensibility of emergence. In Kant's Transcendental Aesthetic we discover space and time as the forms of sensible intuition.* .Bergson takes up Kant's notion in defining space as a homogeneous medium, but in defining duration as a qualitative heterogeneity, he makes duration felt rather than measured. Do we discover in the experience of space and time the sensibility of emergence? Deleuze makes the observation that for Bergson

> Duration seemed to him to be less and less reducible to a psychological experience and became instead the variable essence of things, providing the theme of a complex ontology. But, simultaneously, space seemed to him less and less reducible to a fiction separating us from this psychological reality, rather, it was itself grounded in being and expressed one of its two slopes, one of its two directions.[9]

These ontological aspects of space and time, are more properly revealed two years later upon the publication of Deleuze's *Difference and Repetition*. Virtuality is no longer solely mnemonic but a condition prior to the world, difference in itself, its actualisation taking 'place in three series: space, time and also consciousness.'[10] But consciousness, sitting at the end of this process, with its desire to, as Bergson puts it, delight in 'clean cut distinctions. . . with well defined outlines' . ."[11] cannot come to know this sensibility. It always trails after it, knowing it only by the homogenised qualities of space and time. Thus as Bergson developed the notion of duration, Deleuze develops the notion of depth

> Depth as the (ultimate and original) heterogeneous dimension is the matrix of all extensity, including its third dimension considered to be homogenous with the other two.
> The ground as it appears in a homogeneous extensity is notably a projection of something 'deeper': only the latter may be called *Ungrund* or groundless.[12]

Depth is the first dimension from which consciousness abstracts homogeneous dimensionality. But this first dimension in its groundlessness requires a radical shift in our sensibilities. Might we then begin to consider depth and duration as the trace of the virtual in the world? Deleuze talks of depth which ' bears witness to the furthest past and to the coexistence of the past with the present.'[13] Are we not describing some kind of resonance, coaxed into being by the artist, that shifts our sensibility – a singular moment – beyond a moment – so thick, so *wide*, that it defies time – that we stand in awe – its reverberations felt through every pore and fibre – until we feel time as we know it, in its everyday passage – the shadow that follows our movements – snatching at our heels again – drawing us back into the eye of the storm that we call the present.

The most precious moments of looking and making are, when I encounter the fugitive qualities that I seek, in the wonder of making and looking; the world's disruption is its revealing, a sensibility which, being beyond the object, having the quality of the virtual, the immaterial, finds a radical home in the immateriality Virtual Reality.

In conclusion can we say that in discovering the radicality of Virtual Reality we can fold back this discovery and recognise a quality that is possessed by aesthetic objects. That our habit of identifying the qualities of our experience through matter has marginalised the experience we call the affectivity of the aesthetic object. And which is in reality the sensuousness of pre actualised space and time, *before* it tumbles into consciousness, and is thrust into the instant.

Notes
1. Gibson, W. 1996..*idoru*. London: Viking p.25.
2. ibid p.237.
3. Serres, M. 1995. *Genesis*. Michigan. The University of Michigan Press, p.13.
4. Bergson, H. 1988 *Matter and Memory*. New York: Urzone, p.34.
5. ibid p.103.
6. Proust, M. 1996. *In Search of Lost Time*. Swann's Way. London.Vintage, p.50.
7. Bergson, H.1988 *Matter and Memory*. New York: Urzone, p.133.
8. ibid p.148.
9. Deleuze, G.1991. *Bergonism*. New York: Urzone, p.34.
10. Deleuze, G.1994.*Difference and Repetition*. London:Athlone, p.220.
11. Bergson, H.1910. *Time and Free Will*. London:Swann Sonnenschein & Co., p.9.
12. Deleuze, G 1994 *Difference and Repetition*. London. Athlone, p. 229
13. ibid, p. 230

For those not familiar with philosophy, two things are worthy of note here

i. Intuition in philosophy *does not* mean its usual use as an indirect insight *but a direct relationship* between the mind and an object.

ii. In the Transcendental Aesthetic as *forms* of sensible intuition, space and time are not sensible in themselves.

Mark Palmer – Research Scholar, Staffordshire University
M.W.Palmer@staffs.ac.uk

Selected Group Shows
1997	*Borders, The National Museum of Wales*
	Sculpture in the Park, Festival Park Ebbw Vale
1996	*Trans-formations-The Artists Museum, Lodz, Poland*
	Site-ations 96-The Artists Project, Cardiff
1995	*Words-Aspex Gallery, Portsmouth*
1994	*Cardiff/Berlin Exchange-Galerie edition+caoc, Berlin*
	Site-ations Festival-Cardiff
	Small Works/Small Works-Galerie edition+caoc, Berlin
1993	*Construction in Process Festival (invited artist)-Lodz, Poland*

Sites and Sensibilities-Chapter Arts Centre, Cardiff.

One Person Shows
1997   Closeness and Distance-Flaxman Gallery, Staffordshire University
1996   Transience-Swansea Arts Workshop Gallery
        The High Tower-Henry Thomas Gallery, Carmarthenshire College of Art and Technology
1995   Making Space-Dare Valley Country Park

Collections
        Northern Arts and private collections in the UK and Germany

Papers
1998   'The Sensibility of the Virtual–Art & Technology in the Age of Information',TateGallery, Liverpool
        'Implicit Time – Painting and Time, University Of Lincolnshire & Humberside'
1997   'The Foregrounding of the Issue of Space'.ISEA, Art Institute of Chicago
        'Redefining Spaces,their Phenomenologies and Consciousness'. University of Wales College, Newport
        'Between the Visible and the Invisible', Shades of Time, Staffordshire University.

We are the Consciousness Musicians – Electronic Art, Consciousness and the Western Intellectual Tradition
Mike King

Introduction
At the second 'Towards a Science of Consciousness' conference in 1996 I sat next to a brain surgeon who remarked that most of the speakers were on the wrong track: *he* could tell them how consciousness is produced. As a brain surgeon, he knew the precise combination of chemical and neurophysiological conditions in which a patient was conscious and those in which they were not. This is a materialist, non-dualist viewpoint which regards consciousness as an emergent phenomenon. Previously, consciousness was regarded by the materialists as an epi-phenomenon, or by the dualist philosophers such as Descartes as independent of matter and brought into relationship to it via God. Since the ascendance of science, such dualism became intellectually untenable and a scientific investigation of consciousness thought unfeasible until the late 20th century. The 'new' physics (primarily quantum theory) has made it possible to re-open the debate, because of a new emphasis on the observer in scientific experiment.

However, at the heart of the debate is the question whether consciousness is open to scientific investigation at all. The difficulty can be expressed as the 'zombie problem' where the zombie is defined as 'a behaviourally indiscernible but insentient simulacrum of a human cognizer'[1]. The zombie has been a useful tool in consciousness studies, partly as a theoretical point of reference and partly because engineers can go ahead and build one in the hope of analysing its point of departure from the human. Others believe that it is only a special property of the human brain that gives rise to consciousness, perhaps related to quantum mechanical effects in the microtubules.[2] The reverse-engineered zombie and the quantum investigations are just two approaches to the understanding of consciousness but whatever the results, they will tell us nothing about consciousness *as we know it from the inside*. It is an awkward but irreducible fact that only one's own consciousness is available for direct investigation — in other words every one else *is* a zombie. All the usual reasons for attributing consciousness to others — empathy, common sense, even love — fail the criteria for acceptance as scientific evidence.

My thesis is that consciousness is a suitable subject for a *first-person* science, a science of the subjective, and not for a third-person science.[3] To explore this proposition I shall examine the Western intellectual tradition from an Eastern perspective, using the concept of the *jnani*.

Background concepts

In ancient times and up to surprisingly recently, the Western intellectual tradition was bound up with religious thought. The key distinction in religious matters I am introducing for the purpose of this discussion is between the devotional and the non-devotional orientation, and I will use the Indian terms *bhakti* and *jnani* to cover these. The British medieval mystic Richard Rolle is an example of *bhakti*, while the Buddha is an examples of *jnani*. The characteristics of a *jnani* include an emphasis on knowing rather than loving, on inquiry rather than surrender, on doubt rather than faith, on will rather than abandonment, possibly non-theistic rather than theistic, and possibly *via negativa* rather than *via positiva*. (The latter terms are used in mysticism to distinguish spiritual paths that respectively deny the material world or embrace it.)

Religions are founded by charismatic individuals such as Christ, Muhammed, Buddha, Krishna and so on. The orientation of such men, either *bhakti* or *jnani*, has a profound effect on the type of religion they leave behind and will influence the kind of religious language that the faithful can use as part of their tradition. Christ was *bhakti*, Buddha was *jnani*. Hence for a *jnani* such as Eckhart, born into a *bhakti* religion, it was difficult to use the Christian language to express his own insights and he ran into trouble with the Roman Catholic church.

When we speak of the Buddha, Eckhart or Krishnamurti as *jnani* we are talking about the geniuses of this orientation. However, I believe that all people, whether admittedly religious or firmly secular, have one or other of the two orientations, *jnani* or *bhakti*. It is the person of the *jnani* orientation who doubts, questions and thinks and, given the right intelligence, education and milieu, will become a contributor to a culture's intellectual tradition. The fully developed *jnani* is very different from the intellectual, however, but at the same time I do want to stress that I consider the *bhakti* to contribute as much, if not more, to society as the *jnani*.

Patanjali and Eastern traditions of thought

Using the previous terminology, Hinduism contains luminaries of both *jnani* and *bhakti* persuasion but we will consider a single important *jnani* text in the Indian tradition, the *Yoga Sutras* of Patanjali.

Here are the first five stanzas:

Now the discipline of Yoga.
Yoga is the cessation of mind.
Then the witness is established in itself.
In the other states there is identification with the modifications of the mind.[4]

Patanjali is codifying and summarising a knowledge that already had a long history, and in the first five lines reduces the system to its essence. To condense it even further:

Yoga is the cessation of the identification with the modifications of the mind.

The remainder of the Sutras forms a systematic exposition of this proposition and lays out a method for its practice and achievement of its goal. What is radically different in yoga to any Western tradition is the suggestion that *cessation* of mind is the route to knowledge. In Zen, the doctrine of no-mind is also central to its teachings. However, despite the anti-intellectual implication of this starting point, both *jnani* Hinduism and all the forms of Buddhism, have generated vast literatures which have shaped the intellectual traditions of the far East.

The following sutras are important as they touch on epistemology:

The modifications of the mind are five. They can be a source of anguish or non- anguish.
They are right knowledge, wrong knowledge, imagination, sleep and memory.
Right knowledge has three sources: direct cognition, inference and the words of the awakened ones.

Direct cognition for Patanjali and the sages of the East means a knowing of things from the *inside*, using techniques broadly referred to as meditation. Inference is common with the West; it means any knowledge derived from rational thought. The 'words of the awakened ones' has no credibility in the West outside the traditions of faith, mainly because of the obvious problem: who is to say which speaker is enlightened and which not? However for an inquirer into truth in a tradition such as Buddhism, the seeker is encouraged to test his or her own direct cognitions against those of the 'enlightened ones'. The interplay between Master and disciple in Zen, for example, is an illustration of this.

Jnani and the Western Intellectual Tradition

If we now ask why the intellectual traditions have developed so differently in West and East, we can identify three related points of departure: Greek thought, the dominance of Christianity and the rise of modern science.

Taking the Greeks as the first point of departure from the East, we can formulate this as the difference in emphasis between what we now understand as *philosophy* and what is *jnani*. When we look at Heraclitus or Pythagoras, for example, we find many similarities with *jnani* writings from Hinduism. The central figure in this context is Socrates, however, and I have made a detailed study of the proposal that Socrates was more like a well-developed *jnani* such as the Buddha than a philosopher as we now understand the term.[5] There is not space here to go into the details of the argument, but two pieces of evidence can be mentioned: Socrates's 'fits of abstraction', and his equanimity, or even joy, in the face of death. If we interpret the 'fits of abstraction' from the Eastern perspective as a type of *samadhi*, then much of Socrates's behaviour becomes clear and of a pattern shown in many *jnani's* lives.

Some see the early intellectual debate in the West to be between Plato's mysticism and Aristotle's logic but my analysis emphasises more the divergence between Plato as a philosopher and Socrates as a *jnani*. The West has not previously made this distinction, with the result that philosophy became a formal system of thought divorced from the spiritual but conducive to the rise of science. We can say that Socrates taught from a first-person epistemology, a direct cognition, while Plato developed a theoretical system, partly based on Socrates's teachings.

Some commentators believe that the dominant religion of the West could have derived from Socrates, who is compared to Christ in his teaching manner and in his persecution and execution, though from our analysis Socrates is a man of the *jnani* orientation and Christ of the *bhakti* orientation. There are many contributing factors to the dominance of Christianity but perhaps the most important one was the initial apologist in each case - Plato for Socrates and St Paul for Christ. Paul was probably less intellectually gifted than Plato, but his success in the initial propagation of Christianity lay in his appeal to ordinary people. Plato's diluted Socratism appeals to the rulers and intellectuals, while Paul's Christianity appeals more widely to the poor and suffering and to those whose *bhakti* orientation was touched by the suffering of Christ.

However, the initial *bhakti* nature of Christianity gave an insufficiently broad basis for a religion of the West and the *jnani* element was grafted on from the Socratic source, creating a tension that dictated much of European intellectual history. Plotinus (AD 204-270) is an important member of the neo-Platonist tradition in the West and is considered to be Plato's apologist or successor but on close scrutiny (again from an Eastern perspective) we find that he is a *jnani* in his own right, who happened to use the vocabulary of Plato to express his own ideas. Like Descartes and Spinoza, considered to be the founders of modern philosophy, Plotinus is evaluated for his formal contributions to philosophy but analysing the work of these intellectuals from the *jnani* perspective gives a quite different view of their role.

Going back to Plato, we could say that his impact was to prioritise ratiocination, or the dialectic process, over meditation and so this became the dominant mode for the intellectual investigation of consciousness in the West. Descartes, with his *cogito ergo sum*, locked the West into giving primacy to thought over all other experience. We can say that the West is characterised by ratiocination and mind, whereas the East is characterised by meditation and no-mind. Patanjali considers both to be routes to correct knowledge but the East has prioritised one and the West the other.

– Models –

The Scientific Method

The devotional nature of the dominant religion in Europe could not give free reign to its *jnani*-oriented thinkers, who began to turn to empiricism, that is observations of Nature, as a way of legitimising their instinctive tendencies to doubt and inquiry. One half of the Renaissance was preoccupied with neo-Platonism but the other, epitomised perhaps by Leonardo da Vinci, turned its back on the past and put its energies into a nascent science. Galileo's disagreement with the Roman Catholic church symbolised the parting of science and religion in Europe, which gave the *jnani*-oriented individual the impetus to abandon religion, though the complete secularisation of Western culture took another three centuries.

Third-person science is a consensual one, that is data leading to scientific conclusions are in the public domain, and in principle there is nothing stopping anyone from repeating the experiments that led to the conclusions. In third person science the first person is the object under investigation, the second person is the scientist and the third person is anyone else who can corroborate the measurements and conclusions of the second. When Galileo published his results it was open to anyone with a telescope and some patience to confirm. Although the initial reaction to his discoveries was hostile, it was only a matter of time before 'reasonable' people were convinced, because confirmation was relatively easy.

What then is a first person science? At this point it is not much more than a suggestion by the British mystic Douglas Harding,[6] but clearly the work of the great *jnanis*, East and West, can be examined for a basis. All that can be said now is that in principle it simply replaces the object of the third person enquiry with the subjective world of the investigator.

Conclusions

Insofar as the arts are a systematic inquiry into what is, they are more like a first person inquiry than a third person inquiry, because, although the theme of the artwork may be external and material, the real inquiry is into the subjective response of the artist to the subject matter. Practitioners of the electronic arts are in a good position to engage with a first person science of consciousness because, though artists, they are naturally disposed to science and technology. (I have explored some of these issues in arts and science[7] and also cyberspace.[8])

The brain surgeon suggested that the brain produced consciousness. If one was to suggest that a flute produces music, because a skilled flute-maker can give the precise mechanical conditions under which music can or cannot arise from the instrument, it would be absurd. Yet what if consciousness was as independent of the brain as the music is of the flute? After all, can you remember ever having been unconscious?

References

1   Moody, T.C., 'Conversations with Zombies', *Journal of Consciousness Studies*, Volume 1, No. 2 1994
2   Penrose, Roger, *Shadows of the Mind - A Search for the Missing Science of Consciousness*, Oxford University Press, 1994
3   King, Mike, 'From Schroedinger's Cat to Krishnamurti's Dog — Mysticism as the First Person Science of

Consciousness' in *Consciousness Research Abstracts*, proceedings of the "Tucson II" conference (Journal of Consciousness Studies) Arizona: University of Arizona, 1996, p.141
4   Rajneesh, B.S., *Yoga - The Alpha and the Omega*, Rajneesh Foundation, Poona, 1976.
5   Master's dissertation, University of Kent at Canterbury, 1996, unpublished.
6   Harding, D.E. *The Near End - The Science of Liberation and the Liberation of Science*, Shollond, Nacton, Ipswich IP10 OEW
7   King, Mike, 'Concerning the Spiritual in 20th C Art and Science' *Leonardo*, Vol. 31, No.1, pp. 21-31, 1998
8   King, Mike, 'Concerning the Spiritual in Cyberspace', in Roetto, Michael (Ed.), *Seventh International Symposium on Electronic Art*, Rotterdam: ISEA96 Foundation, 1997. p. 31-36

Mike King is Reader in Computer Art and Animation at London Guildhall University. He is currently working on a computer animated film about Krishna and researches the intersection between the artistic, the scientific and the spiritual. Email: kingm@lgu.ac.uk

Immersive Computer Art and the Making of Consciousness
Laurie McRobert

Introduction
Recently Japan found itself combatting TV `Pocket Monsters' sickness after a video game show mysteriously sparked epilepsy-like seizures in more than 700 viewers, mostly children. The children suffered these seizures after watching an explosion scene with red and blue flashing lights. Medical experts suspect that a scene featuring rhythmic bursts of brightly coloured light was so rapid and intense that it interrupted normal brain function.[1] TV Tokyo is cancelling the program until the cause of the reactions becomes clear and it has urged video stores to stop renting all episodes of the show.

A video game show is not immersive art, of course, but it can focus us on the making of consciousness and computer art which is what this paper is about. I am a philosopher who happened by chance to experience *Osmose*. Before this immersive art experience I was working on iconic dynamics, which led to my investigation of virtual reality dynamics. The dynamics of vision has now led me to turn my attention to the biological aspects of consciousness, particularly the effects of light on the making of consciousness.

Osmose, a Computerized Immersive Work of Art
It has been said of immersive computer art like *Osmose* that something quite different (often explained as a transcendent experience) happens to the person experiencing it. (I have referred to it elsewhere as being an ultimate work of art fulfilling the sense of *techne* that Heidegger talked about.)[2] We have also come a long way from Heideggerian-like philosophising about 'the work of art'. There will always be works of art hung on museum walls but immersive computer art forces us to examine the biological aspects of what it is

that happens to brainwaves because of the colour and sound and motion the eyes are receiving.

Light as Controlling Factor

If light is the controlling factor in viewing a work of art, then the immersive work of art, because one is surrounded by three-dimensional space, and because its lighting effects are produced by an electronic secondary light source, would evoke quite a different response in the subject (distinct from the work of art hanging in a museum that we look at in a lit room, often today supplemented by daylight). This is a rather complex subject. But I hope even after this short inquiry into daylight and computer light and their wavelengths that if nothing more, we can at least put to rest the mundane philosophers' arguments about what the colour red is or isn't and whether we all see the same colour red or not.

Natural versus Computer Light

The first question we must deal with is how would the effects of artificial light differ from natural light? How would the waves of light produced within the computer work on the visual brain and ultimately how might it affect consciousness? We have already noted above that artificially produced light such as that which the children experienced in Japan can, indeed, control vision and the brain and hence supposedly, in extension, control or influence consciousness. There is, for example, EMDR (Eye Movement Desensitisation and Reprocessing) a method which uses a light bar flashing horizontally to treat post-traumatic stress syndrome. There is, also, brainwave entrainment (BWE) or photic stimulation, which uses flickering light and pulsating sound as therapeutic tools.

The knowledge that flickering light can cause mysterious visual hallucinations and alterations in consciousness is something that humans have known ever since the discovery of fire. We have just to sit in front of a fire and gaze into its fractal flames to appreciate the state of reverie that it draws us into. There are also other types of natural BWE that can be experienced by a person, such as the flickering of sunlight through leaves or on water, or the dotted lines that flick by us when we are driving a car. These kinds of rhythms, we ought to note at this point, create alpha and/or theta rhythms in our brain which tune into our dream frequency wavelengths. Not so good when one is driving a car.

Many experiments have been performed to examine the brain's response to photic stimulation including an experiment at the Tohoku University School of Medicine in Japan in 1976 which should be of interest to the Japanese today. Researchers published the influence of colour on the photo-convulsive response (PCR). In measuring the effects of white, red, yellow, blue and green photic-stimulation on the photo-convulsive response they noted that the colour red at a frequency of 15 Hz was most likely to cause a PCR. Interestingly, they also noted that this stimulation could be inhibited by introducing low levels of blue light at the same time.[3]

Wavelengths and Consciousness

In order to establish how light can affect and/or create different orders of consciousness, I shall consider the difference in dynamics between natural photons hitting the eye and artificially produced photons hitting the eye. I think it is apparent from the foregoing that photons which are produced artificially to initiate a photic response do not initiate the same response in a subject as natural photons that hit the eye in a normal state of consciousness. Ostensibly, photic response can and is manipulated. We know the range of wavelengths natural colours produce. Most of us receiving photons in a normal uncontrived daylight state would be experiencing beta or alpha states of consciousness because of the natural range of lightwaves and thus, ideally, receive a comparable amount of energy through our visual systems. It is this energy that drives consciousness biochemically.

The lower brain region, which I think we are dealing with in immersive computer art, includes the brainstem's limbic region. It is a region that is not only involved in the automatic control of basic vital functions but is revealing itself to be the locus of instinctive consciousness not only of humans but of animals. Theta rhythms associated with meditative or dreaming states in human beings but with *conscious* states in animals originate here and it is these theta rhythms, connected with a special order of instinctive consciousness that, I propose, an immersive work of art such as *Osmose* evokes.

Immersive Computer Art and Dream-like States

We are ready to examine why an immersive experience of art such as *Osmose* plunges one into what can be called a different dimension of being or state of consciousness. I believe that this dimension of consciousness is closer in reality to dream-like states of consciousness and I shall attempt to explain all this, briefly, from a physiological/biochemical point of view. The argument hinges on endogenous light, a secondary light, that is produced in the body through the energy of sunlight when photons enter the visual system in a normal way. Lightwaves of different colours control the intensity of energy that enters into the brain. Not very much is known about the effects of lightwaves on the making of consciousness although scientists are deeply ensconced in trying to find out. So this is a relatively new field to be studying.

You are all familiar with colour, of course. Colour is light which travels in waves. Each colour's wavelength strikes the retina of the eye uniquely and is converted into different electrochemical impulses accordingly. These pass into the brain and then into the hypothalamus, which governs the endocrine glands, which in turn produce hormones, and so on. In the middle of the hypothalamus lies that dried-up prune, Descartes' pineal gland, which has highly modified photo receptive cilia that look very similar to retinal photoreceptors.

The energy of visible light must be stored by the brain and body to be reused by the endogenous system of light within the body that I believe energises or lights up the unconscious when we are dreaming or in reverie. Light enters the eye and is instantly gone, transmitted through a series of dynamics that in the eye begins with the vitamin A family and rhodopsin. We can simplify things, here, by stating that mitochondria cells located between rods and cones in the eye are known emitters of photons within the body and, then, jump to the claim that cilium or microtubule-type structures probably transport endogenously created light throughout brain/body.[4] It should be noted at this point that I distinguish between the making of consciousness and the transporting of consciousness. The kind of consciousness, for example, that Stuart Hameroff and Roger Penrose propound is to my mind very much a digital or left-brain consciousness . . . however non-linear they make it out to be. It is not the kind of consciousness that an immersive work of art would evoke.

Photons, as Natural and Artificial

It can be argued that the photons hitting the eye, which are created by the electronic light in an immersive work of computer art, can and do evoke a realm we generally refer to as an "unconscious" or reverie state of mind. In the case of *Osmose*, most people described the immersive experience as evoking a `transcendent' state of consciousness. So far as I know, no actual neurological testing of subjects' brainwaves has taken place while a person (s) was experiencing *Osmose*.

Because the eyes' response in *Osmose* is to an artificially produced environment which the subject sees through the computer's synthetically produced light and colours, the dynamics would, according to my thesis, play a significant role in creating the kind of consciousness a subject was experiencing. Since the subject in *Osmose* experiences a kind of floating through space created by his/her bodily-harnessed movements and since a transcendent, timeless experience is reported by immersants, we could hypothesise at this point that the subject is also experiencing alpha and/or theta rhythms. This can be adduced because she/he falls into a dream-like state of consciousness quite different from the cognitive processes of the left hemisphere even when the brain is exposed to Char Davies's text world.

Instant 3D Images

Another reason this "unconscious" dimension of consciousness is accessed by immersive computer art is because the photons that hit the eye in an immersive work of art relay three dimensional images to begin with. This leaves the eyes/brain with much less to do—a person more or less simply observes the image that emerges all around him/her. The eye does not have to translate two dimensional input into three dimensional input . . . this is already done for it by the computer. i.e., consider how difficult it is to achieve the stereoscopic effect by manipulating one's eyes and how easy it is for the eyes when one is simply immersed in 3-D.

Watching Dreams Unfold

The effect of immersion in 3-D is similar to what happens in a dream-like state where one is also immersed in three-dimensional space, viewing images of a symbolic nature. I.e. the hypnagogic state just before passing over into slow-wave sleep in which holographic images in glorious colours arise. During one of the four periods of REM sleep one *watches* the dream unfold - images are ready made - in the same way that they are in an immersive work of art.

Endogenous light

The link, then, between immersive art and the dream-like state lies in the effects of secondary light on the brain, not unlike what occurs in brain entrainment. Dream light is endogenous - produced and stored by the cells of a human being possibly in the pineal gland - that 'third eye' in the middle of the brain that receives light. Computer light is also 'endogenous' artificially producing colours and thus controlling wavelengths reaching the subject. The images produced by dreams and by the computer, I submit, are a result of a different order of consciousness based on an internal dynamic spawned by endogenous light. In neither states do the eyes respond to light and/or the image in the same way that they would if the eyes had first processed natural light and natural images that we see two-dimensionally and then translate into three. If we cannot yet prove that the brain is induced into a brainwave state of theta dream-like consciousness in a work of art such as *Osmose*, we can, I think, on the evidence we have of brain entrainment which utilises wavelengths to produce soothing alpha and or theta rhythms in the brain, state that something similar to this happens.

Inducing Instinctive Consciousness

The upshot of all this is a curious one. It seems that immersive computer art has a way of stimulating the visual system biochemically because, in the end, lightwaves are reduced to biochemicals that evoke a deep-rooted sense of archaic or instinctive consciousness similar to dream consciousness. Jung would have agreed because,curiously,this has to do with his astute claim, many years ago, that "instincts are archetypes". Recent research with animals who, unlike us, evidence theta rhythms *consciously* while engaged in predatory behaviour substantiate this Jungian insight. Researchers have confirmed that instinctive theta rhythm exists by comparing it to the animal's theta rhythms in the REM dream-state. This is done by carefully studying the response of the animal's neurons in a conscious state and in REM sleep.[5]

Unlike photic stimulation which purposefully attempts to control brainwaves, *Osmose*, I suspect, has innocently transgressed, with its flickering lights, pulsating sounds and use of subdued colour, into the realm of the 'making of consciousness'. In the case of this immersive work of art and the reports of hundreds of immersants, the results appear to have been beneficial in respect to the 'making of consciousness'. It elicits a dream-like effect in the subject that is capable of allowing herself/himself to transcend the logical, sequential-like progressions of the left-hemisphere and to enter into a *simultaneous* sense of consciousness. This is something that artists who do not use technological means are not always successful in doing, or sometimes, as with modern and postmodern artists, deliberately strive not to do because they want their art to be abstractly 'intellectual'.

Conclusion

Immersive computer art is, indeed, employing digital means that are being described as being post-biological. Yet, ironically, the computer it seems is turning out to be a 'consciousness-heightening' instrument in the hands of some talented artists. Immersive computer art handily reveals its complicity with underlying biochemical human nature, not only with the consciousness of human beings, but with cosmic consciousness itself.

Notes

1. The incident took place in December, 1997. More than 200 viewers with symptoms including blackouts, nausea and convulsions had to be kept at the hospital overnight.

— *Models* —

2. McRobert, L. 1996. 'Immersive Art and the Essence of Technology', in *Explorations, Journal for Adventurous Thought*, Fall 1996, Vol. 15, No.
3. Siever, D. 1997. 'The Rediscovery of Light and Sound Stimulation', excerpt found at: http://www.comptronic.com/history.htm
4. Simanonok, K. 1997.'A theory of physiologically functional Endogenous Light and a Proposed Mechanism for Consciousness", p.4, http://www.dcn.davis.ca.us/go/karl/consciousness.html
5. Winson, J. 1990. *The Mysteries of the Mind*, A *Scientific American* special issue, 1997, 'The Meaning of Dreams', pp. 62,63.

Laurie McRobert, Ph.D., is a freelance philosopher, associated with Thomas More Institute for Adult Education in Montreal, Quebec, Canada. A past director of the Institute, she is still a discussion leader. She has also taught philosophy at McGill University. Studying the nature of the dynamics of consciousness brought her first to consider the radical Evil of Holocaust. In recent years she has focused her attention on the iconic imagination, virtual reality and holographic dynamics. She is presently studying light, its effects on the visual system and hence consciousness. She has published many essays in books and articles in journals on these subjects. Some of the published work can be found at http://idnet.qc.ca/~mcrobert. E-mail mcrobert@intlaurentides.qc.ca

2 Memory

Casablanca and Men in Black
Consciousness, Remembering, and Forgetting
Michael Punt

'I remember my first date, not the girl or even her name but the movie. Not *Casablanca* but Sergei Eisenstein's 1924 film *Strike*. I remember the scene which obliged her to clutch my hand, I can play it over: a Bolshevik caught beneath a fallen door as a Cossack horse, shot from the victim's point of view, trampled it onto the man's chest, his eyes, shot from the horse's point of view, wide in terror as the breath was forced from him. I forget if we kissed, if we saw each other again; went on another date to another movie. Instead I remember the glamorous audience-earnest young men and women in black polo neck sweaters smoking in the foyer after the film-I remember the strike in 1912 broken by the brutal forces of the Tsar when Sergei was only 14 and remembered by him with crystal clarity 12 years later. I remember understanding montage editing, and that this history was told from the point of view of the masses and not the false continuity of individuals so favoured by the bourgeoisie. Sergei made sure I would remember only the heroic struggle of the workers. This movie wrote my history as it wrote the history of anonymous revolutionaries. This partial story as full of absences as presences helped me forget and become fully conscious of my own presence in the histories playing on the screen and off'.

Art, Technology, Consciousness-the coalition of these three terms in recent years has been inescapably linked with the proliferation of electronic communications networks, which blur the boundaries between history, memory and self. It is a common place that users of Muds and Moos apparently change their personal details with impunity. Virtual environments, according to many researchers, allow quite normal healthy people to construct new personal histories and live with them satisfactorily. A few keystrokes and mouse clicks are enough to launch a relatively complete other self. Multiple mailboxes also apparently allow for distinct identities which can provide the scope for 'acting out' different aspects of a single person. Again it is a received wisdom that people frequently adopt different genders and personas associated with different usernames. Less dramatically perhaps, institutional identities are often at variance with those expressed by the same individual when they are using email accounts established on private servers. It is as though the particular apparatus constructs the individual subjectivity of the user.

The idea established in Europe in the late 19th century by, among others, Sigmund

Freud, that unfettered desire meets social convention to produce a coherent (if fractured) character is now under strain from new means of social interaction. Electronic media, however, enlarge the possible varieties of voice since there is no ecriture, the personal is no longer regarded as necessary or is even expected. Moreover, the dominant critical discourses around Muds and Moos especially the 'acting out' personalities which in the normal circumstances of social intercourse would either be unsustainable or even socially transgressive. In this sense we are currently on the cusp of an historical rupture in which there is a discontinuity in the social deportment of the individual relative to orthodox psychoanalytical explanations. Whereas a brief survey of the biographies of the famous reveals the remarkable frequency of men leading double-even treble-lives, with mistresses and even established families in the same city, now the opportunities for acting out multiple personalities are much more socially accessible.

Unlike the Freudian idea of the ego, the multiple personalities facilitated by Muds and Moos emerge from vicarious experiences acquired through television, novels, cinema, etc. as well as personal memories and histories. The tendency has been criticised as encouraging a distorted view of the self in which the trivial is amplified and the significant trivialised. On the other hand, such a skewed self-image, it has been argued, can have a therapeutic effect by, for example, empowering the disenfranchised and giving them a sense of voice in their own affairs if not the actuality of the wider world. Adolescents can rehearse possible social and psychological solutions to problems in order to realise satisfactory outcomes, and for those for whom youth is a memory, electronic networks allow them to redeploy the problem-solving strategies of their teens. Such gains however, are not without costs. To envisage oneself as a multiple personality is to legitimise discontinuity in the very place where we have come to value continuity if not wholeness, that is in consciousness. How can such a split subject be reconciled with health? How can a false consciousness, an imagined relationship with reality which does not correspond with the reality of that relationship, be useful? Unless, of course, you happen to be an old fashioned Marxist theorist brought up in the 1970s in which case Muds and Moos are, as has been suspected all along, trivial distractions which are merely the extension of the ideological project as capitalism responds to new technologies of interaction.

Multiple personalities are thought by some mental health workers to be clinically significant. Primarily they are recognised as one of the consequences of repeated trauma and a means of defending the self against damaging experiences by building false memories. In some cases in later life, for one reason or another, these false memories become unsustainable and the individual loses personal coherence and becomes sick. In particular, in the last decade it has become common for patients to have their real memories recovered through therapy in order that a 'proper' reconciliation of the self with history can take place. With alarming frequency these memories, recovered with the help of MPD (Multiple Personality Disorder) therapists have concerned repeated sexual trauma and ritualistic satanic abuse in childhood. This is to such an extent that there is some suspicion that memories of abuse are cultivated and even implanted under hypnosis. In 1992 a grass roots response to the frequency of these allegations led to the foundation of the False Memory Syndrome Foundation. On the other hand, the increased frequency of recovered memory may simply reflect the permission that society grants from time to time to express one's inner self in particular ways. It is well known for example, that hysteria was axiomatically a female condition until the mid 19th century when men too were allowed to be hysterical and it became a generalised condition. It is equally well known that after about 1910 hardly anybody was diagnosed as hysterical, as other ways of expressing self were legitimised-for example, in neurasthenia-and that during the First World War soldiers were once again forbidden access to certain psychological conditions. Similarly, in more recent times, the range of discrete phobias-arachnophobia to xenophobia-gave way to the more generalised 'panic attack'. Like MPD, all of these conditions have their fashions and their professional advocates and institutionalised treatments.

For Ian Hacking, at stake in this high profile dispute about the existence of multiple personalities, recovered memories and false memories, is not so much the individual who suffers but a professional dispute over memory and psychotherapy. As he plots assiduously in *Rewriting the Soul: Multiple Personality and the Science of Memory*, mental conditions are historically a convergence of competing discourses which achieve some temporary stability in a social consensus. If there is a generally publicised condition which patients know about then they feel able to experience the symptoms associated with it. Through a variety of means, he argues, we acquire new understandings of the term sickness. In the struggle for power in the psychiatric branches of medicine, a well-publicised scepticism of Freudian psychoanalytic theories and therapies first gave advantage to pharmaceutical treatment (psychopharmacology). The considerable successes in treating the most severe conditions of schizophrenia and depression placed doubt on Freud's theories of mental illness based on a failure of the normal processes of the repression of desire through the interdiction of the social. With it came the collapse of psychoanalytic therapies. In this professional dispute, the intellectual investment in MPD and recovered memories restores some of the ground lost to pharmpsychotherapy by reintroducing the idea of the repressed.

Hacking suggests that the recent phenomena of Multiple Personality Disorders, virtually unheard of until a decade ago and now diagnosed in one case in 20, might be a direct result of late 19th century science of memory. Nineteenth century science incorporated culturally sanctioned expressions of the soul into its own rationalist discourses. On the side of science in this project was the rich enterprise of technology which too was rewriting the human and developing body prosthesis either to amplify muscular power-as for example bicycles, steam shovels and cranes, or to replace the scientific observer with apparently less subjective instrumentation such as the chronophotograph. Hacking's position is that in the face of this professional struggle, the very idea of soul and indeed the sick soul has been abandoned.

Machines, which could somehow store memories, were in great demand from both scientists and entrepreneurs in the 19th century. The railroad and telegraph facilitated access to global markets. To manage this corporate expansion and benefit from economies of scale, new kinds of data storage devices were required. In the latter half of the century the various versions of the phonograph and the cinematograph were developed alongside the photograph. Each of these apparatuses made claims to a superior human facility of recollection. The transience of sound and the ephemerality of the moment were subjected to technological transformations to make them permanent. The photograph, unlike the painting and print which were subject to the interpretative hand of the artist, was considered a superior device for recording the world as it was at a given moment. Its regulated mechanical and chemical processes offered a guarantee of a contractual link with the world that no artist could match. Underpinning these technologies was a bourgeois investment in an aggressive materialist realism which saw the potential of these devices to cheat death. It promised to recover from the dead a lasting presence which did not, like their flesh, wither and change. Almost from the outset of the diffusion of the technology, post-mortem photographs were in great demand. Photographers first manipulated corpses into allegorical poses from the great history paintings, but quickly discovered that a naturalised pose, usually of restful sleep or contemplation was preferred by clients who wanted realism. The phonograph and the cinematograph too were enlisted in the project of remembering the dead as though they were living, much to the disgust of many poets and artists who saw the conceit of realism and the reduction of the human being to a material presence. And as has been well noted, 19th century science, especially medicine, also intersected with literature in dynamic ways. Consequently, as Charcot, Bauer, Freud, etc. were developing a science of memory, their ideas about human recollection were entering the public domain through the romantic novel. For example, the causes of Madam Bovary's social transgression are attributed (through careful character delineation) as strong sexual urges combined with an excess of religious devotion. Notions of irrational behaviour and of amnesia became commonplace through the novel but as this became identified with malaise, so highly sophisticated machines appeared in amusement parks as white knuckle rides which temporarily sanctioned excess and the loss of self and disorientated the rider from reality. Paradoxically, these periodic excursions to the theme park with their institutionalised opportunities to forget oneself were intended to furnish unforgettable moments.

For the Freudian and MPD advocates we can never escape our own historical construction; at best we can come to terms with it. Implicit in this is our historical determination of what it means to remember. In 19th century practices of science, technology and entertainment, a consensus around the idea of memory built up which both depended on precise detail (usually of a materialist nature) and simultaneously demanded that substantial elements of experience were forgotten. The way in which we learnt to describe memory-or the symptoms of memory-meant that those forgotten areas of touch, smell, affect and for want of a better word, soul, were consigned not only to the technologically unrepresentable but also the repressed and unremembered. From the end of the 19th century personal memory, as for example in the recollection which begins this paper, is understood as authentic only if it acknowledges its own forgetfulness. Those recollections, are minutely catalogued in detail from other categories-from the morbid and fetishistic to the specialised professionalism of the spy and the card sharp. Real memories must incorporate forgetfulness. As new storage modes have developed, however, this has become formalised in the intellectual enterprise of what has become known as post-modernist histories, which insist on the partiality of institutionalised recollection. It is now acknowledged that constructive forgetfulness is necessary for the purposes of building a national identity-especially in coming to terms with the unspeakable atrocities of a nation; for example, the genocide of Tasmanian aboriginals, the holocaust or the horror of Vietnam. These can never be narratives of repression but in their place are the multiple personalities of individual histories, idiosyncratic perspectives, in the multifaceted project of post-modern history.

The effectiveness of post-modern histories cannot be denied. With them new generations are able to live with the perpetrators of horrific actions without necessarily understanding the 'causes' of them. But this historical methodology must increasingly confront its own repressed. The conceit of electronic communications is that nothing can ever be forgotten. Digital storage systems insist that even discarded trash in the wastebasket is recoverable. Electronic memories do not admit to forgetfulness and their recollections seem too real to be authentic, they form another class of memory and need interpreting. This places the user in an unsustainable schizophrenia: he needs to forget to remember but the new apparatuses which have emerged from the scientific and technological projects of the 19th century have impossible memories. The cinema and the amusement park built the new reality of the moment of its own consumption as in: 'do you remember when we saw *Casablanca*? The film was so touching that I can't remember where it was, but I loved the ice cream, wasn't it a great time', etc. Part of the cinema's enduring appeal is that it constructs a subject which can forget as well as remember. Muds and Moos and multiple email personas do not; everything is recoverable at a mouse click and without the aid of psychotherapy. The user cannot re-enact a Freudian scenario of recovery, repression and disavowal with a technology which will not forget; instead he must confront everything. If wholeness is still regarded as an important feature of the human condition, as the *Men in Black* insist, and if it is still an important aspiration for poets and artists as well, then perhaps only a sceptical distance from the seductions of technology might save us from the incipient sickness and schizophrenia of Muds and Moos and help us to use electronic media to rewrite the soul rather than invent new and impossible personalities.

Virtual Space and the Construction of Memory
Andrea Polli

STM, LTM and Chaos
In 1989 I was fortunate to work with George Lewis, a jazz trombonist and pioneer in human/computer musical improvisation. He counselled me in the creation of *Chaotic Systems in Musical Improvisation*. This project, programmed in IRCAM's Max software, was a system of improvisation for musician and computer based on the Lorenz attractor.

A lorenz system is described by the solution of three simultaneous differential equations:

$dx/dt = -ax + ay$
$dy/dt = -xz + rx - y$
$dx/dt = xy - bz$

In *Chaotic Systems in Musical Improvisation*, the Lorenz attractor took an initial value input from a midi keyboard and generated a series of X and Y coordinates. These values were used on a macro scale as time and duration values. That is, the program captured chunks of numerical data (midi information) in real time based on the notes played by a musician. The size of these 'chunks' and when they were captured was determined by the Lorenz attractor algorithm. The system would then play back these chunks in time using the same algorithm.

The resulting improvisation, which for performer and listener felt very much like an improvisation between two human musicians in performance, was inspired by Robert R. Snyder's work on memory in musical perception. In his book *Music and Memory: A Brief Introduction*, (1996), cognitive concepts are used to analyse musical structure. Short term memory (STM) including chunking and long term memory (LTM), including non-declarative, declarative, episodic and semantic memory are discussed at length in relationship to musical concepts such as rhythm and meter.

Snyder's analysis of musical structure based on short term memory (STM) is the basis of the design of *Chaotic Systems in Musical Improvisation*. Short term memory is the second stage in the memory chain, a temporary memory which holds its contents for 3 to12 seconds and has a limited capacity (5-9 items). This area of memory is the location of awareness of the present. In my system, groups of notes played by the live performer were selected using duration values within the limits of STM generated by the attractor, as were durations of delays (pauses) between these groups. What was unexpected and especially interesting about the system in performance was that the live performer was able to use his/her STM to anticipate and manipulate the reactions of the system. I believe that this predictability was possible because the Lorenz attractor was used to generate values. The attractor established a waxing and waning pattern that a live musician could anticipate and respond to.

Interior/Exterior

White Wall/Black Hole, shown in 1993 at Artemisia Gallery in Chicago, was one of the first pieces I created to address the issue of memory in gallery installation. The piece consisted of a flour-coated gallery wall shaken by live sound from the area outside the gallery using a police radio scanner. The flour on the wall slowly fell to the floor of the space owing to the vibrations of the sound. The flour-coated wall visually referred to a topographical map, changing slightly with each new bit of audio information.

Police radio in the piece acted on two forms of the listener's memory. On the one hand, the comprehension of speech is the basis of acoustic memory. Comprehension of

Installation view, The Twins

the text depended on the listener's short term phonological memory which serves as an 'articulatory loop' which helps preserve order and allows the listener time to process continuous streams of speech. The listener's long term semantic memory was also engaged, that part of the mind which contains schemas or generalisations about world order.

The Observatory was an international site specific collaborative project which took place in the last weeks of April 1997 in Vilnius, Lithuania. The observatory building, established in the centre of Vilnius by Tomas Zebrauskas in 1753, is one of the oldest observatories in Europe. In 1876 the western tower of the observatory burned and in 1883 the observatory was closed, inaccessible and almost forgotten until 1997. During the 18th century, the observatory was an international centre of scientific research with significant discoveries in the orbit of Mercury and the nature of light itself. In the present day, technology has made the observatory once again internationally significant. *The Observatory* is metaphoric, an interface created as a means to understand the world on a global scale.

In the observatory, through the use of robotics, light, interactive computer technology and human interaction, I designed a system in which a structured set of rules created a complex and unpredictable event. Complex patterns in black chalk were created by two performers controlling line-tracker robots. The cylindrical tower space served symbolically as interior mind space and in this performance the chalk marks left as a record of an event could be seen as a map of the mind.

Storage/Retrieval

> One explanation is that throughout one's lifetime, experiences with common objects are stored in permanent memory-not as singular instances, but as items organised around a central theme . . . We recognise and classify a variety of disparate objects (cups and saucers) as members of a class by rapidly comparing them with an 'idealised' image of the class . . . It is the idealised image, or prototype, of an object, person, feeling, or idea that is stored in our long-term memory

(Gilinsky, Alberta Steinman. *Mind and Brain, Principles of Neuropsychology* 1985).

Research into the concept of desire led me to consider that I along with many others have multiple layers of possessions. We have possessions in physical space and we have possessions in virtual space: images, sounds, and texts in analogue and digital media. Porcelain plates suspended on the walls of a space contained actual materials symbolic of personal desires. A cellular phone, for example, referred to protection, i.e. the idea of being untouchable; keys

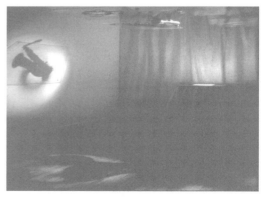

Installation view, Fetish

— *Memory* —

referred to power and control . Each material on the plates was photographed in its 'ideal' state, lit like a commercial product. Objects of desire in the virtual world exist in a visually heightened state to compensate for the lack of physical touch. Remote visitors could access the desires in the virtual world through the World Wide Web.

The idea of possessions in virtual space explored in this work, *Appetite*, installed in the summer of 1995 at Here in New York City, led me to the conscious realisation that virtual possessions are actually an integral part of non-digital life. Every human being has a storage bank of virtual possessions, memories. In fact, the computer storage bank is understood in human terms only through the metaphor of the memory.

Fetish, in 1996 at NAME Gallery in Chicago, further explored the issue of memory in virtual and physical space. The work consisted of twelve objects suspended over the heads of the viewers on a glass drop ceiling. A computer in the space provided a virtual replication of the objects. In positioning the objects, I was attempting to create a physical correlate to an emotional state. For example joy is experienced as a physical buoyance, and in contrast, grief is experienced as physical weight. When trying to remember, often humans will move their eyes up and to the side. I wanted to create a physical space that might refer to the virtual space in the mind when remembering events and objects. Certain events have prominence in the mind and the physical metaphor of size/importance was referred to in the space through oversized shadows, which are foggy reproductions of the actual event/object.

The objects were selected as signifiers of personal experiences. Viewers could access the computer using an interface sensitive to touch. A visitor could select each object to obtain a personal story related to the memory of the object and a sound which was used as a signifier of the emotional quality of the story. The stories were selected for their prominence in my personal database of memories and described in a way that left the reference to the object somewhat ambiguous and/or cross-referenced to more than one object. This structure referred to the fluid nature of the experience of a memory. One memory leads to another in unpredictable ways.

The sounds on the system were an effective means to evoke memories in the viewers. They were chosen for their familiarity: a door knock, a car door slamming, birds calling, ice clinking in a glass, etc.

May I Help You, *storefront view*

May I Help You, *second floor*

Another manifestation of this work was created for the Nylistafnid Museum in Reykjavik, Iceland . One glass plate, filled with wine, was placed above a large speaker which emitted the sound of a heart beat. With each beat, the surface of the wine would distort to a series of circular ripples. Light, bounced of that plate and onto the wall, created a pattern of motion that was reproduced on the wall. This work used material, sound and movement as a direct reference to an emotional state.

Fetish, May I Help You was a collaborative version of *Fetish* created in 1997 with students of Alfred University. The students collected their personal fetish objects: some were found and some were made and the students wrote short stories related to the objects and desires. These objects were then photographed and placed with the stories on an interactive CD-Rom. The objects were displayed in glass cases on the first floor of a storefront .

On the second floor, a dark 'cabaret' atmosphere was created with a projection of the interactive application and live performance. Visitors were invited to enter a tiny 'confessional' and record their private fetishes on video tape which was then broadcast onto the street of the town.

Since beginning this series, there had been a fluid exchange of objects in each work. Objects gain and lose importance in memory, and that is reflected in the objects evident in each work.

Motion and Perception

In Milan Kundera's *Slowness* (1997), the relationship between body motion and memory is discussed. Kundera describes a man walking. When he tries to recall something, his walking pace slows down. When he tries to forget, he speeds up the pace. New work attempts to change viewers' physical body perceptions and alter their expectations of control through inconsistencies in control and response in interactive installation.

Tight, shown in 1997, was a collaborative work between myself, Louise McKissick and Barbara Droth. An interactive computer application was displayed through an antique stereoscope. Sound, stories and images could be accessed through a standard mouse interface but the way the images were seen created a false sense of perspective. The application used stereoscopic 3-D modelled objects as well as

2-D graphics and video. The text was from a conversation with a stranger stored in memory.

The performative aspect of the exhibition consisted of two performers suspended from the ceiling of the gallery in climbing harnesses. Visitors to the gallery were then invited to suspend themselves as well. This process was my first experimentation in combining altered physical sensation with interactive media. This has recently been expanded to performance events.

The most recent project, *Gape*, was created using a simple eye tracking device created as part of the diploma project of An Reich at the Academy of Media Arts Cologne under the supervision of Dr Seigfried Zielinski and Phillipp Heidkamp (http://www.khm.uni-koeln.de/~an/imagery/). The device, which determines the position of dark or light pixels, uses input through a video capture card to control an interactive application. In *Gape*, a live performer uses the eye tracking device as a mode of communication. A grid of nine regions on a computer screen output the sound of eight words (one of the nine regions was inactive and used to create the effect of a pause or breath in the spoken words). The performer worked with the device and several sound processors to create a sound composition. At first the viewer believed that the performer was trying to speak a complete sentence but was unable to control her eye movements enough to tame the sensitive technological device; but then the viewer began to listen to the soundscape created by the overlapping words and to appreciate the complexity of the combination and repetition. The eight words I (You) (don't) want to be young (old) in combination created conflicting statements about the human body while the viewer watched a performer locked into an unmoving position, limited by the same technology she was controlling.

Conclusion

Art can be viewed as parallel to memory. Many of the same terms are used to describe the two. For example: art and memory both employ and integrate the senses, both are representations, and both refer to a sense of timelessness. Art can evoke memory and vice versa.

There are a number of metaphors in use today to help us understand how memory functions. I have concentrated on three major schemas in my work and used these schemas in the organisation of this article: the spatial metaphor, the computer database metaphor, and the temporal metaphor. None of these schemas completely define memory with all its complex and inexplicable behaviour. I have come to believe in the course of this research that at the present time there are many aspects of memory that, like art, are not quantifiable.

Andrea Polli is an independent artist and educator. She is currently chair of the Robert Morris College Institute of Art and Design where she has directed large scale interactive media projects such as the Great Chicago Cultural Center Adventure with the Chicago Department of Cultural Affairs and Live Live!, an international collaborative public art project with the Museum of Contemporary Art and the Chicago Transit Authority (http://www2.rmcil.edu/~live). She also teaches interactive multimedia at Columbia College and the School of the Art Institute of Chicago and has been a guest artist at Alfred University in New York and the University of Illinois at Chicago. She has performed and exhibited interactive work internationally, most recently at the Imagina 98 conference in Monaco.
http://homepage.interaccess.com/~apolli
apolli@interaccess.com

Memory Maps and the Nazca

Bruce Brown

I would like to discuss the notion of cultural identity in digital space. As communications networks are embedded within the global fabric of daily life, the notion that cultural identity is tied to geographical location, a nation or a village, is changing. It is likely that the geographical frontiers circumscribing cultural identities will give way to electronic boundaries. In the future, cultures and identities will exist in digital territory, across space and time, and so need specific strategies to help maintain social cohesion amongst a group of dispersed individuals so defining the electronic frontiers of their collective memory.

To address this I will first look back to the future and research that I started 20 years ago. Then, in 1978, I was fortunate to win major research awards that allowed me to study the Pre-Columbian civilisations of Peru. This enabled work in Lima, Cuzco, the Nazca desert and other sites at Sacsuahaman, Pachacamac and Tiahaunaco. My task had been to research the underlying design principles of beautiful objects and artifacts produced throughout many centuries of Pre-Columbian civilisations in Peru and how these principles had been transported through space and time, from one place and from one generation to the next, so ensuring both the social cohesion and cultural continuity.

Indeed, when in 1532 Francisco Pizarro rode into Peru and conquered the Inca empire he encountered a highly advanced civilisation. But, though the Inca empire was the most well known, it was by no means the only culture to flourish in Peru; others such as Chavin, Chimu, Chancay, Nazca, Tiahuanaco, Paracas and Moche all produced beautiful examples of textile weaving and design, ceramics and architecture. It is difficult to discuss the quality and grandeur of Peruvian designs without being suspected of exaggeration. For example, Peruvian craftsmen employed practically every known method of textile weaving including several special processes peculiar to Peru and perhaps impossible to produce by mechanical means. These achievements extended to ceramics, architecture and road building along with the arts of medicine and healing. Three aspects of these achievements first engaged me.

Firstly, when looked at dispassionately today, the design of all artifacts and buildings in Pre-Columbian civilisations seemed to be circumscribed by a cultural continuity that existed for over 2,000 years. The historian Victor Von Hagen observed that '. . . structures along a wide and lengthy geography show that we are dealing with a master plan. . .' the persistence of design throughout thousands of miles of varied terrain proves the point that "All the best known monuments of Inca architecture were constructed not by individuals but according to careful plans".

Secondly, and surprisingly, all of these high achievements were made without the invention of two very basic technologies, the wheel or any known system of writing. Both of these technologies for moving things (for transporting materials and knowledge, usually thought essential in the creation and conservation of any high culture) were absent from the Pre-Columbian civilisations of Peru. How could individuals, for example, have the knowledge to weave specific textiles with very complex structures and designs when there were no records to assist them; and how were they able to transport this knowledge across distances and through time from one generation to the next in order to ensure the continuity of their cultures?

Finally, and more specifically, I like many others wondered why Nazca cultures had inscribed gigantic drawings of animals on the desert pampas and lines that went dead straight across the land, then up hills and across valleys, for miles without deviation. Since they are only visible from an aeroplane, many people have wondered what motivated production of these gigantic lines and drawings in the desert and who they thought might have looked on them.

— Memory —

To cut a long story short, some of the answers to these puzzles lay in the extraordinary ability Pre-Columbians showed for weaving and the memory systems needed to support such work. The complicated textiles structures they evolved were, essentially, based on numerical pattern systems stored in memory.

But the Pre-Columbian cultures of Peru had no artificial memory technologies (such as writing) to assist them. Instead they invented a system of number patterns that could easily be held in their own biological memory systems. This was the digital mnemonic of their oral culture, its manifestation resting in a device called the Quipu. This a fibre construction with many strands and knots. Each Quipu being constructed of a main fibre spine from which ran a network of sub-fibres, each of these carrying an array of knots. These knots represented numbers and the numbers themselves represented knowledge stored in biological memory. So this was knowledge coded in digital form, stored in biological memory and transported by fibre from one place to another (via a Quipu).

So, official translators known as Quipucamayocs were trained to decode these memories stored in the digital code. And to transport this digital knowledge via the fibre of Quipus a spectacular road system was constructed. These superhighways ran dead straight throughout the empire. The coastal highways alone ran for 6,000 miles with such networks creating a spectacular information system that covered the whole empire. At two mile intervals along each highway post-houses were located. And each post-house housed a courier who, in relay, would transport the fibre Quipu onward, from one point to the next, until it had reached its destination to be decoded.

The transport of such knowledge into people's memories not only came via these official translators the Quipucamayocs but also through objects and images of daily life. Pots, ponchos, jewellery, furniture, statues and buildings were all designed to carry on their surfaces the symbols and stories of culture. And each design was given numeric proportions corresponding to the digital mnemonic upon which their culture was founded. So, on a daily basis people encountered objects that carried both the knowledge of their culture and the digital template on which it was built, this to be transported from the object itself into each person's landscape of mind so creating a collective memory of cultural identity and a social cohesion that worked over 2,000 miles and 2,000 years, the individual tribes of Peru gradually shaping a common cultural framework of mind that developed the most sophisticated empire of the old world.

The most dramatic example of this process can be seen in the gigantic drawings made by earlier Nazca cultures on their desert pampas. These enormous images of birds, animals and fish were not intended to be looked at by some other being. They were intended to be transported from the desert surface into the memory landscape of each person. A closer look will show that each drawing is made from one continuous line. Each a processional route to be walked, not to be looked at or seen. And the proportions of each line conformed to the digital mnemonic of all other structures in the culture. By walking over the surface of each symbol, the digital information it contained would be transported into the memory of each person with the images finally being held within the landscape of memory.

Though Pre-Columbian civilisations of Peru were similar to other oral cultures in that they harnessed biological memory to store information, what distinguished them from all others was the design of visual systems and languages intended to create a collective memory in the minds of people who were dispersed over vast tracts of territory. To achieve this they developed the skill to abstract things into digital form; these digital abstractions were then recorded as strings of information on fibre devices , the Quipus , that were transported via superhighways from one place to another.

This perhaps sounds like digital culture as we know it. We create abstractions of things, say photographs, as digital code, transfer them via networks to remote locations, then translate the code back into its original form, reconstituting the image in print.

But the difference rests in the use of memory technologies. In Pre-Columbian civilisations technologies were invented to virtualise knowledge from the world of objects into the landscape of biological memory. In our cultures technological invention has been to reactualise knowledge in biological memory back into the world of objects. The transport of information has gone in opposite directions and so we have come to elevate technological memory systems over biological ones.

Pre-Columbians never invented the technology of writing (though they presumably could have done so if necessary). Our invention of writing marked a shift in culture away from such oral traditions and their reliance on biological memory. One person saw the implications of such a shift, though, typically, Socrates committed nothing to writing himself, relying on his younger pupil Plato to do so. Famously, in the *Phaedrus*, Plato records Socrates' suspicion of writing which he believed would place outside the mind of each person that which should rightly be within it. And he has Socrates say in the *Phaedrus* that the effect of this upon us would be threefold:

1 we would lose our memories;
2 we would cease to be private individuals;
3 we would change the way we educated ourselves.

History has proven Socrates right on each of these counts, though the shift was gradual until around 500 years ago with the invention of printing from movable type. Preceding this, written manuscripts were produced at great expense in very limited editions, more often than not only one copy would exist of manuscripts such as *The Book of Kells*. And people would travel many miles to look upon its pages, transporting the images into memory for subsequent use throughout a lifetime. So manuscript cultures still operated from the premise that knowledge was to be virtualised into, and then re-actualised from, biological memory. In these cultures manuscripts were still considered as aids to biological memory rather than its substitute outside the mind; they were specifically designed to assist the process of virtualising things into memory in a way easy to later re-actualise.

This all changed with the invention of printing from movable type, the widespread distribution of standard texts and the growth of literacy. This started the suppression of visual language in our culture and the crucial role it had played in helping people of earlier times to virtualise things into their private landscapes of memory.

The invention of the printing press also saw the end of a tradition whereby objects and buildings were seen as texts. For example, it was no longer necessary for the side of a bowl or the wall of a church or cathedral to visualise a story from the Bible when it could be read in book form. So the role of objects and images as carriers of knowledge began to

— *Memory* —

decline. And this gradual loss of our capacity to handle visual language resulted in a progressive erosion of our ability to design and navigate biological memory.

Our technological culture has become supreme in its ability to externalise knowledge through the fragmentation, freezing, packaging and distribution of our memories outside the mind of each person-these being actualised and cloned in the mass produced artifacts of culture-its books, photographs, films, videotapes and CD Roms. These artifacts are no longer aids to memory but its substitute outside the mind of each person. This has eroded our control of visual language and our ability to design and navigate biological memory so causing us to forget the need to remember.

And as digital networks increasingly link people from remote locations in electronic space, each person will be the recipient of fragmentary information. This process of fragmentation and isolation could increase. But information is not knowledge and cultures exist through coherent self identities that connect people with others of a like mind. This turning of information into knowledge and the creation of coherent pictures from many fragments will only be achieved through the power of memory, not digital memory but the biological memory system located inside each person. As a vital resource, for the future we must resensitise ourselves to the power and flexibility of our own biological memory system ; we must repossess it and reestablish our natural ability to virtualise things into our own landscape of memory. In this we have much to learn from earlier oral cultures, such as the Pre-Columbian civilisations of Peru, and their need to design biological memory systems so to maintain social cohesion and create cultural identity through space and time.

In this sense Cyber Space is really Psychic Space, the inner landscape of biological memory made visible by technology and globalised. And tomorrow's cultures will constitute groups of individuals who, though they have no geographical coherence, will each share a collective memory defined by the electronic frontiers of their shared digital landscapes.

Professor Bruce Brown is Dean of the Faculty of Art, Design and Humanities at the University of Brighton. Educated at the Royal College of Art in London, he has worked as a graphic designer for many years, undertaking commissions for important clients. He was, for some time, art director of Crafts magazine and his work has been included in British Visual Communication Design 1900-1985 (published by 'Idea' in Japan). He has published and lectured widely, at the Pompidou Centre in Paris, in Chicago and most recently in Germany, Norway and Portugal on the theme of 'Graphic Memory'. In the past he has won research grants to study the Pre-Columbian civilisations of Peru and spent time in South America working in the field — in Lima, Cuzco and the Nazca desert where he researched the Nazca line drawings with the guidance of Maria Reiche. He has worked on many juries and committees in the past including the selection juries of the Designers and Art Directors Association and been invited to advise on issues of academic management by institutions in the UK and abroad, most recently in Israel, Japan, Hong Kong and New Zealand.

3   Transcendence

Parallel Worlds: Representing Consciousness at the Intersection of Art, Dissociation and Multidimensional Awareness[1]

Kristine Stiles

Traumatic dissociation, USA military projects in 'remote viewing,' and psychic phenomena have been extensively studied in terms of consciousness. Yet few researchers have noticed more than a passing correlation amongst them. In this speculative provisional report, I shall point to an overlooked interpenetrating arena of consciousness where dissociation and psychic phenomena may share common mechanisms in the production of mind.

    Dissociation is often described as 'the compartmentalisation of experience, identity, memory, perception, and motor function'.[2] In traumatic dissociation, aspects of consciousness are truncated from normative experience and memory, only to reappear in altered forms. Dissociation is frequently the key instrument for survival from trauma, providing a form of cognitive homeostasis (however fragile) that enables severely traumatised individuals to live with experiences otherwise too painful for consciousness to acknowledge. Dissociation includes derealisation, depersonalisation, amnesia, confusion and alterations in identity where various parts of the subsystems of mind 'disconnect in terms of information exchange or mutual control'.[3] This is especially true of dissociative identity disorder (DID)-formerly called multiple-personality disorder (MPD) - a psychological behaviour that is primarily attributed to childhood sexual abuse.[4] In DID, consciousness and identity integration is fractured and subsystems of mind appear able to communicate dissociatively across a terrain of consciousness. I am particularly interested in DID because of the high incidence of paranormal abilities reported in those suffering from it.[5]

    Researchers on the paranormal are familiar with the capacity of artists to test high in psychic abilities; the very process of producing art has sometimes been described in dissociative terms.[6] For example, the idea that artists 'lose' themselves in the work which comes to 'speak' for them is stated by Rimbaud, Nietzsche and Rilke, all of whom accounted for their artistic abilities as if they had been 'spoken by another'.[7] Moreover, there are 'complex relationships amongst hypnosis, dissociation, absorption, fantasy-proneness, somatization, and paranormal experiences' in both artists and traumatised people.[8] Indeed, it appears that a traumatised individual may be the best subject for testing of, and experimentation on, psychic abilities, especially if one observes closely the recruitment practices of the United States military for 'remote viewing', even though many developers of this relatively new field distance themselves from the stigma of the occult.

Ingo Swann, a remote viewer for the US Army Intelligence and Security Command programme (INSCOM c.1972-95), a CIA and DIA (Defence Intelligence Agency) project,[9] argues that the American intelligence community adapted the term remote viewing precisely because it was *not* a concept intended 'to replace the term "psychic" as a stereotyping label,' but rather because the CIA was interested in 'whether the bio-human possesses additional receptors for organizing information that exceeds the local limits of the five physical sense'.[10] Lyn Buchanan, also a former remote viewer, defines it as 'the structured, scientific use of natural human potential for the acquisition of real-world information without depending on the normal five senses, or equipment such as photographs, electronics or other devices, or logical deductions'.[11] But Joseph McMoneagle, a 17 year veteran of remote-viewing for the US government, admits that remote viewing is 'a method of identification . . .*knowing details about something which could only be known through extrasensory perception or ESP*'.[12] McMoneagle adds that remote viewing is "always done within scientific or approved research protocols'.[13]

What intrigues me about remote viewing is the fact that many recruits had themselves experienced trauma. McMoneagle was conscripted 'as a result of many unusual events in his life, a near-death experience (NDE), a UFO sighting, and numerous spontaneous out-of-body experiences'.[14] When asked whether trauma was integral to the ability to do remote viewing, McMoneagle replied in the affirmative, explaining that the US government recruited its remote viewers from a group of Vietnam veterans who had endured 'impossible situations during long-range reconnaissance missions where they were dropped many miles behind enemy lines and returned unharmed'. He added that the government reasoned, 'They could not have survived unless they were psychic'.[15] David Morehouse, another remote viewer for the army's elite Airborne Ranger company commanders, also had a near-death experience, visions and nightmares.[16] One of the students that Buchanan trained in remote viewing described experiencing something akin to dissociative-identity-disorder: '[I was] floating on a different personality foundation' and undergoing 'a slight-but-noticeable personality change while . . .working'.[17] When more documents about this project are declassified, it will be possible to scrutinise recruitment practices of the US government to determine the accuracy of these preliminary observations.[18]

I think that the ability to dissociate may be linked to psychic capacities through hyper-vigilance, a classic symptom of, and response to, trauma. Hyper-vigilance is a 'phenomenon representing excessive attention to external stimuli *beyond* that called for given a realistic appraisal of the level of external threat'.[19] Hyper-vigilance plays a vital role in protecting trauma victims from an environment of risk and is central to their sense of well-being. It may reflect not only a healthy but *normal* response to the exigencies of survival. Hyper-vigilance is also connected to the development of greater powers of concentration. Hypnotic and dissociative states have long been associated with unusual effects on the body. They are mental acts in which the 'cognitive resources are fully allocated to the central task [of perception], with little in the way of distraction, [becoming] vehicles for increased control over neurophysiological and peripheral somatic functions'.[20] For example, in McMoneagle's attempt to grapple with his newly-acquired psychic abilities after his near-death experience, he remembered: 'A psychologist suggested that coming out of the NDE made me more sensitive to other forms of detail', and he described this ability as 'spontaneous knowledge . . .[a] new-found psychic functioning'.[21] Dissociative hyper-vigilance literally may block out 'noise' interfering with a signal of focus, enabling consciousness to receive in remote viewing and psychic acts. 'Noise equates to inaccuracy', Swann has pointed out, and must be eliminated if perception in the 'bio-mind sensorium' and 'mental information processing grids' are to understand the signal.[22]

Evidence is beginning to suggest that the environment of the electronic interface and virtual reality are themselves encouraging, if not augmenting, conditions of consciousness associated with what was once considered anomalous states of mind. Roy Ascott has theorised this hybrid state as 'cyberception'.[23] Swann has described it as 'being WIRED into alternative realities, cross-dimensions and multidimensional awareness'.[24] My hypothesis is that hyper-vigilance may well be a prime feature in the connection between trauma and the capacity for multidimensional forms of awareness, which may also account for why both meditation common to Eastern philosophy and the development of Western technologies that enhance concentration are increasingly important to such processes as remote viewing.[25] Telematic culture may contribute to *both* dissociative abilities and expanded consciousness. But with its exponential growth, the grasp of these otherwise evasive forms of cognition may again be crowded out by the increasingly redundant and meaningless 'noise' of electronic media.

Even as dissociative and multidimensional modes of consciousness continue to evade scientific method, they parallel the behaviour of matter described by many physicists. For example, David Albert, a professor of theoretical physics and sub-atomic particles at Columbia University, in collaboration with Bryan Loewer, a philosopher from Rutgers University, discussed conditions of non-locality in the 'the micro-world of sub-atomic particles'. Describing the behaviour of electrons in 'super-position', they stated that an 'electron is located somewhere, but it is not located in any particular place', a behaviour that is 'paradoxical and strange'.[26] If the properties of an electron in super-position are thought of metaphorically as a question of colour, they note, then to say that it is either blue or green would be false. For in super-position electrons require an entirely different mode of thinking, a model that would identity them as 'nor both nor neither'. Albert and Loewer add: 'What is going on here is that we are being forced to confront odd (and metaphysically odd) claims about the structure of the world by the mathematical structure of our best scientific theory of how the world operates'. Similarly, the Princeton Engineering Anomalies Research (PEAR) programme calls for a 'generously expanded model of reality, one that allows consciousness a proactive role in the establishment of its experience of the physical world', and one that 'regards many of the concepts of observational quantum mechanics, most importantly the principles of complementarity and wave mechanical resonance, as fundamental characteristics of consciousness, rather than as intrinsic features of an objective physical environment'.[27]

I suggest that traumatic dissociation and hyper-vigilance may become processes that filter mental 'noise', thereby enabling consciousness to exercise multidimensional modes of awareness. As the parallel worlds of mind and matter both appear to function paradoxically, trauma may be key to understanding conditions of consciousness that functions as 'nor both nor neither'.[28]

Notes

1. Aspects of this paper have been presented in two longer unpublished versions at 'The Incident: An International Symposium to Examine Art, Technology and Phenomena', organized by Rob Le Frenais and Kathleen Rogers in Fribourg, Switzerland, 1995, and at 'From Energy to Information: Representation in Science, Art, and Literature', an international conference organised by Linda Dalrymple Henderson and Bruce Clark, sponsored by the Center for the Study of Modernism and the Center for Interdisciplinary and Digital Arts Studies at the University of Texas at Austin, 1997 (where my paper was cut from the published proceedings).
2. Spiegel, D. 1994. 'Introduction', In Spiegel, D., ed. *Dissociation: Culture, Mind, and Body*. Washington, DC and London: American Psychiatric Press, Inc., p. ix.
3. Erdelyi, M.H. 1994. 'Dissociation, Defense, and the Unconscious'. In *Dissociation*, p. 3.
4. Braun, B.G. 1993.' Multiple Personality Disorder and Posttraumatic Stress Disorder: Similarities and Differences'. In Wilson, J.P. and Raphael, B., eds. *The International Handbook of Traumatic Stress Syndromes*. New York and London: Plenum Press, pp. 35-6, 583.
5. I began to notice these connections over two decades of research on Performance Art where traumatic and dissociative representations and behaviours are so regularly displayed that I have theorized dissociation as the phenomenological centre of this medium. Many performance artists are interested in, or state that they possess, various forms of non-local consciousness traditionally associated with psychic phenomena. short list includes: Marina Abramovic, Joseph Beuys, Michel Journiac, Paul McCarthy, Linda Montano, RaphaelMontanez Ortiz, Ulricke Rosenbach and Carolee Schneemann. On these subjects, see my, 'Uncorrupted Joy:International Art Actions', in *Out of Actions: Between Performance and The Object, 1949-1979*. Los Angeles: Los Angeles Museum of Contemporary Art, 1998, pp. 226-328,and my 'Shaved Heads and Marked Bodies: Representations from Cultures of Trauma' (1993), reprinted in O'Barr, J., Hewitt, N., and Rosebaugh, N., eds.1995. *Talking Gender: Public Images, Personal Journeys, and Political Critiques*. Chapel Hill: University of North Carolina Press, pp. 36-64.
6. Bem, D.J and Honorton, C. 'Does Psi Exist? Replicable Evidence for an Anomalous Process of Information Transfer', in *Psychological Bulletin* 115:1:4: pp. 14-18.
7. Rimbaud observed, '*Je est un autre*'; Nietzsche's '*es denk*'" names something '*thinking* in' him; and Rilke wrote, 'Where there is a poem, it is not mine but that of Orpheus who comes to sing it'. In de Certeau, M. 1988. *The Writing of History*. Trans. Tom Conley. New York: Columbia University Press, p. 257.
8. Ross, C.A. 1994.' Dissociation and Physical Illness', in *Dissociation*, p. 173.
9. INSCOM'S remote viewing project began in the early 1970s and ended in 1995. See McMoneagle, J. 1997. *Mind Trek: Exploring Consciousness, Time, and Space Through REMOTE VIEWING*. Charlottesville, Va.: Hampton Roads Publishing Company, Inc.
10. Swann, I. 1995. The emergence of Project SCANATE the First Espionage-Worthy Remote Viewing Experiment Requested by the CIA (Summer 1973). In http://www.ameritel.net/lusers/rviewer/ingo07.htm
11. Buchanan, L. 1996. 'Controlled Remote Viewing: A Seminar and Mini-Workshop',8 September , 1996, in http://idt.net/~jwc/cvi1.htm/
12. McMoneagle, *Mind Trek*, p. 22.
13. ibid, p.21
14. Inside back book jacket of McMoneagle's *Mind Trck*.
15. McMoneagle in conversation with the author 22 June 1998. McMoneagle also stated that he grew up in a dangerous area of Miami where he had learned special survival skills.

16. Morehouse, D. 1996. *Psychic Warrior: Inside the CIA's Stargate Program: The True Story of a Soldier's Espionage and Awakening*. London: St. Martin's Press.
17. Turell, S. A. 'Write Up'. In http://www.amritel.net/lusers/rviewer/skye.htm
18. LSD and other mind-expanding drugs are well-known as vehicles for opening consciousness. Some suggest that drugs enhance remote viewing. See Constantine, A. 1995. *Psychic Dictatorship in the U.S.A*. New York: New York: Ferel House. Constantine wrote a chilling account of the mind-controlling aspects of remote-viewing projects, and claims that many of its founders (Harold Puthoff and Ingo Swann) came from the Church of Scientology. See, Constantine, A. 1995. 'The Constantine Report No. 1: "Remote Viewing" at Stanford Research Institute or Illicit CIA Mind Control Experimentation?' in http://www.redshift.com/~wmason/inreport/articles/const1.htm/
19. Weiss, D.S. 1993. 'Structured Clinical Interview Techniques', in *International Handbook of Traumatic Stress Syndromes*, p. 183.
20. Spiegel, D. and Vermutten, D. 1997. 'Physiological Correlates of Hypnosis and Dissociation', in *Dissociation*, pp. 185, 186, 202.
21. McMoneagle, *Mind Trek*, p. 37.
22. Swann. I. 1996. 'Remote Viewing: The Signal-to-Noise Ration, a.k.a.' "The Noisy Mind — Dirty Data" Issue. In http://www.ameritel.net/lusers/rviewer/stereo1.htm
23. Ascott, R. 1994. 'The Architecture of Cyberception'. In http://www.sat.qc.ca/~sat/isea/intersoc/sym/archives/isea94/pr413.html
24. With several others, Swann coined the term at the American Society for Psychical Research in 1971. Swann, I., 1995. 'Hey, Guys! What Are We Talking About?' In http://www.ameritel.net/lusers/rviewer/ingo02.htm
25. See the 'Hemi-Sync' Audio Technology, an auditory guidance system manipulating Theta brainwaves developed and improved by The Monroe Institute (also associated with the US government's remote viewing projects).
26. Albert, D. and Loewer, B, 1989. *Physics and Philosophy with Wayne Pond on Soundings, National Public Radio. 1989*. All quotes by Albert and Loewer are from this conversation unless otherwise cited.
27. Statement by Princeton Engineering Anomalies Research: Scientific Study of Consciousness-Related Physical Phenomena. In http://www.princeton.edu/~pear/
28. Similarly, recently researchers have drawn credible links between chaos theory and epileptic seizures. See 'In Brief', in *Scientific American* 274: 2 (August 1998), p. 26.

Sacred Art in a Digital Era *or* The Internet and the Immanent Place in the Heart

Niranjan Rajah and Raman Srinivasan

With my body become His temple,
my sense driven mind His willing slave,

With truth for purity
and my one-pointed mind the Deity within;

– Transcendence –

With my love as offerings for his sacred ablution,

I have performed a virtual 'pooja' for 'Ishwara'.

Thirunavukarasar [1]

Post-industrial Reality
As the millennium approaches, the materialist trajectory of Western civilisation seems to have arrived, perhaps unwittingly, at a technological threshold, one that promises to yield an ontological evolution. In the modern conception of reality, based on the empirical confirmation of the sensible realm, the world was understood in terms of its material form. Today, the post-modern, post-industrial 'information' technologies that form our interface with the world multi-user dimensions, telematic communications, immersive virtual environments and bio-electrical interfaces,are leading us to a less materialistic conception of reality and even of our very human 'being'.

Multiple Identities
In his keynote address to the Seventh International Symposium on Electronic Art, Jos de Mul (1996, p3) applied a phrase from Michel Foucault's 'Technologies of the Self' to the discourse on virtual communitiy. He proposed 'that the computer is a technology of self which permits individuals to affect on their own bodies and souls, thoughts, conducts and ways of being, so as to transform themselves in order to attain a certain state of happiness, purity, wisdom, perfection, or immortality'. De Mul foresees a future in which human beings will realise them'selves' as multiple identities. What is today approached as pathology will, in the future, become the norm in the construction of the self. Indeed, the forerunners of this ontology are the innumerable *'avatars'* that inhabit the multi user dimensions of today's Internet.

The Cyber Avatar
The 'gurus' of virtual reality have , of course, borrowed from religious ideas and terminology in order to produce their secular 'theology' of cyberspace. The cyber 'avatar' is the epitome of the confusion that prevails in this technocentric theosophy. To Hindus, an *avatar* is a manifestation of God eternal in a particular time and space. For instance, Krishna is a particular form of Vishnu, of whose sacred 'principle' there have been and will be other *avatarams*. All these manifestations are known to belong to the realm of physical and psycological being-*maya*. The cyber avatar is an illusory persona, a second order projection, emanating from the realm of *maya* itself. Conferring upon these projections a sense of being, however disembodied, and then applying sacred Hindu terminology to index them is sheer delusion or, in Hindu terms, *moha*.

Traditional Ontology
In Hindu philosophy the corporeal sphere of extension, *maya*, is a relative reality that is not best understood by its material forms and objects but in terms of the 'manner of its operation'. As Ananda Coomaraswamy (1935, p58) observes, '*Maya* is not properly

delusion, but strictly speaking creative power, *sakti*, the principle of manifestation'. Ultimate reality or *satya* is the eternal principle that generates and sustains the derivative realm of spatial and temporal phenomena that constitutes the empirical world. *Satya* 'subsists there where the intelligible and sensible meet in the common unity of being . . . as knowledge or vision, that is, only in act' (Coomaraswamy 1935, p.11). The role of art in this sacred paradigm is to symbolise the spiritual 'action' immanent in the world.

Sacred Art
Sacred art is a technology or 'calculus' with which to transcribe universal truths. As Titus Burckhard (1967, pp7-10) insists, 'No art merits that epithet unless its forms themselves reflect the spiritual vision characteristic of a particular religion . . . in a rigorous analogy of form and spirt.' Indeed, sacred art imitates the 'Divine Art' of creation.

'What must be copied is the way in which the divine spirit works. Its laws must be transposed into . . . artisanship.' This sacred art does not seek to evoke feelings or communicate impressions. It generates symbols for 'supra-formal spiritual realites'. Its purpose is to demonstrate the 'symbolical nature of the world . . ', thereby 'delivering the human spirit from its attachment to crude and ephemeral 'facts'.

The Hindu Temple
In the Vedic 'Hymn of the Cosmic Man' or *Purusasukta*, the gods sacrifice the giant *Purusa* to create the physical universe. '*Purusa* is this all, that has been and that will be . . . From his navel was produced the air; from his head the sky was evolved; from his feet the earth...' (Gombrich 1975, p115). The human body is the source or, in linguistic terms, the root metaphor for the universe, which in turn, is what is 'modelled' or represented in the Hindu temple. As Bruckhardt (1967, p17) observes, 'that which is in ceaseless movement within the universe is transposed by sacred architecture into permanent form'. Just as the 'Vedic sacrificer identifies himself spiritually with the altar, which he builds to the measure of his body . . . the architect of the temple identifies with the building and with that which it represents . . . ', conferring 'upon his work something of his own vital force' (Bruckhardt 1967, p33).

King Kadavaraja's Temple
The great King Kadavaraja decided to build the grandest temple to Lord Shiva in Kanchipuram.[2] He commissioned skilled craftsmen from all over the country to build this offering to the Lord. Every day the king would come to the site and offer his own physical labour to raise the temple. In due course, the temple was built and crowds upon crowds came to see this regal temple. And soon the temple was near completion and the king set a date for the ceremonial installation of the deity in the temple. The astrologers were consulted, the date decided and the announcements made. But then one night the king, woke up with a start. He had had a dream. Lord Shiva had appeared and asked to be excused from the ceremonial installation at the king's temple. 'Why?' the king had asked. Shiva said that he had to be present at the inauguration of yet another temple nearby.

Poosalar's Virtual Architecture

'Who has built that temple?' asked the king. 'Poosalar,' said Shiva and disappeared. The next morning, the king began searching for the temple of Shiva built by Poosalar. His men were sent out on fast horses in all the four directions. No signs of a grand temple. Finally one of the men came back with the news that in a distant village there lived an old man called Poosalar. The king rushed to the village and saw no temple. Finally he made his way to the hut of Poosalar and asked him, 'Tell me where is the temple you have built?' Poosalar, a little nervous on seeing the mighty king, said, 'Sire, I have built no temple. I am no King. Only a poor Brahmin. The only temple I have built, brick by brick, is in my imagination.' The king understood and fell at Pooslar's feet humbled by the devotion of Poosalar.[3]

http://www.sarawak.com.my/temple

A VRML temple[4] has been 'erected' in the Internet based on an existing physical structure.[5] Deities have been placed in this temple and soon it will go on-line, inviting virtual pilgrimage. Ultimately a temple is a sanctuary for 'divine being' and as Burchardt (1967, p17) observes, 'Spiritually speaking, a sanctuary is always at the centre of the world'. The Internet is a network of computer networks connected by Transmission Control Protocol/Internet Protocol communication. This protocol has the ability to interconnect networks that use different local protocols, while also allowing networks linked by other internetworking protocols to connect with the ubiquitious metanetwork. (Glister , p22). Indeed, the 'centre' is everywhere in the virtual geography of the Internet.

Hindu Sculpture

In the Hindu tradition, sculptural form is realised by way of meditation or *Yoga*. As Ananda Coomaraswamy (1935, pp5-6) explains, the image of a *devata*, latent in canonical prescription, is inwardly visualised by the icon maker, or *stapthi*, in a meditative act of 'non-differentiation'. From this inner image he then proceeds to execute the sculpture in a chosen material. In the great medieval bronzes of the Cholas, the 'lost wax' method was used and this tradition is still alive today. In an enlightened comparison, Peter Gardener (1982, p478) notes that 'the Hindu visual arts bear more resemblance to the performance arts of the West than to its visual arts'. The textual codification of Hindu iconography is analogous to the score in Western music. The sculptural image is realised in an interpretive 'performance' of sacred text in which the performer imparts his 'vital force'.

Ganesha! [6]

A master *stapthi* will be invited to produce a wax model of Lord Ganesha in the traditional meditative method, with all the attendant rituals intact. The topological coordinates of this model will be mapped digitally and the deity finally rendered in VRML. The only departure from traditional sculpture will be the incorporeal realisation of the icon no longer cast in bronze but in bytes. The finished Icon of Lord Ganesha will then be installed on-line. If there are no theological objections, the deity will be consecrated by a priest. A Cuseeme reflector will be set up to serve as the Lord's temple. There will be regular on-line 'poojas' during which the networked congregation will worship in virtual communion. [7]

The Immanent Place In The Heart

In the proper adoration of a deity the physical or 'gross' form or the icon is echoed in an analogous inner or 'subtle' image, in whose radiance 'harmony or unity of consciousness' is achieved. Just as Poosalar had done, the worshipper applies his or her own 'imaginative energy' to its image, 'realising' the *devata* within the 'immanent space in the heart'

(Coomaraswamy 1935, p26). This ancient understanding of the relativity of realities might help us comprehend more completely the implications of the Internet for human consciousness. Indeed, it is crucial that computer mediated communication is not construed, simply, as the technological manifestation of post-modern theories, and that virtual reality is not understood as the metaphysical backlash of a materialistic Western civilisation. As Internet technology becomes universally available, its potential must be realised in terms of a multiplicity of cultural paradigms, particularly in terms of the living sacred paradigms of the East.

Notes

1. Thirunavukarasar is a Saivite saint from the 7th century ad. The Tamil verse was remembered and translated by Sathiavathy Deva Rajah (author's mother).
2. Kadavaraja of this legend may be the Pallava King Rajasimha (Narasimhavarman II) who built the great Kailasanatha temple at Kanchipuram (Coomaraswamy, 1965,p101-4)in the 8th century ad.
3. The story of the virtual temple of Poosalar as told and retold to Raman Srinivasan by Saraswathi Raman (author's mother) and Dr.Karan Singh and confirmed in Cekkilar's (1993, p616-17) 12th century narrative poem called the *Periya Puranam* or Great Legend.
4. Before returning to Madras from the USA in the winter of 1994, Raman Srinivasan had been 'visiting' a series of dream temples and writing devotional poems. At that time his central desire was to build a virtual temple and when he joined Orpheus Multimedia in 1997, one of the first things they did was to embark on building a VRML temple.
5. Email from Srinivasan to Niranjan. Sat, 28 Mar 1998 08:28:42 +0000: 'Let me tell you what has been happening here. My American family from Philadelphia are here in Madras to attend my wedding. They wanted to see as much of India as possible in ten days. So two days ago, we drove to a small obscure temple in the village of Tirunindrayur about 20 miles from Madras. We chose this temple, because of our staff artist, Prabhakar. Tirunindrayur is his home village and he had been doing sketches of it for our virtual temple. We had driven for an hour and a half and arrived at the temple at the most sacred hour of Lord Shiva, *Pradhosham*. As I enter the temple, I see the priest approach us, and after saluting him, I ask him for the *Sthala Purana* or history of the temple. He says, much to my shocked delight, that this is the temple that was imagined by Poosalar! Unknowingly, entirely in ignorance, we had been sketching the temple of Poosalar for our virtual temple'.
6. The Ganesha! project was concieved by Niranjan Rajah and finalised after discussions that followed the presentation of 'Prosthetics For The Mind' at 'Consciousness Reframed' 1997. One of the questions asked was why not 'sculpt' the icon directly in digital media. It is indeed central to this project that the transmission of sacred form from text to the 'heart' of the worshipper, is achieved with the traditional visualisation intact. It is also essential that the 'subtle' imaging technologies are developed in the new media.

7   Email from Srinivasan to Niranjan, Sat, 28 Mar 1998 12:19:43 +0000: 'I want to alert you- the Ganesha section looks difficult this year. Ganesh is still not pleased with us and I think we need to do the this part with greater seriousness of intent next year. Somehow Lord Ganesha is placing obstacles'.

References

Burckhard, T. 1967. *Sacred Art in East and West*, Middlesex: Perennial Books Ltd.
Cekkilar. 1993. *Periya Puranam. Arumukanavalar* (ed.) Kuala Lumpur: Saiva Sithantha Nilayam.
Coomaraswamy, A. 1965. *History of Indian and Indonesian Art*. New York: Dover Publications, Inc.
Coomaraswamy, A. 1935. *The Transformation of Nature in Art.*Cambridge, Massachusetts: Harvard University Press.
De Mul, J. 1996. 'Networked Identities', In Roetto M. B., (ed.) Seventh International Symposium on Electronic Art Proceedings. Rotterdam: ISEA Foundation.
Gardener, P. 1982. 'Creative Performance in South Indian Sculpture: An Ethnographic Approach', In Onias, J. (ed.) *Art History, Journal of the Association of Art*
Historians. Vol. 5, no. 4. December Issue.
Gilster, P. 1995.*The New Internet Navigator*. New York: Wiley.
Gombrich, R. 1975. 'Ancient Indian Cosmology', in Carmen, B.and Michael, L. (ed.) *Ancient Cosmologies*. London:George Allen & Unwin Ltd.

Niranjan is an Internet artist and theorist. He has an MA in Fine Art from Goldsmiths College, University of London. He lectures in Art History and Theory at the Faculty of Applied and Creative Arts, Universiti Malaysia Sarawak. Niranjan is concerned with the relationships between post-modern theory, traditional metaphysics and the new ontology engendered by Internet technology. Please visit 'La folie de la Peinture' at http//www.kunstseiten.de/installation/   Email: niranjan@faca.unimas.my
Srinivasan is an elephant trainer by imagination and an engineer by training. He also holds a doctorate in History and Sociology of Science from the University of Pennsylvania. He currently heads the Madras Foundation at Chennai. Prior to this, he worked at the Rockefeller Foundation. Srinivasan is also on the board of Orpheus Multimedia, a firm specialising in visual simulation and modelling. Email: srini@mf.org

Negotiating New Systems of Perception: Darshan, Diegesis and Beyond

Margot Lovejoy and Preminda Jacob

At the end of the optical age when images produced in the visual field were the privileged measure of reality-and at the beginning of an age when interactive information systems will be relied upon more than light for communicating images-we need to reexamine theories of perception and vision we have been relying on for more than five centuries.
   Clearly the new medium of computer technology increasingly extends our perception to broader territory where we are forced to deal with greater contextualisation of information and complexity of memory. The artist now becomes a designer of interactive navigational concepts and choices of directions; simultaneously the viewer, positioned on the interstice

between intent (the artist's programming) and interpretation (the interactive, digital artwork) becomes an explorer and participant-instead of a passive onlooker. The new pluralistic condition of interactivity moves us away from the subject-centred concept of passive/receptive positions to the rebirth of active/receptive modes.

As we struggle to find theoretical frameworks that will enable us to understand the new systems for perception becoming available to us, we might profitably (re)turn to concepts about visuality (that is, the socio-historical dimensions of vision) developed in contexts culturally and /or historically distinct from our present moment. We believe this cross-cultural, trans-historical montage of concepts is in keeping with the increased awareness of global cultural exchange today that fosters negotiation of extremely disparate systems of perception and communication.

The two theories of interactive seeing we discuss in this paper are '*darshan*' and 'diegesis'. The former is a spiritual experience through the effect of the gaze in Hindu practice and philosophy and the latter refers to a mediated experience through technological construction of reality which can bring about suspension of disbelief. In both *darshan* and diegesis the participant, the locus of an intense experience, may surrender human agency and experience overpowering feelings and loss of control. Moreover, both experiences occur through a form of seeing that is deeply interactive and which may transform consciousness. Extrapolating from these insights we suggest that the experience of the participant-spectator in these contexts illuminates, in a unique way, the positioning of the viewer in an interactive digital media artwork.

Darshan, or the process of viewing an iconic image or superior being, is an everyday practice in Hinduism. Although the roots of this concept can be traced to Vedic philosophy of circa 1500 BCE, the practice of *darshan* and associated ideas actively informs modes of visuality in the new media-dominated contexts of contemporary India. The second concept we discuss, diegesis, is a Western theory that references the technological image structure of the cinematic medium in the modern era.

Perhaps the most common and ubiquitous practice of Hindu religious ritual, *darshan*, which translates literally as 'seeing,' can be defined as the act of exchanging gazes with a divinity. The power given and received occurs on an individual basis and it occurs through the gaze. Further, *darshan* is also imagined to be a way of touching, of making actual contact solely through the gaze. And finally, by absorbing, in this manner, some of the superior powers of deity, *darshan* is a way of knowing, of arriving at a superior state of consciousness.

This concept is interesting because it is based on a theory of the eye as an active transmitter rather than a passive receptor. In this and related concepts, the eye is the locus of tremendous power and energy that radiates outwards in all directions. This energy can also be focused on a specific target with beneficial or destructive effects.

Indeed, this belief in the extramission capabilities of the eye was the most widely accepted theory about vision in Western culture until the 18th century. (Brennan,1996). However, with the ascendancy of the rationalist mentality of the Enlightenment era, this understanding about vision was largely discredited as scientifically unverifiable. Divested of its power of extramission, the eye was reduced to a passive receptor much like the hole in the *camera obscura* through which light travels.

– Transcendence –

The practice of *darshan* habitually extends from spiritual to secular realms. Moreover, electronic media, rather than curtailing the practice has multiplied the avenues for *darshan*. Documenting the contemporary practice of *darshan*, anthropologist Lawrence Babb discusses two religious movements, the first dating back to the mid 19th century and the second to the early 20th century, wherein the eyes are a chief instrument in attaining a higher state of consciousness (Babb,1981). The relation between the devotee and the deity or guru (teacher, superior being) pivots on seeing and being seen. As Babb describes, in such instances, the visual sense is trained to function as a physical manifestation or extension of the mental and spiritual state of an individual. The process by which this acquisition of power and knowledge occurs is circular. The individual who gazes upon the icon or spiritual teacher has to fully acknowledge the superior powers of the latter. The individual gazes on the deity in adoration and the deity gazes back in benevolence. By gazing in the right way the individual can begin to absorb the power of the deity-in other words, the individual acquires the capacity to see him/herself as the deity sees him/her. One goes outside oneself to see oneself-a means of self-realization and hence transformation.

In our opinion, there are unmistakable parallels between the philosophy of *darshan* as it is practised in contemporary India and theorisations of the gaze in contemporary critical theory generated in the West. In the post-Enlightenment era in the West, the most influential theories about gazing have been based in Freudian and Lacanian psychoanalysis. Until recently, theorists of the new media have focused on the hegemonic, binaristic aspect of the gaze which was characterised as voyeuristic, fetishistic and objectifying. Such theorisations of perception dismiss any interaction between the viewer who is unequivocally fixed in the position of a spectator, and the viewed, who is, for all intents and purposes, essentially inert.

Recent scholarship on Lacan, however, pays closer attention to his later lectures wherein discussions about the gaze shifts to a more complex, triangulated interaction of subject, other and community (Melville,1996). Scholars trace this shift to Lacan's interest in Merleau-Ponty's phenomenological theories about the invisible realm that pervades the visible world as a constant, watchful presence. However, Lacan interprets the invisible presence not as a fixed, essential gaze but as a mutating, unstable gaze of culture and community. In Lacan's later theories, then, the moment of gazing changes from a hegemonic, voyeuristic, guilt-ridden moment to a moment of self-realisation. By looking out at others, one looks back at/into oneself in the way one is seen by others. In this version not only has the gaze become more interactive but the fixed positions that subject and object occupied in earlier discussions are now destabilised as the viewer becomes both the subject as well as the object of the gaze. Similarly, that which is viewed is no longer rendered powerless and inanimate,but actively gazes back.

Clearly the emphasis of both theories, *darshan* and Lacanian psychonanlysis, is on interaction and the fluid or unstable positioning of subject and object. Moreover, in both theories the practice of gazing or seeing extends to knowing, self-realisation and empowerment.

Film relies on the well-known phenomenon of 'persistence of vision' (where the eye fills in the blanks between images shown at 24 frames per second) to create the illusion of reality. It also relies on how moving images are constructed-film as a constructed language consonant with the idea of images acting as phrases and sentences to create a narrative image structure. The construction of a film's represented instance is termed its diegesis. The narrative structure or diegesis of a film is usually designed to cause an intense mediated experience which can bring about 'suspension of disbelief.'

The flow of moving images permits analysis from different points of view. Through slow motion, rapid panning, gliding close-ups, and zooming to distant views, space, time, illusion and reality fuse in a new way, representing a deepening of perception by revealing entirely new structures of a subject than is available to the naked eye. With film, the individual usually becomes part of a collective audience whose attention is controlled by the images moving in time which cannot be arrested. These constant sequenced changes of images are designed to affect emotions and dominate the thinking process through juxtaposition of edited views of subjects, close-ups and distant views, in edited, timed sequences.

Eisenstein, in his pioneering editing experiments, regarded montage as an essential structuring principle where meaning could be found to exist in the collision of disparate images to create a dynamic, active form of viewing. Montage involved an active spectator whose understanding could be affected as a result of this new form of visual juxtaposition. Though transposing vivid alternate views of reality, Eisenstein sought to transform consciousness of his viewer through active viewing-by stimulating thought as well as emotion to make it a 'carrier of meaning'. Montage as a unique interactive method of structuring images creates an aesthetic experience of being and life, capable of revealing deep and hidden elements of reality.

Walter Benjamin commented on the "optical unconsciousness" triggered through the structure of film grammars and image flow (its diegesis) as connected to hidden emotional drives of the spectator. The dialectics of this form of mediated image construction could release powerful forces all the way from an epiphany of rapture and beauty to the most debased forms of aggression and violence-as an active response to hidden personal drives and new understanding of one's times.

The dialectics of the more technologised conditions we are experiencing in our century where optical vision has been dramatically extended by so many new tools for seeing-the camera and photography; the mediated images of cinematography and of television and video brings us to the point where the average psyche is deeply affected as the result of a media-driven culture. How we see is influenced by deep transformations of consciousness brought about by major advances in technical standards.

Now that we rely more and more on the mathematical modelling systems of the computer rather than on light and film as the basis for making an image, the direct connection to the visual world is broken. With digital cybernetic technologies we have moved beyond the copy as image of the visible world. The computer chip can simulate an infinitely manipulable model of this world. We can no longer believe in the truth of our own vision when all our captured images can be invaded and changed at will. This new condition makes it more difficult to "suspend disbelief". We are at a point where we can determine that different theories of visual reception may have been more true for their times than they appear to us today. What had been fixed poles of visual reality, now collapse. 'Simulation displaces any antecedent reality, any aura, any referent to history.' (Nichols,1988).

Instead of the conventions and signs of other languages of representation such as cinema, new forms of social practice are countered by simulation. Computer-based systems are primarily interactive rather than one-way; open-ended rather than fixed. They reject hierarchies and rely on linked messages-in-circuit, rather than on a single text. Now, with digital linkages leading to possibilities for interactive dialogic exchange, the concept of a single 'text' slips away and takes on a continually variable form, de-emphasising the concept of single authorship. Instead, the interactive takes on a form both addressed to and addressable by us.

Up to now, the primary structural form, including the visual ones of photography and film traditionally based on a narrative, have been subsumed by the analogue 'flow' of electronic television transmission. But, 'like face-to-face encounter, cybernetic systems offer (and demand) almost immediate response. The temporal flow and once-only quality of face-to-face encounters becomes embedded within a system ready to restore, alter, modify or transform any given moment to us at any time. Cybernetic interactions can become intensely demanding, more so than we might imagine from our experience with texts, even powerfully engaging ones.' (Nichols,1988).

While diegesis describes and explains the actual technological process of constructing an architecture of meaning in cinema, cybernetic simulations with their linkages and interactive potential bring such processes under closer, sharper focus. Interactive technologies force adaptation to new ways of communicating which involve also new questions about direct, individual social interaction, communication and organisation. The shift is one from the cinematic narrative, where image structuring created meaning, to an interactive one based on linked iconic images and concepts.

Interactive implies pluralistic omnidirectionality-something which lies between people, is inter-mediatory, and therefore can relate and bind them together. It is a 'speaking-between' similar to the power of montage and *darshan* to actively transform thinking where one goes inside oneself to seek one's own response as a means of self-realisation. It offers new possibilities for transgressing old boundaries and limits and systems of control imposed by social systems which have had the effect of stifling dialogic communication.

The politics of the new cybernetic world we have entered may in turn work to impose limits to this interactive dialogue-in the form of copyrights and powerful methods of commercial control, but the challenge is to use the democratic, participatory, apperceptive powers it has brought into being. Thirteenth century Persian poet Jalauddin Rumi reminds us: 'New organs of perception come into being as a result of necessity-therefore, increase your necessity so that you may increase your perception'.

References

Babb, Lawrence. 'Glancing: Visual Interaction in Hinduism', in *Journal of Visual Anthropological Research* 37:4, Winter 1981, p387-401.

Benjamin, Walter. 'The Work of Art in the Age of Mechanical Reproduction', in *Illuminations*, Shocken Books, N.Y., 1973.

Brennan, Teresa. 1996 'Conclusion.' in *Vision in Context. Historical and Contemporary Perspectives on Sight.* eds.Teresa Brennan and Martin Jay.London and New York: Routledge.

Melville, Stephen. 'Division of the Gaze, or, Remarks on the Color and Tenor of Contemporary Theory'. in *Vision in Context. Historical and Contemporary Perspectives on Sight*. (op.cit.).
Metz, Christian,1974 'Film Language, Semiotics of the Cinema'. London and New York, Oxford Univ. Press, 1974.
Nichols, Bill. 'The Work of Culture in the Age of Cybernetic Systems', Screen 29:1, 1988, p22-46
Viola, Bill,1995 *'Reasons for Knocking at an Empty House, Writings,1973-1994'*, Boston,MIT Press.

Seeing Double: Art and the Technology of Transcendence

Roy Ascott

The new telematic adventure in art, currently played out in the Net but swiftly migrating to the 'smart' environments of ubiquitous computing, has brought questions of distributed mind and shared consciousness to the definition of a new aesthetic. This *Technoetic Aesthetic* recognises that technology plus mind, tech-noetics, not only enables us to explore consciousness in new ways but may lead to distinctly new forms of art, new qualities of mind, and new constructions of reality. What is more ubiquitous than consciousness; what is less understood than mind? In both art and science now, the matter of consciousness is in the forefront of research. Science, in trying to explain consciousness, faces the most intractable of problems. In this endeavour many disciplines are brought into play. For the artist, although consciousness is more to be navigated than mapped, and more to be re-framed than explained, it too poses an overarching challenge. The mysterium of consciousness may be the final frontier for both art and science, and perhaps where they will converge. Certainly, objectivity and subjectivity intermingle there. It may be the space in which art's classical concern with representation and expression gives way to processes of construction, emergence and evolution; a space in which both the self and its reality can be redefined and transformed.

As for conscious experience in itself, while there is nothing we know more intimately than our inner sense of being, there is nothing we can experience with less comprehension than the conscious states of another. It may be only the profound empathy of mutual attraction, 'love' if you will, that can break this barrier[1]. Telematic connectivity may provide the 'sympathetic technology' for such a breakthrough, but neither reductionism nor postmodernism would countenance such a possibility. Fortunately there are signs that, in some important respects, science is becoming more subjective and that, in the temples of high art, cultural pessimism is on the wane. There is no doubt that both scientists and artists are curious about the ways that advanced technology can aid in the exploration of mind.

Life in cyberspace can be seen as essentially technoetic. Our experiments with the technology of being, involving for example VR, telepresence, hypermedia, may be the prelude to our eventual migration from the body into other forms of identity. Unlike the material body, the mind cannot be contained; it leaks out everywhere. It is as if our destiny is to make intelligence ubiquitous. Migration from the body does not imply its disappearance but the emergence of the multiple self, the distributed body, whose telepresent corporeality creates its own field of being.

Artists are beginning to recognise the primacy of consciousness in both the context and content of their art, and as the object and subject of their study. Indeed, the very provenance of art in the twentieth century, through its psychic, spiritual and conceptual aspirations, leads towards this technoetic condition. It may be necessary only to point to the work of Duchamp, Kandinsky, Klee or Boccioni, in the earlier part of the century, to indicate the roots of this tendency. It is equally clear that the impact of technology can have the effect of reducing art to a form of craft in which the spectacle of special effects and dazzling programming alone can replace the creation of meaning and values. A more optimistic view is that our concern in interactive art with whole systems, that is, systems in which the viewer plays an active part in an artwork's definition and evolution, may express an ambition to embrace the individual mind by a larger field of consciousness. But as much as interactivity may be emblematic of this desire for a shared consciousness, it is problematic in its assumed resolution of the object/process, observer/participant dichotomy.

Before examining this further, the notion of double consciousness and its relationship to art must be considered. By double consciousness is meant the state of being which gives access, at one and the same time, to two distinctly different fields of experience. In classical anthropological terms this is to describe the shamanic 'trance' in which the shaman is both in the everyday world and at the same time navigating the outermost limits of other worlds, psychic spaces to which only those prepared by physical ritual and mental discipline, aided often by plant 'technology', are granted access. In post-biological terms, this is mirrored by our ability, aided by computer technology, to move effortlessly through the infinities of cyberspace while at the same time accommodating ourselves within the structures of the material world. Confronted by an array of technological devices that offer us a pathway into virtual worlds we are invited on the plane of prosthetics to enact the shaman's journey. Immersion in such noetic simulation may induce real changes of consciousness and eventually real transformations of self.

To research this apparent parallelism between shamanic realities and virtual realities, psychic space and cyberspace, and the double consciousness that seems to be a part of both fields of experience, I have spent time immersed both in the virtual reality of advanced computer systems and in the traditional reality of an indigenous Brazilian tribe, that is under the influence of the computer and of the plant, albeit an extremely powerful computer and a particularly potent plant (ayahuasca, the 'vine of the soul'). My access to virtual reality was at locations on both sides of the United States, at the Human Interface laboratories in Seattle and at the University of North Carolina at Chapel Hill. My introduction to the psychic world was in the very heart of Brazil, with the Kuikuru pagés (shamans) of the Xingu River Region of the Mato Grosso, and through my initiation into the ritual of the Santo Daime community in Brasilia.

The shaman is the one who 'cares' for consciousness, for whom the navigation of consciousness for purposes of spiritual and physical wholeness is the subject and object of living. Consciousness occupies many domains. The pagé is able to pass through many layers of reality. In his altered states of awareness he engages with disembodied entities, avatars and the phenomena of other worlds. He sees the world through different eyes, navigates the world with different bodies. In parallel with technologically aided cyberception, this could be called psi-perception. In both cases it is a matter of the double gaze, seeing at once both inward realities and the outward surfaces of the world.

The double gaze and double consciousness are related. In my experience of ingesting the ayahuasca I entered a state of double consciousness, aware both of my own familiar sense of self, and of a totally separate state of being. I could move more or less freely between these two states. Similarly with my body: I was at one and the same time conscious of inhabiting two bodies, the familiar phenomenology of my own body sheathed as it were in a second body which was made up of a mass of multicoloured particles, a million molecular points of light. My visual field, my double gaze, alternated, at choice, between the coherent space of everyday reality and a fractal universe comprising a thousand repetitions of the same image, or else forming a tunnel in space through which I could voluntarily pass with urgent acceleration. I could at any point stop and review these states, moving in and out of them more or less at will.

Many shamanic tribes not only enhance their psi-perception by drinking the ayahuasca on a regular basis, but their culture, by adoption, has given rise to a ritualised practice known as Santo Daime which has spread to most parts of Brazil, not least in its urban and metropolitan areas. In addition to the ritual drinking of the ayahuasca, Santo Daime has precise architectural and social codes. The design of the building that houses the ritual, the ordered placing of participants in that space, the rhythmic structure of the music, the pungency of the incense, the repetitive insistence of intoned phrases, punctuated by extended periods of absolute silence, the recurrent demand to stand or sit, one's own inclination to move into and out of the new field of consciousness that the ceremony and the drink together induce, leads one's awareness to fluctuate between the two realities. It raised the question, of course, of the way in which specific protocols and conditions control or construct a given reality, and leaves unanswered the question of where or how or indeed if a ground of reality might be identified or even be said to exist.

This immersion in a controlled environment, affecting sight, touch, taste, smell, and hearing respectively, conferring on the mind the ability both to induce and create new conceptual and sensory structures (in philosophical jargon new 'qualia'), while at the same time giving the freedom to step aside from the visionary experience, back into the 'normal' field of experience, is mirrored to an extent in our artistic aspirations in cyberspace. Here double consciousness, is engendered by digital technology, as for example by Virtual Reality, hypermedia structures and, the fast developing field of Augmented Reality[2] with its superimposition of cognitive schemas on real world situations. In both cases, esoteric and technological, there is a kind of rehearsal of the Sufi injunction to be both in the world but not of the world, although the original context of that phrase is more emphatically spiritual than perhaps many artists would want to acknowledge. Here technology plays an important part in the experience of 'double consciousness', just as it is clearly integral to our emergent faculty of cyberception and the double gaze. It is as if, through our bio-telematic art, we are weaving what I would call a *shamantic* web, combining the sense of shamanic and semantic, the navigation of consciousness and the construction of meaning.

— Transcendence —

Historically, our command of the material world has been such that we have little option but to keep the worlds of our double consciousness in separate and distinct categories, such as the real, the imagined, and the spiritual. The advent of the Artificial Life sciences, in which I include both dry (pixel) and moist (molecular) artificial organisms and the whole prospectus of nanotechnology, points to the possibility of eroding the boundaries between states of mind, between conception and construction, between the internalisation and the realisation of our desires, dreams and needs of our everyday existence. Let me give you an example, which can be found in our cyberception of matter at the atomic level. The scanning tunnelling microscope (STM) enables us to view matter at this level, but to image individual, single atoms. However the real significance of this process does not end there. Not only can we select and focus on individual atoms, but we can, at the same time, manipulate them, one by one, atom by atom, to construct from the bottom up atomic structures of our own choosing[3].

This means that, in an important sense, the prosthesis of vision can be at one and the same time instrumental in constructing what is envisioned. To see in the mind's eye is to realise in the material world. The worlds of the double consciousness, supervenient[4] as they are on the processes of the double gaze, become less distinctly separate. The immaterial and material loose their categorical distinction. Cyberception is as much active and constructive, as it is receptive and reflective. As this kind of double technology develops, and it is doing so at an accelerated rate, artists, no less than the philosophers and neuroscientists, must increasingly turn their attention to what I will call 'techno-qualia', a whole new repertoire of senses, and to a new kind of relationship between the tools of seeing and building.

Within this context the whole question of the control of consciousness arises. Historically, art has served as a subtle, almost invisible but none the less effective, means of shaping social, political, and religious, as well as cultural consciousness. But interactive art has sought to place this power in the hands of the viewer-quite literally through the physical manipulation of the interface. The momentum towards disseminating the overall control of an artwork's meaning is leading to the emergence of self-controlling images and structures, self-assembling forms, with self-governing behaviours. Currently, this finds its expression in the work of artists employing artificial life technology, biomedia, and the strategies of artificial intelligence. Artificial Life is as much a part of our quest for self-definition as it is an instrument in the construction of reality. In exploring the technology of life we are exploring the possibilities of what we might become. The self as an ongoing creation gives rise to essentially non-linear identity.

Let us return to the question of interactive art. Although it may seem to solve the problems associated with the distancing of the observer and passivity of reception that art cast in Renaissance or Euclidean space seemed to imply, interactive art is by no means unproblematic. At the moment, by its structure, placement and presentation (usually in a traditional museum or gallery space), the work of networked interactive art presupposes, in spite of itself, an audience of more or less passive observers, just as much as it proposes a participant in open-ended interaction with at its interface. In this sense, the total system including the participant viewer, however dynamic a process it may be, is actually incarcerated within the very status it despises, that of pure object-an envelope, bracketed in space and time, to be viewed by a second observer. This creates a dichotomy between the aspiration toward open-ended evolution of meanings and the closure of an autonomous frame of consciousness, a contradiction that necessitates the removal of the second observer and the phantom audience from the cannon of telematic art.

Here, by way of contrast, the shamanic tradition may usefully be invoked. All the activity of the pagé, and of those who interact with him in image making, dancing, chanting, making music, are performative but *is not intended as a public performance*. It is never played to an audience, actual or implicit. No one is watching or will be expected to watch what is being enacted. This is not a public performance but a spiritual *enactment*, which entails the structuring or re-structuring of psychic forces. To paint the body elaborately, to stamp the ground repeatedly, to shake the rattle, to beat the drum, to circle round, pace back and forth in unison, is to invoke these forces, to conjure hidden energies. This is an enactment of psychic power not a performance or cultural entertainment. This perspective, although seen at a great distance from our current hypermediated culture, may be of value in our consideration of the function of works of interactive art, thereby avoiding the double observer, the phantom audience. Art as an enactment of mind implies an intimate level of human interaction within the system, which constitutes the work of art, an art without audience in its inactive mode.

Eschewing the passive voyeur, the traditional gallery viewer, this technoetic aesthetic speaks to a kind of widespread intimacy, closeness on the planetary scale. It is the question of intimacy in the relationship between the individual and cyberspace, which must be at the heart of any research into technologically assisted constructions of reality. The quality of intimacy in the relationship between artist, system and viewer is of the greatest importance if a technologically based practice is to engage or transform our field of consciousness.

What then is the role of the artist in an art that increasingly sees its content and meaning as created out of the viewer's interactions and negotiations? An art which is unstable, shifting and in flux; an art which parallels life, not through representation or narrative, but in its processes of emergence, uncertainty and transformation; an art which favours the ontology of becoming, rather than the assertion of being; an art moving towards a post-biological re-materialisation; an art of enactment, without audience. An intimate art, the free-flowing outcome of interaction between actors within networks of transformation. An art, in short, which reframes consciousness, articulating a psychic instrumentality, exploring the mysteries of mind.

I have identified a major aesthetic shift which has taken place in our century, from the art of appearances, classically concerned only with the static order of things, to an art of apparition, concerned with dynamic relationships and processes of coming-into-being. It seems that increasingly, existence is regarded as an aesthetic phenomenon, rather than a moral or religious one, as Nietzsche insisted. His story is that of the gradual rejection of the dialectic of being, and its mystification, in favour of a yea-saying, life-affirmative recognition of the primacy of becoming.

The questions I have raised are those artists are taking into the new century. They may find answers in the deep past, in the remotest parts of the planet, or simply within the double consciousness to which we all have access. They may find them in understanding the role of the shaman, re-contextualised in the bio-telematic culture but re-affirmed in its capacity for the creation, navigation and distribution of mind. They may dedicate themselves to the conservation of what emerges from the complexity of interactions in the Net or from the self-assembling processes of artificial life. One thing seems certain, the technoetic principle will be at the centre of art as it develops, and consciousness in all its forms will be the field of its unfolding.

Notes
1 Ascott, R. 1996. Is There Love in the Telematic Embrace? In *Theories and Documents of Contemporary Art*. Stiles, K and Selz, P eds. Berkley: University of California Press, pp.396, 489-98.
2 see <www.cs.unc.edu/Research/stc/predictive_tracking_html/azuma_AR.html>.
3 For an aspect of this process, see the IBM website <www.almaden.ibm.com/vis/stm/lobby.html>.
4 For the definition of 'supervenience' see Chalmers, D. 1996. *The Conscious Mind*. New York: Oxford University Press.

Jumping Over the Edge: Consciousness and Culture, the Self and Cyberspace
Lily Díaz

Introduction: Altered States of Consciousness
Possession and possession trance have been a part of the religious practices of the Western world for hundreds of years. For example, 'the formal role of the exorcist exists in Catholic priesthood and all possession beliefs of the Christian churches have a justification in the New Testament.' Bourguignon (1991: 3). Most recently, culture-specific information artifacts from pagan religions such as Vodoo and Santeria have found their way into the genre of science fiction. In this latter, the condition of possession trance has been used as a type of interface to represent the idea of multiple states of consciousness the self can access. Furthermore, concepts about the self extracted from non-Western belief systems permeate the design strategies in fields such as avatar development. That is, artifacts from the religious domain provide users with ways of manouvreing through multiple representations of the self.

What is Possession?
Overall, in the social sciences the term possession pertains to ideas or concepts used to interpret behaviour.That is, the behaviur is observed as existing within an explicit social context. Thus, the anthropologist speaks of possession only when (s)he finds such a belief among the people (s)he is studying.

However, in approaching the same phenomena a psychologist may find that the causes of possession reside elsewhere: A clinical point of view will attribute possesion to symptoms of dissociation and other multiple personality disorders. This is why a cross-cultural approach that assesses the origins, recurrence and acceptance (or rejection) of the behaviour over a period of time is used to study these phenomena.

Through the method briefly described above, two basic forms of the state of possession have been identified in anthropological literature. The first type, simply referred to as possession, is used to describe a physical state involving a change in bodily function that is involuntary and undesirable. Such a state of possession not only exhibits a split between the body and the self but also between the will and performance of which the actor is quite conscious. A key feature of this state is the idea of the person acting without willing to do so.

Possession trance, on the other hand, refers to a state in which the person possessed submits and allows his or her body to be taken over. *Glossolalia* (or speaking in tongues) and gifts of the spirit, such as prophecy, are associated with this type of behaviour. The main difference between these two types of states of consciousness is that in the latter the behaviour is seen as desired and is, therefore, intentionally induced and regulated.

Possession as a Ritual Practice

Santería and Vodoo are synchretic religious practices. This means that their belief systems contain elements originating from diverse sources. Historically these sources date back to the slave trade. When the African slaves were brought to the Americas by the Europeans, they carried with them the beliefs and traditions of their tribes. The conversion to Western religions was superficial, so that from the combination of these different components resulted beliefs and practices that include elements from several African cultures and also from Catholicism. In the case of Vodoo, Dahomean practices relating to the cult of the snake have been crystallised into the religious framework. The African elements in Santeria, however, have been found to relate to practices from the Yoruba religion.

Possession trance is an intrinsic component of the ritual practices of these two religions. In both it is viewed as the voluntary surrender of the individual to an external, pre-defined entity, such as an ancestor spirit or deity. Whereas in Santeria the deities are referred to as *orichas*, in Vodoo they are called the *loa*.

The presence of these agents, or deities, via the subject undergoing possession trance is the driving force in many of the rituals of these religions. This is partly because in these rituals the deities perform narratives that re-enact aspects of the cult that transgress boundaries of time. These aspects represent a reality that has been preserved as common collective knowledge in oral traditions. These agents are also symbolic units that contain a multiplicity of representations and co-exist within rationally disparate frameworks. An example of this is the case of Chango. In the African belief system of the Yoruba religion, he is the god of thunder. In the synchretic representation of Santeria he is depicted as Saint Barbara. Yet in the ritual, during the moment of possession trance, a woman possessed by Chango will speak with a deep bass voice and adopt very masculine attitudes (González-Wippler 1976:54).

Both religions differ from Western monotheistic traditions in that there is the belief that nature is an animate being that humans stand in relation to: We are moulded by nature but we also have at our disposal the means to control and define it. Chaos, destruction and evil are not opposites but part of a dynamic process that clears the path for the evolution of life. Therefore, the relationship between the believer and the deities is one in which the former understands, and accepts, that the latter can be as passionate as human beings and in certain cases even unjust.

Culture in Cyberspace

William Gibson's use of aspects from Vodoo in his novels is an example of the appropriation of culture-specific knowledge to represent ideas for which there is no equivalent in the West. In this work, Gibson seizes the notion of possession trance to indicate the difference between the fluid consciousness that can be easily deployed into cyberspace and the physically situated, stationary, body. Gibson also uses this concept as a platform to denunciate the ideology of the technology-enabled consciousness and self (Gibson 1987).

These altered states of consciousness are features of a semi-organic interface to cyberspace. In the novel *Count Zero*, for example, the Vodoo practitioners Lucas, Jackie and Beauvoir have accommodated to the new way of life by using mental structures provided by their cultural mileu. Through the use of cultural strategies they become key players in the spectral marketplaces of the Matrix. As such they represent a cultural adaptation to the post-human condition:

> 'Think of Jackie as a deck, Bobby a cyberspace deck, a very pretty one with nice ankles. . . Think of Danbala, who some people call the snake, as a program. Say as an icebreaker. Danbala slots into the Jackie deck, Jackie cuts ice. That's all.'
>
> (Gibson 1987:114).

Gibson's visualisation of the technology-enabled self juxtaposes the ideal of the augmented mind to the imperfect and replaceable physical body. This is also presented in *Count Zero* through the botched attempts to maintain Joseph Virek's existence. A member of the corporate elite, Virek's body parts are distributed in a series of containers parked somewhere in a suburb of Stockholm. The tragic consequences of the technology-dependent existence reach us through Virek's amputation from the human sensorium: 'I was touched . . . at your affairs of the heart. I envy you the ordered flesh from which they unfold', he says (Gibson 1987:16).

Paradoxically, though Virek can afford diverse forms of presence,or manifestations, he is at the same time forcibly isolated from the life-cycle. Thus lacking a physical body, the cumbersome network that prolongs his existence at the same time symbolically denies him the simple act of death: 'I imagine that a more fortunate man, or a poorer one, would have been allowed to die at last, or be coded at the core of some bit of hardware'(Gibson 1987:16).

Virek also symbolises how the intrusion of capitalist, market-driven forces on the configuration of the body can result in collective malady and disease. Virek himself confesses that rather than being part of one body his cells 'have become autonomous, by degrees; at times they even war with one another. Rebellion in the fiscal *extremities . . .* (Gibson 1987:13).

This ideology of the protean self is also articulated in this novel through the character of Angie Mitchell. The lastest prototype,or Virgin, she demonstrates a synthesis of the machine and the human achieved by tapping into the unconscious: Angie's experiences of the Matrix occur almost always in her sleep, when she is dreaming. The bio-circuitry, or 'tumors' in her head trigger possession-like states during which her body involuntarily spits, acts and vocalises dreams, hallucinations and events from the Matrix. Like a participant in a ritual, she 'sees' things and others in the Matrix sense her as well.

Culture and Design in Cyberspace

Cultural models can be used as paradigms in the field of design. In the case of the new media, they are being used to fashion the 3-D interface environments of virtual worlds. Examples of this can be found in how avatar design is trying to address questions of how to negotiate identity, presence and behaviour in cyberspace.

Its literal meaning being 'reincarnation', the origins of the word 'avatar' are cited back to the Sanskrit language. The term, however, is not restricted to this cultural domain and I would posit that the concept is common to many pagan belief systems. In Santeria, for example, the term is also used to mean the diverse manifestations or 'paths' of a deity. Therefore, Yemaya, who is one of the Seven African Powers, the goddess of the sea, and is represented synchretically as Our Lady of La Regla, also possesses seven different aspects or avatars. In her oldest avatar, she is Yemaya Olokun, the first Yemaya, who is masculine and who is the god of the deepest part of the ocean.

Xenomorphic behaviour, or the quality of survival and adaptability, whichis part of synchretic belief systems, is seen as a desired component in avatar design:

> 'An important design feature, in the development and evolution of an intelligent avatar, is the inherent ability to directly influence, and affect the dynamic 'reconstruction' of an avatar entity. As an interactive process, this process renders a continuously evolving virtual character'(Ostman 1997.)

More than a costume or mask, an avatar aims to provide the displaced self with channels for interaction. One of the ways that this is done is by allowing the user access to a symbolic field of action through pre-programmed behaviour and synthetic *persona*. Touted as the next level in the evolution of the interactive experience, avatars are supposed to offer the user experiential, emotional engagement asa reward for participation. Underlying the development of this technology is the profit-driven agenda that views cyberspace as the new frontier for commercial development (Ostman 1997).

Religion and Ritual as Interface

The diverging views and the philosophical issues raised by the introduction of culture-specific discourses are merely signals of how volatile a territory grounds this notion of the self. Figueroa-Sarriera's remarks on the self as a socially and historically constructed category centre the current dialectics on a post-Cartesian emergence of the body and soul as two substances that are mutually dependent on each other (Figueroa-Sarriera 1995:129). Allucquere-Stone focuses on how larger shifts in cultural beliefs and practices that have been brought about by the emergence of new technologies, such as the telephone and the computer, are redefining both the notion of self but also of presence (Stone

(1995:16). However it is approached, the issues are about the accommodation that must take place if we are to accept the promised reality of a global networked digital environment: Who am I in the virtual domain? What am I and how do I represent myself in cyberspace? Is my self where my body is?

The interface can be described as a 'surface forming a common boundary, a meeting point or area of contact between objects' (Figueroa-Sarriera 1995:129.) As the 'consensual hallucination', cyberspace has much in common with altered states of consciousness, such as possession trance. The equilibrium of cyberspace as a symbolic system is based on the immaterial presence of others and their ability to represent and to communicate amongst themselves.

As mentioned earlier, possession trance does not occurr in the context of collective frenzy. It is a premeditated response that is built to close a cognitive gap between the physical and non-physical realm inhabited by practicioners and spirits. In both belief systems mentioned, the ritual and performances of the actors possessed by spirits serves as a collective communication strategy that enables a 'rational' exchange between living and non-living entities. Inherent in these practices is the notion of communication between entities from both the material and immaterial realms. This is a dynamic that is easily ported to cyberspace. It also creates a framework that allows for the transfer of human consciousness from the realm of the organic into the space of artificial cognition.

Then there is the ability of belief systems that feature the practice of possesssion to rationally explain how diverse, and even contradictory, entities can inhabit the same body at different points in time. A similar phenomenon is experienced upon entering cyberspace where in the time span comprising several 'logins', or 'immersions', we can assume many identities.

Conclusion

Who will I be when I log into separate IRC channels as Alma, Fatima or Perla? Where will the real me be, during my morning shopping in the virtual plazas of cyberspace? Will my avatar like me? Will we share the same thoughts? Will I be like the Vodou practitioner whose *ti-bon age* abandons her body when mounted by the *loa* and share my body with other avatars? These are some of the questions that we should ask ourselves when contemplating the future life in virtual worlds.

Most probably, those who practise Santeria and Vodoo do not harbour such questions. Why? For one, their belief system provides rational boundaries between the ritual and everyday life. The lack of ritual in most of the secularised societies of the West may prompt us to reconsider whether to take the jump over the edge into new, non-corporeal, modes of existence. Is the challenge merely based on technological development? Are there parameters that deal with cultural adaptation?

References

1. Bourguignon, E. 1991. *Possession*. Illinois: Waveland Press.
2. Gouldner, A.W. 1976. *The Dialectic of Ideology and Technology*. London: MacMillan Press.
3. Gonzales-Wippler, M. 1973. *African Magic in Latin America*. NewYork: Julian Press.

4. Figueroa-Sarriera, H. J. 1995. 'Children of the Mind with Disposable Bodies: Metaphors of Self in a Text on Artificial Intelligence in Hables Gray, C., ed. *The Cyborg Handbook* . New York: Routledge Press.
5. Gibson, W. 1987. *Count Zero*. New York: Berkeley Publishing Group.
6. Ostman, Charles 'Synthetic Sentience on Demand'. MONDO 2000 Archives
    <http://www.mondo2000.com/left%20brain.html>
7. Stone, A. R. 1995. *The War of Desire and Technology at the End of the Mechanical Age*. Cambridge: The MIT Press.

Lily Diaz is an artist and a researcher. She is currently working at the Media Laboratory of the University of Art and Design in Helsinki, Finland. <lily@mlab.uiah.fi>

Space-time Boundaries in the Xmantic Web
Maria Luiza P. G. Fragoso

Space and time are concepts always present in any artistic manifestation. Present in their absence, present in their representation, present in the sole existence of the manifestation, the act of creation. The two concepts are inherent to our existence as humans. The representation or re-creation of these two concepts is an eternal challenge for all artists. The definition of a specific space and of a specific time, in the creation of a work of art, determines the existence of the work and determines paradigms. Technology, materials and instruments will be chosen from the needs of space and time.

Space
Space defines intentions, feelings, cultural heritage and many other aspects of our world . The space we perceive is the space we know within us. The space we need around us is the negative space of what we have inside us. The visualisation of space is directly related to our understanding of space (Bachelar 1978). Herbert Read in his book 'Icon and Idea' (Read 1955) explains that the intuitive manifestation of the image precedes the conscience of why and how the image came about. When the need for space comes from within the being, this vital need will manifest itself in all creations. If it represents a social necessity, it will be studied, analysed, measured and will then be incorporated in the social system as a paradigm. In this way, different perspectives are disclosed and we are able to understand them and, consequently, they will change our lives and our consciousness. Just as an example, today we can easily imagine a suggested three-dimensional space in a bi-dimensional painting, or we can have an idea of the size of a city block by looking at a picture where it is situated in perspective. The second example is based on an exercise of simple proportion and basic topologic reasoning (Abbot 1991).

The idea of space is so important to us that we preconceive measurements, we imagine territories in order to create conditions to interact with our world and with others. From the moment we were capable of, in our minds, conceiving the world as a whole, we engaged in the adventure of overtaking distances. Large spaces were related to large distances which were related to large periods of time. For example, the Kuikuru, one of Brazil's native tribes, use the stars not only to guide them through space but mostly to determine time periods in relation to space.[1] Scientifically, with the advent of telecommunications, distances were diminishing, and now, with telematics, seem to be conquered. In an inverse proportion, space is growing, expanding, becoming infinite and this is a new paradigm, a change in collective consciousness.

Artistically, during the Renaissance period, the representation of three-dimensional objects in perspective, brought to a conscious level the notion of space as a source of human experiences (Fraga 1995). In visual representation, artists re-created space in Cubism and Surrealism, and transported locations from one place to another, consequently taking one part of the world to another part. The simulated representation of space was replaced with its actual re-creation and transformation. Works of Duchamp, Beuys, Pina Bauch, Calder, Man Ray, Kazuo Ohno, Cage, Cunningham, Nyman, Christo and many others, expressed in different artistic languages, were not only re-creating spaces and defining non-spaces, but were expressing the need of a conscience of a totality, the break down of boundaries. Concepts such as multidisciplinary, 'transdisciplinary', global, multi-cultural were introduced decades ago. From land art to telecommunication experiences, from Bill Viola's videos to installations in art galleries , artists are using science and technology and are expanding the concepts of space and time.

Time

Time as a concept is as difficult to define as is space. As artists, we speak about these two concepts in a different way from scientists such as Galileo, Newton, Einstein or Prigogine, but we can introduce you to a universe of sensitivity, sensibility and perception, to a universe of beauty and harmony, emotions and feelings. In this sense we engage in the adventure of speculating about space and time, referring to the findings of these scientists.

In time, back to time: a closed system that has the primordial function of organising intervals, rhythm, biological functions, physical needs for this life, this world. We need time to have an orderly life. A time measured by our watches but not part of our living bodies. The entire time system is part or our developed logic system. In the view of Ilya Prigogine, 'we are in the presence of two times: there is the time of the watches, of classical dynamic's trajectory, of communication; and, in the other hand, there is the structural time, the internal time, marked by fluctuations and the by the irreversible'(Pasternak 1993).

The work of art is testimony of its time, its moment. When we observe an art work, we can recognise time, pointing out a moment lived by a social group (external time) or exposing the artist's moment (internal time). Time may also be trapped within the work. Movies, videos, dance, theatre, performances and computer animation have the power of re-establishing time periods, seducing the viewers' attention and transforming 20 minutes in hours, days, years, centuries. As spectators, we are capable of living in n-time systems at the same moment. Our minds travel in time, through time. Time travelling is no mystery for those who allow their minds to flow.

If the world is our imagination and if we imagine that each one of us is dreaming a world and that we pass from one thought to another without even feeling it, we may suppose that there are no pre-established sub-divisions in time or space (Borges 1996). Scientists and artists developed intelligent machines, computers and software and consequently came up with a wonderful gift for mankind. A gift, accessible to all, that offers the opportunity to physically, mentally and sensorially interact with multidimensional space and time. For a few of us, this type of experience is not new; rather, it is very old. Ancient Tibetan practices (Yeshe 1987), passed on from generation to generation, offer the experience of a rupture in time and space. Recent anthropologic studies on primitive societies, such as Brazilian Indian tribes (Langdon 1996) are disclosing shamanic practices that deal with the experience of several time-spaces, acknowledge by their community and present in their everyday life. The artist himself, truly involved with the act of creation, reaches a state of equilibrium, defined by Hannah Arendt (Arendt 1993), where the opposite forces of past and future meet at the present. This encounter defines an original point, from which the mind is propelled into infinite time and space: a state of bliss in the act of creation.

On the World Wide Web, or should we say, in cyberspace, time is unlimited, it challenges the arrow of time, time's paradox of irreversibility. Our imagination runs free in an unfolded universe of unimaginable possibilities. Telematic presence happens anywhere, any moment, past or future, like a moving pendulum. Once involved with this new environment, artists are again using science and technology to expand the concepts of space and time. But instead of expressing ourselves through media subdued to the external space and time, we are creating with VRML (Virtual Reality Modelling Language) in cyberspace, using our intuitive and sensitive capability to experience a new dynamic in a multidimensional world.

Xmantic Web

The advent of digital technology initiated meditation on the changes occurring in the field of visual arts: the influence of telecommunications and globalisation mediated by on-line systems, the phenomena of multicultural interchange on the Web, the transposition of space-time boundaries and the apparition of an interfaced being. This threshold lies in the questionable capability of transforming the machine, not only into an extension of the living body, butas an extension of the mind.

The project of a Xmantic Web proposes the study and the experience of ecstasy techniques, to arouse the complementary sensation between reality and virtuality. The experience of five artists, either in the VR environments and the multicultural connection, or by being in a native tribe in Xingu (Brazil) has already begun the process of powerful personal transformation within the group. Cultural transformation is pursued by these artists working with VRML, telepresence and dialogic systems, dealing with different states of mind in search of different states of consciousness. These transformations made evident the possible transposition of boundaries and the potentiality of reconstructing myths, of new paradigms, of new behaviours and of a expanded human conscience.

Conclusion

'Einstein considered that the distinction between past, present and future was an illusion, and that it did not belong to the domains of science'(Pasternak 1993). Our research into shamanic rituals is based in the possibility of producing immersions into virtual environments. Shamanism, understood as 'any ecstatic phenomenon and any magical technique ' (Eliade 1989) uses rituals which abolish daily space-time boundaries. A shaman is a mediator between worlds, one who cares for consciousness. The Xmantic Web will stand as mediator, as an environment that will offer the navigator an opportunity to engage in three fundamental movements: the mind's emergence from its limited concrete reality into the desire of realising it's unlimited potential in a multi dimensional environment- cyberception;[2] the openness and determination to share with others, to create and to expand with others-connectivity;[3] and dissolution, being freed from a dualistic and individualistic vision of the universe- hypercortex.[4] One of the positive functions of art is to stimulate integration between individuals and cultures. A new paradigm, the advent of computer technologies, based on the capability of communication and interchange, will question techno-scientific tendencies and an aesthetics based on rationalism and mathematics. The attempt to subvert aspects of technology into a part of the being, present in the work of contemporary artists, and the concern with the construction of reality, will help build new moral and ethical values for a new society. The beauty of it all lies in the understanding that we may now express externally the infinite space-time that exists within us.

References

Abbot, E.A.1991. *Flatland*. New Jersey: Princeton University Press.
Arendt, H. 1993. *A Vida do Espírito*. Rio de Janeiro: Relume Dumará, p. 157.
Borges, J.L. 1996. *Cinco Visões Pessoais*. Brasília: Editora Universidade de Brasília, p. 46.
Eliade, Mircea. 1989. *Shamanism*. New York : Penguin Books, p. 375.
Fraga, T. (unpublished) 1995. *Simulações Esteroscópicas Interativas* (PhD thesis). São Paulo: PUC, p.32.
Langdon, E.J.M. 1996. *Xamanismo no Brasil*. Florianópolis: Ed. da UFSC, p.91-92.
Yeshe,T. 1987. *Introduction to Tantra: A Vision of Totality*. Boston: Wisdom Publications.

Notes

1   Kuikuro Indians, native Brazilians living in the Park of Xingu, bordering the Amazon jungle, explained briefly to some of the members of the reconnaissance expedition that they had an organised system which allowed them to determine 'lengths' of time. This system was based on the reading of the position of the stars, their movement or dislocation from one point of the sky to another. They use this kind of observation to guide them spatially and to determine how long it takes to reach different locations.
2   Term included in 'Technoetic Aesthetics' A glossary of definitions and terms coined by Roy Ascott.
3   ibid
4   ibid

Graduated from the Licentiate programme in Fine Arts at the University of Brasília (Brazil) in 1984. Obtained a Masters degree in Fine Arts at the George Washington University (USA) in 1993. Admitted as Professor at the University of Brasília (Brazil) in 1993. Member of the research group Corpos Informáticos since 1993, developing research in Contemporary Art and Multimedia, which includes photography, experimental video, computer graphics and performance. Member of the research group working on the conception of the Xmantic Web since January 1997.

Part II – Body

4   *Post-biological Body*

5   *Space and Time*

4 Post-biological Body

Recreating Ourselves
Romeo & Juliet in Hades by Future Movie
Naoko Tosa, Ryohei Nakatsu

Introduction
When humans dream during sleep, we are centrally involved in the events of our dreams. We are the lead character and walk, talk, feel happy, feel sad and actually even sweat. Is our consciousness active at this time? Moreover, is our deep psyche working at this time? When a novel or a film touches our emotions, our mind enters into its fantasy world. However, no matter how much we empathise with this world, we can never be more than just an outsider. I have thought what it would feel like to break down this barrier, and have started researching interactive theatre using the latest technology. Interactive theatre is an activity whereby one enters the film world and experience the narratives of the movie world not as an outsider but as a subject. One then tries to influence the actors and change the storyline depending on the conversations and actions that have taken place.

Empathy and Catharsis
In theatre and film, the most difficult but valuable effect is catharsis. Defined by Aristotle, it is the soothing release of the emotions. In theatre, however, not all emotions aroused by theatre are happy ones. Pity, anxiety and fear are mainstays of the theatrical style. It is the release of feelings that are thought to be pleasant rather than the emotions themselves. In the specific context of dramatic activities, the arousal and release of emotions have the power to purify our soul or inner world. The deeper the empathy we feel towards the narratives of the theatre and film, the higher a purified state we can get. Therefore both empathy and catharsis are essential for these media.

The Film Media
The events told by a film is the narrative. This tells of the relation among humans and feelings between one human being and another. These emotions are depicted by the film maker in his/her own special way. The major difference between film and theatre is the camera work, the way scenes are expressed, in an ordinary life way, and the fact that the time axis can be edited when working with film. Because of these factors, film may be called the expression of memory.

The Theatre Media
Theatre styles are visualized through the patterns of emotional tension coming from the audience. In the typical form, tension increases as the play progresses and a climax in the action is reached. This is then followed by a period of calm. The climax of a play is both universal and necessary and marks the moment when all other possible story directions are eliminated. It is for this reason that the climax is the pinnacle, not only in a emotional sense, but also in an informative sense. There is a direct connection between what we find out and how we feel. The control of information establishes fate and universality and triggers the release of tensions and emotions just at the moment of catharsis.

Theatre Does Not Describe
1. Theatre is performance. It is not reading; it is the expression of action. Both what we directly feel and what we understand have relevance to theatre performances. Story consist of description, while theatre consists of action.
2. Theatre does not express events as they occur in our daily life. Only significant or symbolic actions/utterances are selected and presented to the audience in an exaggerated form. In other words, theatre presents each event in a condensed form, thus giving strong impressions to the audience. On the other hand, film expresses events as they occur in our daily life. This means that events are expressed in an diffused way in the case of film. On the other hand, film has the capability of controlling the time axis more freely than theatre. Film gives the audience a feeling of empathy by utilizing this time control function.
3. Theatre has consistency in its action, while storytelling has an episode-like structure. For example, as each scene of film is a clip extracted from our everyday life, film is a sequence of 'episode'. Therefore to give the audience a strong impression, it is necessary for a director to utilise the contextual and temporal relationship among each scene. On the other hand, theatre can show each scene as a different one from our everyday life or even as an abstract one. Therefore it is easier for a theatre director to construct tight relationships among each theatre scene and as the result to create the stronger effect of climax and catharsis.

Future Movie as a New Media
The proposed interactive theatre is a medium that mixes film (prerecorded material) and live theater, and brings together the viewpoints of both those watching and those being watched (the actors). Remote performances of this type of theatre across different time zones and in different languages will become possible through the use of networks.

Avatar
To create the impression that you are really living in the fictitious world of film, an alter ego that can reflect you as you are and can be controlled by you is necessary. The proposed avatar design concept adopts human silhouettes. However, using silhouettes alone will not produce a strong representation when shown in a three-dimensional space. Therefore, wooden marionettes on shafts have been added. The hearts of these marionettes are made

of three metal cogs which only rotate when the avatars are being controlled by members of the audience. The head, body, arms and legs of each avatar are controlled by a audience member who is fitted with magnetic sensor ware. In order to make you aware of your movements, the avatar is devised to make the noise of these movements by hitting wood against metal. This is designed to make you feel like you are moving like a robot. *(Fig.1)*.

Interactive Story

A style has been adopted that brings a story to life, in a way similar to how we dream, through communication between the actors and with the avatars in a film-like production. We have put together an interactive scenario that deals with the fate of Romeo and Juliet after their deaths. The theme is the sentimental relationship between people in an imaginary world.

It goes like this. Humans, like other animals, keenly pursue other lives. Despite this pursuit, however, there are many circumstances that are out of our control. Encounters and separations, whether good or bad, are strange events, and cannot be accurately predicted by any person. When a computer stands between such living relationships, we must ask ourselves whether it is possible that this intervention may change these relationships.

The scenario is devised by the people playing Romeo and Juliet who select lines depending on their mood. How the drama unfolds depends on this selection process. Figures 2 to 4 show a number of scenes.

Figure 1

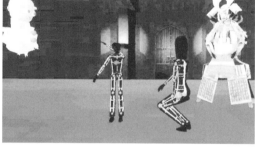

Figure 2

Figure 3

Figure 4

Conscious Interaction

Let's now take a look at the possibility of feeling empathy while undertaking conscious interaction. Our aim is to bring out catharsis, namely the seemingly pleasant release of emotions. However, if the techniques used for interaction do not work properly, this release will stop midway, producing unpleasant feelings. The interaction technology used and the combinations of these patterns can greatly affect good and bad consciousness.

Software Realising Conscious Interaction

1. *Anytime interaction*

Anytime interaction is the design function that allows players to interact with other characters or objects in the scene at any moment during the story whether prompted by a sign, obstacle, conflict, discovery, change of heart, success or failure.
* Follow interaction: This function enables the characters in the movie system to follow or accompany the avatar of a player in an appropriate situation. For example, if the player who plays the role of heroine Juliet is in a serious situation, an autonomous agent Shin, an angel, recognises her mental situation by emotion recognition, accompanies, and takes care of her.
* Background speech interaction: This function allows you to talk from the background to an character about your feelings and doubts concerning the performance running in the foreground.
* Touch interaction: This function is used when, for example, the lead actor Romeo becomes very angry and hits his old enemy Paris. Paris then pulls back in fear.

2. *Emotional recognition*

If the people playing the roles of Romeo and Juliet speak words of anger or of happiness, depending on the context of the play, autonomous characters recognise the emotions from the tone of voice and express their emotional reactions by speech and animations.

Emotional recognition technology allows a neural network to learn emotional speech and through this learning, the network can create personalities, such as an angry character or a cheerful one.

3. Voice recognition

The function whereby the lines chosen by a player to go with the scene are said, and the autonomous character recognises the meaning of these lines and reacts to them by their utterance and gestures.

4. Gesture recognition

The reaction of the character to action, contact or a pose in any location. Take for example if the impassioned Romeo was to try to kill his former friend, Macutio, with a pistol. Macutio would recognize this behaviour, escape and condemn Romeo.

Hardware

The system uses nine computers. In creating the computer graphics, SGI Onyx2 is used. Workstations are utilized for speech and gesture recognition. Also, PCs have been adapted for voice input and output.

Future Work

Future topics include describing interactive situations to improve the real-to-life feel of each scene. Here I do not mean a realism that blurs the line between what is real and what is not. What I envisage is the generation of unknown pleasant imagination that lies between people. This experience does not necessarily have a happy ending. The independent actors may betray the humans, while the humans may grumble about their problems, be afraid or feel pity. After this experience, however, we will have a digital catharsis in which a sense of fulfilment can be felt. To improve this cathartic feeling, we will need to use the most advanced technology to express the context of each scene. Ultimately this is linked to research on the unknown elements of humans such as the boundary between consciousness and unconsciousness.

Conclusion

Machines become most beautiful when they most resemble living forms. Even computer graphic actors are touched when something human-like is felt. In our age of machines, visual experiences using the most advanced technology are probably the closest to this feeling in machines. We have already come across unimaginable new man-made beauty in films using computer graphics. If we move from the era of machines to the era of images, it can be said that we will bring forth a new world of consciousness deeply related to the imagination. This will generate a new consciousness for humans that communicate with machines. This may even be experienced through the up and coming Future Movie.

Naoko Tosa Director of Interactive Cinema Project at ATR Media Integration & Communications Laboratories. Address: 2-2 Hikaridai, Seika-cho, Soraku gun, 619-02 Kyoto, Japan E-mail:tosa@mic.atr.co.jp UTL:http://www.mic.atr.co.jp/~tosa Naoko Tosa is also guest Professor in the graduate school of Science and Technology Kobe University. Her major research area is Art and Technology where she is working on the creation of Film, Video and media arts. Her recent work includes the Neuro-Baby project. Her work was exhibited at the Museum of Modern Art (New York),the Metropolitan Art Museum, SIGGRAPH, Ars ELECTRONICA and other locations worldwide. In addition, her works are collected at The Japan Foundation, American Film Association, and other institutions in Japan.

Ryohei Nakatsu received his B.S., M.S. and Ph.D. degrees in electronic engineering from Kyoto University in 1969, 1971 and 1982, respectively. After joining NTT in 1971, he mainly worked on speech recognition technology. Since 1994, he has been with ATR and currently is the president of the ATR Media Integration and Communications Research Laboratories. Recently he has become interested in the recognition of non-verbal information such as emotions in speech. In 1995 he met Naoko Tosa, a media artist, and started collaboration.Since then they have developed several computer characters which are able to communicate with people based on emotions. Their works were exhibited at the National Museum of Art in Osaka, O Museum in Tokyo and other museums and art exhibitions.

The Body as Interface
Jill Scott

Body Memory
Many inbuilt components of interaction are naturally present in the human body, for example the skin is interactive, the nervous system, the qualities of sight and sound, and all of these form the memory of the body's location in a given space and its navigational ability. Biologist David S.Goodsell describes the role of memory cells in the immune system which register whether/if so, the human body has been in touch with a virus and , call upon the same antibodies for resistance.[1]

Philosophical extensions of this idea include the beliefs of Vilem Flusser, who thinks that the human body has a memory composed of both a genetic memory and a cultural memory combined, 'as the human body stores and transmits equal amounts of both acquired and inherited information over successive generations'.[2]

Interskin, one of the three components of Digital Body Automata, explores the idea that according to alternative forms of medicine, genetic memories are also affected by cultural memory. For example, in Chinese acupuncture we find the opinion that old, bad habits of the mind can become stored in the organs of the body and block the energy flow.[3] In relation to the concept of seeing the human body as interface the computer, which is usually programmed to be logical and predictable, attempts to mimic our genetic memory, while the human elements of reflection and unpredictability consequent while engaging with the computer, are mainly culturally based. That is, they are learned from education or subject to elements of associative interruption and adapted in social exchange. As Flusser goes on to point out, there maybe a disadvantage in the fact that 'Electronic memories can forget more efficiently than human memories'.[4]

Navigation
In artificial surgery, the hardest thing a surgeon and computer can simulate is the simplest mobility of a three-year-old human action. Navigation is an inbuilt memory mechanism which deals with the sense of the body in space and this concept was used in *Digital Body Automata* to explore the genetic and cultural difference between two- and three-dimensional environments and images as well as the body's process of moving between these dimensions. As such, the three components provide a study for the location and orientation of the viewer inside a virtual environment and hence they explore the idea that the body can be seen as an interface. In *Interskin,* by using a torch in free space, to probe the alternating two- and three-dimensional representations of the human body and its separated parts, the viewer is confronted by a navigational challenge. as 3-D tends to be a genetic experience while 2-D tends to be a cultural one, connected to the primal centre of the human mind and body.

Interaction

In *Interskin*, the viewers are made aware of how their belief systems may affect their own health in the future but also about their own 'body-sense' in a confined space. The spatial design was based on the French telephone booths from the 50s [5], when intimacy and privacy were a major concerns and were influenced by the comments of Sandy Stone.[6] The *Interskin* telephone booths extend the idea that certain technologies can conceal identity while communication takes place.

The opposite occurrence can be found in another part of *Digital Body Automata*, called *Immortal Duality*. Here, the viewers move along a wall, where they can see a shadow and mirror of themselves upon a set of monitors. By using reverse key methods, they see a digital image reflected inside these shadows of their own bodies which display a micro-biological story about genetic transformation into a future-body state and a list of the desires of science. Here the viewers bodies become public identity, a live interface, incorporating what Mryon Kruger suggested, that the body is already the best form of gestural interaction can use in any artificial reality environment.[7] *Immortal Duality* extends the body into a virtual realm of communication which not only reflects the representations of the body it also allows the viewer to play with the navigation of movement to reveal the identities of future bodies.

Smart Sculptures

As the name *Interskin* implies, interactive skin experiments with the idea that immersive virtual environments offer the potential to enter through skin, which both Marc Lappe.[8] and Barbara Maria Stafford[9] imply. Myth, laboratory and clinic can be intimately interwoven, with the visual history of human skin being 'dissected, abstracted, marked, conceived, sensed and inscribed' particularly via scientific visualisation which will allow us 'see' inside the body. However, there may be only a little difference between the machines which we create to enter 'see' ourselves and the way we 'see' the machine as extensions of the skin of ourselves. As Donna Haraway says, 'The Machine is not an it to be animated, worshipped or dominated. The machine is us, our processes, an aspect of our embodiment. We can be responsible for machines, they do not dominate or threaten us[10]. This blurring of the edges between machines and us calls for a redesign of imput devices, a device which, as Sally Pryor has suggested, could contain our senses of smell, touch and taste[11]. Can artists improve the relationship to machines by altering their design to be touch capable, soft and more organic?

With this in mind, I am in the process of converting sculptures into impute devices which I call Smart Sculptures. These incorporate technology and require the touch and substance of the human body to be made smart. Our bodies, for example, are composed of 95 percent water; it is not

Overview

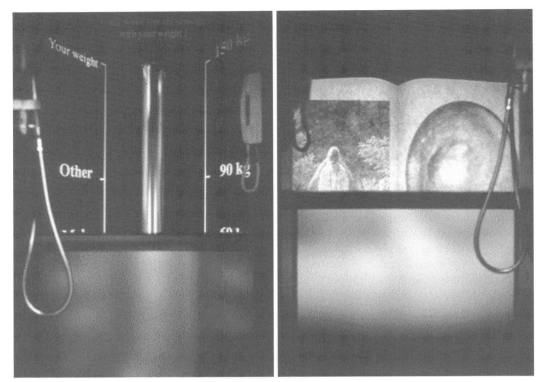

Interskin – inside the telephone booths

dangerous to pass an electronic charge of approximately 2-5 amps through the body's skin. In *A Figurative History*, one of the interesting things about art objects is their potential to be metaphorically and historically pertinent or associative for the viewer, particularly if they are placed out of context as in the Readymades of Marcel Duchamp[12].

Within this theory, which I call *smart sculptures*, the participant can actually enter the content via these objects to become an immersed part of the artwork, even though the content or associations can cause dislocation. By touching a smart sculpture, a symbiotic meeting of two worlds places the viewer in the role of a performer, increasing the role of the body to be an interface. Thus through participation, the viewer becomes a 'modern cyborg', literally re-enforcing Haraway's statement 'We are all modern Frankensteins', cyborgs, fabricated hybrids of machine and organism'[13]

Here, the concepts of local or/and foreign body interaction are enhanced by two important issues, ease of use and access to symbolic association, which in this case are animated and transformed body types. Within each sculpture local or foreign connections can occur. While the local connection allows the viewer to talk to virtual characters privately, the foreign role of the smart sculptures is to instruct the players to join hands with each other and touch the objects simultaneously. In this way some of Brenda Laurel's ideas are confirmed about extending the word 'interface' to include the body. While opening

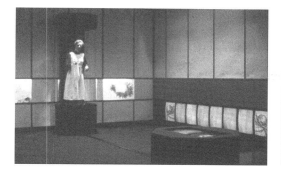 

Immortal duality *A figurative history*

up new ways of experience in terms of 'bundling functionality' could encourage the idea of a group of bodies becoming one giant interface as attempted in *Immortal Duality, A Figurative History*, which brings conductivity into physical reality. The Smart Sculptures cause the breakdown of the fine lines between the organic body and the mechanical sculpture. Thus the viewer's body becomes an accomplice to the artificial in order to access virtual characters.

As suggested in this paper, the three parts of Digital Body Automata attempt to integrate the body and the machine into a closed symbiotic relationship on a practical and therefore theoretical level. *A Figurative History* does this by incorporating new interface design (smart sculptures) with both cultural and genetic memories and desire for body transformation, while *Interskin* places the viewers in the centre of the virtual human body, tossing them between the interior and exterior metaphors of the body in order to discover new identities. *Immortal Duality*, on the other hand allows for the direct reflection of the future of molecular transformation in the shadow of the viewer's body, without encumbering the body with any extra devices. Consequently, these explorations extend the study of body as an interface, as well as help in the analysis of machines which we ourselves create as extensions of our bodies. Therefore, they tend to shed light on the way we 'see' and 'interpret' the direct role of the body, graduating it from being an accomplice to an impute device, to becoming an essential part of the actual interface.

Notes

1. Goodsell, David S.1993 *The Machinery of Life*. New York: Springer Verlag, p.29.
2. Flusser,Wilem.1990 'On Memory.Electronic or Otherwise' in *Leonardo Magazine of Art and Science*, vol.23,p.397.
3. For a social discussion on the effect of memory on disease from different Eastern viewpoints see Sheldrake,R. 1990. The Prescence of the Past-Morphic Resonance. Los Angeles.Park Street Press.
4. Flusser.op.cit. p.20.
5. In the early fifties, French Telecom conducted a set of research tests on the levels of intimacy and concealment of identification, in order to design a new telephone box.
6. Stone, Alucquiere Rosanne.1995.*The War of Desire and Technology at the Close of the Mechanical Age*. Cambridge, Mass. 'The MIT Press. Allucquiere Rosanne Stone says that the telephone is in fact an extended part of

ourselves, and the continuing popularity of the telephone on our daily lives 'shapes us, structures our ways of seeing and changes us'.
7   Kruger was one of the first artists to advocate using the body as an interface. For further discussion, see his book 1992. Artificial Reality. Los Angeles. Addison Weetley. p.23.
8   See Marc Lappe.1992. '*The Body's Edge, Our Cultural Obsession with Skin.*' Plus, as cited in Jan Holt's book review of the same title in *Nature Magazine* vol.20 U.S.A (1996) for example, praises the skin as the ultimate interface. Lappe says that the skin is much more than a boundary of the physical self; instead, skin is a metaphor, it is 'a living screen of unmatched efficiency, a focal point of anesthesia, the conscious faculty that contributes to physical preservation. It engages with the forces of sexuality and the level of our spiritual condition and it is a sensitive radar alert to the subtlest pertubations of the environment'.
9   Stafford, Barbara Maria. 1995. '*Body Criticism*' Mass.:MIT Press,p.15.
10 Haraway, Donna. 'A Manifesto for Cyborgs.Science,Technology and Socialist Feminism in the 1980's. *Australian Feminist Studies*, vol.5,(Autumn 1987)p.30.
11 Pryor, Sally.1991. 'Thinking of Oneself as a Computer', in *Leonardo*, vol.24, no.5,pp.585-90.
12 Block, Rene.1990.'Duchamp and the Readymades', *Art is Easy*. Sydney Biennale:Sydney Press.
13 Haraway, Donna.1990. *Simians, Cyborgs and Women, The Re-Invention of Nature*. (Autumn 1987) p.30.

Art at the Biological Frontier
Eduardo Kac

Working with multiple media to create hybrids from the conventional operations of existing communications systems, I hope to engage participants in situations involving biological elements, telerobotics, interspecies interaction, light, language, distant places, time zones, video conferences and the exchange and transformation of information via networks. Often relying on contigency, indeterminacy and the intervention of the participant, I wish to encourage dialogical interaction and to confront complex issues concerning identity, agency, responsibility and the very possibility of communication.

Biotelematics
'Teleporting an Unknown State' is the title of my biotelematic installation which linked the Contemporary Art Center in New Orleans to the Internet (August 4-August 9, 1996). This piece was part of 'The Bridge', the Siggraph '96 Art Show. 'Teleporting an Unknown State' combined biological growth with Internet (remote) activity. In a very dark room a pedestal with earth served as a nursery for a single seed. Through a video projector suspended above and facing the pedestal, remote individuals sent light via the Internet to enable this seed to photosynthesise and grow in total darkness. The installation created the experience of the Internet as a life-supporting system.

As local viewers walked in they saw the installation: a video projector hung from the ceiling and faced down, where a single seed lay on a bed of earth. Viewers did not see the projector itself, only its cone of light projected through a circular hole in the ceiling. The circularity of the hole and the projector's lens flush with it are evocative of the sun

'Teleporting an Unknown State', Eduardo Kac, 1996.

breaking through darkness. At remote sites around the world, anonymous individuals pointed their digital cameras to the sky and transmitted sunlight to the gallery. The photons captured by cameras at the remote sites were re-emitted through the projector in the gallery. The video images transmitted live from remote countries were stripped of any representational value and used as conveyors of actual wavefronts of light. The slow process of growth of the plant was transmitted live to the world via the Internet as long as the exhibition was up. All participants were able to see the process of growth via the Internet. The computer screen, i.e. the graphical interface on which all the activity could be seen, was dematerialised and projected directly onto the bed of earth in a dark room, enabling direct physical contact between the seed and the photonic stream.

This piece operated a dramatic reversal of the regulated unidirectional model imposed by broadcasting standards and the communications industry. Rather than transmitting a specific message from one point to many passive receivers, 'Teleporting an Unknown State' created a new situation in which several individuals in remote countries transmitted light to a single point in the Contemporary Art Center in New Orleans. The ethics of Internet ecology and social network survival were made evident in a distributed and collaborative effort.

During the show, photosynthesis depended on remote collective action from anonymous participants. Birth, growth and death on the Internet formed a horizon of possibilities that unfolded as participants dynamically contributed to the work. Collaborative action and responsibility through the network were essential for the survival of the organism. The exhibition ended on August 9, 1996. On that day the plant was 18 inches tall. After the show, I gently unrooted the plant and replanted it next to a tree by the Contemporary Art Center's front door.

Biorobotics

'A-positive' was a dialogical event realised by Ed Bennett and myself on September 24, 1997, at Gallery 2 in Chicago, in the context of the ISEA 97 art exhibition. This work probes the delicate relationship between the human body and emerging new breeds of hybrid machines that incorporate biological elements and from these elements extract sensorial or metabolic functions. The work created a situation in which a human being and a robot had direct physical contact via an intravenous needle connected to clear tubing and fed one another in a mutually nourishing relationship. To the new category of hybrid biological robots the general epithet 'biobots' is ascribed. Because of its use of human red blood cells, the biobot created for 'A-positive' is termed a 'phlebot'.

In 'A-positive', the human body provided the robot with life-sustaining nutrients by actually donating blood to it; the biobot accepted the human blood and from it extracts enough oxygen to support a small and unstable flame, an archetypal symbol of life. In exchange, the biobot donated dextrose to the human body, which accepted it intravenously. In 'A-positive', oxygen is extracted by the phlebot and used to support the erratic flame. The conceptual model created by this dialogical work is far from conventional scenarios that portray robots as slaves that perform difficult, repetitive or humanly impossible tasks; instead, as the event unfolds, the human being gives his own blood to the biobot, creating with it a symbiotic exchange.

This work proposes that emerging forms of human/machine interface penetrate the sacred boundaries of the flesh, with profound cultural and philosophical implications. 'A-positive' draws attention to the condition of the human body in the new context in which biology meets computer science and robotics. We can no longer regard the body as isolated from firm contact with the technoscape or protected from the biological surveillance of biometrics. Not even DNA or blood are immune to the invasion of the body by technology. A DNA computer has been successfully demonstrated through the joint effort of a biologist and a computer scientist. Instead of electrical impulses, it employs deoxyribonucleic acid, or DNA, to control the commands a processor gives to a computer and uses nucleotides, the basic units of DNA, to replicate the actions of a processor. The technologies that condition our imagery and sensibility at the end of the century (including nanotechnology and genetic engineering) also penetrate our skin-our blood stream, even-enabling new forms of therapy. Miniaturised electronic devices and new chemical compounds are invading (and cohabiting) the physical structure of an organism. For example, a new technology aptly called 'electroinsertion' proposes to increase a drug's effective potency manifold by binding specific drug molecules directly to the red blood cells, rather than adding a drug to the circulatory system. This and other related developments clearly reveal that technology has already permeated the body in subtle ways. The dialogical situation created in 'A-positive' quite literally 'wires' the human being to the robot, with four connection points in a prototypical biological LAN (Local Area Network). Once extracted and released inside the sealed chamber, the oxygen supports the minuscule glowing mass of burning gas, the symbolic 'nanoflame'.

Bioimplants

'Time Capsule' was a work-experience realised on November 11, 1997, at Casa das Rosas, a cultural centre in São Paulo, Brazil. The piece lies somewhere between a local event-installation, a site-specific work in which the site itself is both my body and a remote database, and a simulcast on TV and the Web. The object that gives the piece its title is a microchip that contains a programmed identification number and is integrated with a coil and a capacitor, all hermetically sealed in biocompatible glass. The temporal scale of the work is stretched between the ephemeral and the permanent, i.e. between the few minutes necessary for the completion of the basic procedure, the microchip implantation and the permanent character of the implant. As with other underground time capsules, it is under the skin that this digital time capsule projects itself into the future.

When the public walked into the gallery where this work took place, what they saw was a medical professional, seven sepia-toned photographs shot in Eastern Europe in the 1930s, a horizontal bedstead, an on-line computer serving the Web, a telerobotic finger and

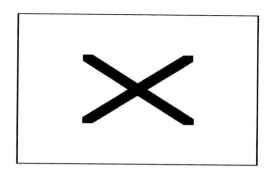

'A-positive', Eduardo Kac and Ed Bennett, 1997.

additional broadcasting equipment. I started (and concluded) the basic procedure by washing the skin of my ankle with an antiseptic and using a special needle to insert subcutaneously the passive microchip, which is in fact a transponder with no power supply to replace or moving parts to wear out. Scanning the implant remotely via the Net generated a low energy radio signal (125 KHz) that energised the microchip to transmit its unique and inalterable numerical code, which was shown on the scanner's 16-character Liquid Crystal Display (LCD). Immediately after this data was obtained I registered myself via the Web in a remote database located in the United States. This is the first instance of a human being added to the database, since this registry was originally designed for identification and recovery of lost animals. I registered myself both as animal and owner under my own name. After implantation, a small layer of connective tissue formed around the microchip, preventing migration.

Not coincidentally, documentation and identification have been one of the main thrusts of technological development, particularly in the area of imaging, from the first photograph to ubiquitous video surveillance. Throughout the 19th and the 20th centuries photography and its adjacent imaging tools functioned as a social time capsule, enabling the collective preservation of memory of our social bodies. At the end of the 20th century, however, we witness a global inflation of the image and the erasure by digital technologies of the sacred power of photography as truth. Today we can no longer trust the representational nature of the image as the key agent in the preservation of social or personal memory and identity. The present condition allows us to change the configuration of our skin through plastic surgery as easily as we can manipulate the representation of our skin through digital imaging, so that we can now embody the image of ourselves that we desire to become. With the ability to change flesh and image also comes the possibility of erasure of their memory.

Memory today is on a chip. As we call 'memory' the storage units of computers and robots, we anthropomorphise our machines, making them look a little bit more like us. In the process, we mimic them as well. The body is traditionally seen as the sacred repository of human-only memories, acquired as the result of genetic inheritance or personal experiences. Memory chips are found inside computers and robots and not inside the human body yet. In 'Time Capsule', the presence of the chip (with its recorded retrievable data) inside the body forces us to consider the co-presence of lived memories and artificial memories within us. External memories become implants in the body, anticipating future instances in which events of this sort might become common practice and inquiring about the legitimacy and ethical implications of such procedures in the digital culture. Live transmissions on television and on the Web were an intergal part of "Time Capsule" and brought the issue closer to our living rooms. Scanning of the implant remotely via the Web

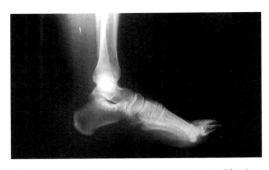

'Time Capsule', Eduardo Kac, 1997. X-ray of kac's left ankle indicating the position of the microchip implant

revealed how the connective tissue of the global digital network renders obsolete the skin as a protective boundary demarcating the limits of the body.

The need for alternative ways of experience in the digital culture is evident. The wet hosting of digital memory-as exemplified by 'Time Capsule'-points to traumatic but perhaps freer form of embodiment of such proposition. The intradermal presence of a microchip reveals the drama of this conflict, as we try to develop conceptual models that make explicit undesirable implications of this impulse and that, at the same time, will allow us to reconcile aspects of our experience still generally regarded as antagonistic, such as freedom of movement, data storage and processing, moist interfaces and networking environments.

Eduardo Kac is an artist and writer who works with electronic and photonic media, including telepresence, robotics and the Internet. His work has been exhibited widely in the United States, Europe and South America. Kac's works belong to the permanent collections of the Museum of Modern Art in New York, the Museum of Holography in Chicago and the Museum of Modern Art in Rio de Janeiro, Brazil, among others. He is a member of the editorial board of the journal Leonardo, published by MIT Press. His anthology 'New Media Poetry: Poetic Innovation and New Technologies' was published in 1996 as a special issue of the journal Visible Language, of which he was a guest editor. Writings by Kac on electronic art as well as articles about his work have appeared in several books, newspapers, magazines, and journals in many countries, including Argentina, Australia, Austria, Bolivia, Brazil, Finland, France, Germany, Holland, Hungary, Mexico, Paraguay, Portugal, Spain, Russia, Uruguay, The United Kingdom, and The United States. He is an Assistant Professor of Art and Technology at the School of the Art Institute of Chicago and has received numerous grants and awards for his work. Eduardo Kac can be contacted at: ekac@artic.edu. His work can be seen at: http://www.ekac.org

Images and Imaging:

Amy Ione

Introduction

Images are everywhere today and when we speak about them with others it often becomes clear that many people assume that the invention of photography moved us away from language and toward the iconic (e.g. see Barry 1997; Dondis 1973). This characterisation of image proliferation fails to acknowledge that non-optical images differ significantly from photographic ones. In this paper I demonstrate that the difference is an important one because non-optical images are capable of piercing through physical surfaces. Thus they allow us to bring information generated from non-visible wavelengths into a visible form. This is not a trivial point. Being graphic renderings of invisible domains, the non-optical images are maps capable of placing 'something' in the portion of the spectrum used by our eyes, but something that *could not be seen without* the new technologies.

The visible domain is shown in Figure 1. It includes wavelengths ranging from red light (at about 700 nanometres) down to violet light (at about 400 nanometres). This is the section where we find *all* of the colours of the rainbow. The self-propagating waves that comprise the *entire* spectrum include a variety of electric and magnetic fields which, like visible rays, travel at the speed of light. These other waves, however, have different frequencies from visible light and it is frequency that determines whether we characterize waves as radio, microwave,infrared, visible light, ultraviolet,X-rays or gamma rays.

Briefly, the non-optical portion of the spectrum could be said to have become a visual part of our lives in November 1895, when Wilhelm Conrad Roentgen became the first person to see the bones of his hand through his skin. Recognising he had glimpsed something that should not have appeared before his eyes, Roentgen investigated and quickly discovered how the use of 'invisible' radiation can produce and record a 'visible' image of an 'invisible' object property (Beck 1994). One critical element here is that this kind of non-algorithmic insight cannot be classified as something that was *directly* perceived , directly *received* or analytically conceptualised. Another important point is that the visual quality allowed people to quickly grasp that we could make the skin, as well as other physical and opaque surfaces, transparent.

Images and Consciousness

Overall the event had a varied and nonetheless momentous impact on life and culture in general. Scientists like Roentgen were enthusiastic about opening a door to another way of seeing, while lay people like Frau Roentgen were often less enthusiastic. After confronting her skeletal hand in her husband's laboratory, she was convinced it was an omen of death and never returned. Some people were also taken aback to discover electromagnetic waves had been surrounding and bombarding humans throughout history-without their knowledge or awareness of them.

This is not to say that the idea of rays piercing the body was itself totally novel. To the contrary, ideas of rays had been around for centuries. Even as far back as the 13th century the philosopher Roger bacon had noted that no substrate is so dense that it can prevent rays from passing through and he pointed out that the walls of a vessel of gold or brass show this

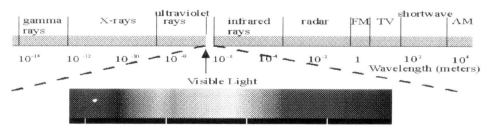

Figure 1. Electromagnetic Spectrum.

when they heat up (Kevles, 1997). There was also the belief that spiritual rays emanated from the body to the outside world and these were often portrayed in paintings by halos around the heads of saints and religious figures. But nowhere did artists, mystics or philosophers suggest that any rays, whether passing from the inside or emerging from the soul out through the skin, could reveal everything beneath the skin to a human eye, even with technology. There was also no suggestion that the images could leave an impression on something else, like a shadow on a wall or a permanent imprint on glass or film.[1] Yet these rays could be tamed to do so.

Philosophical concerns were also formulated as people realized only a tiny fraction of the waves are directly perceived by us as visible light and thus we needed to re-consider what 'seeing' and empirical evidence in general are. Reactions also varied in regard to how the waves should be integrated. Biological and physical scientists were attracted to the evidence that the generated data had a testable, replicable and tangible quality. They were also captivated by the ability to chart possibilities never imagined to exist! Artists, on the other hand, were captivated by how newly revealed information offered new perceptions of how the mind, imagination and body exchange with the environment. In addition, the spiritually inclined were immediately attracted to the images because they revealed unseen worlds (Henderson 1983). Virtually unfazed by the fact that this new domain did not confirm the so-called spiritual truths recorded by psychics, mystics and others, who had made forays into the 'invisible' domains of 'spirit' throughout history, artists quickly responded to the metaphoric qualities being suggested.

Some, like the Russian mystic Pavel Tchelitchew, saw the X-rays as a door to another dimension. This new dimension was not time but a fourth dimension echo of spiritualism. His perception is evident in his painting *Hide-and-Seek*, produced in 1942. Purchased by New York's Museum of Modern Art, it has been the most visited painting in that museum for four decades (Kevles 1997). Gerald D.Fischbach, a Professor of Neurobiology, described Tchelitchew's rendering as one that

> [C]aptures the interplay between the mind and environment that influences the brain's development as well as it's architecture. Hidden forms are embedded figures, a delicate test of mental function. Roots, branches and vines suggest neuronal arborisation and the ability of such structures to change (Fischbach 1992, p.49).

Thus the images Tchelitchew chose, quickly authenticate the painting as a 20th century artifact. More specifically, the image displays a chorus of X-ray images of children's see-through heads arranged in puzzle-like patterns as parts of a growing tree. The veins and bones of the children merge with the roots and bark of the tree in a now-you-see-it, now-you-don't puzzle pattern of arteries and landscape. Looking at the piece we see the interior of the bodies include an accurate physiological rendering kind of skeletal structure and internal elements that only become transparent with the imaging technologies (Kelves 1997).

Cognitive neuroscientists have produced another quality of image using digital advances. Their research is especially intriguing because it represents a multi-dimensional perspective on how digital images have joined with human consciousness. What is perhaps most noteworthy in probing these contributions is that the scientists themselves continue to respond to the images of consciousness as if they are photographs illustrating the mind, overlooking the insight that made the non-optical investigations possible.

For example, Michael Posner and Marcus Raichle begin their 1996 book *Images and Mind* by noting that the study of imagery forms a part of a more general enquiry into the relationship of mind and brain that is now producing real insights into the workings of the brain. They note that

> [R]ecent advances in technology have made it feasible to create pictures of our brains as we think. For the first time, scientists are able to render certain aspects of thought visible, by recording its physical effects on brain activity; these records . . . can show which areas of your brain are active, and which dormant. One of these streams is the description of mental processes in terms of component operations that can be precisely specified . . . A visual image is not created as a whole but rather formed over time by an orderly set of operations that includes, for example, placing the parts of the images in their proper relationship and scanning the content for specific features. The measurement of these operations have been the task of cognitive science for the past 30 years. The second stream of research, called neuroscience, is the study of how brains are constructed (Posner and Raichle 1997, p.2).

This points to how brain-imaging methodology allows system-level analysis of human cognition by providing an image of complex tasks. This technological link to biological tasks-such as visual attentional control, memory storage and language interpretation-allows for collection, characterisation and cataloguing. None of these modes, however, explain insights like the X-ray itself. As I explained earlier, the realisation of non-optical image formation that began with the X-ray, was not created over time by an orderly set of operations that included things like scanning, placing the parts of the images in their proper relationships and seeing how the form is constructed. These operations followed after the visual anomaly was noticed. To be sure, the subsequent investigation included operations like placing the parts of the images in their proper relationship and scanning the content for specific features. Nonetheless, image was needed to *begin* scanning, searching for relationships, drawing conclusions, etc. Thus, what Posner and Raichle fail to include is how *discovery, creativity, innovation* and *invention* brought the data-driven non-optical technologies into the arena where mapping and the study of the brain take place.

Conclusion

Once non-optical images helped us glimpse information long unknown to human eyes, human minds and human artistry began to map the once invisible terrain. As I have shown, both artists and scientists have re-framed questions in regard to how much we really see and what we can actually 'see'. One link between artists and scientists at this initial stage is that scientists are using some measure of artistry to evolve digital image formation, interpretation and clarity. On the other hand, artists have used the digital advancements to personalise their human reflections on what the scientific ideas have to say to us in a more general sense.

Subtle issues have also emerged. For example, we have slowly discovered that the images do not just 'look' into other domains. Exposure to X-rays and other forms of radiation can easily *damage* the body-and the specific nature of the damage might not be evident for years.

This dangerous aspect of what had at first seemed to be a miraculous discovery was initially as startling as the rays themselves. Having been discovered at a time when science meant progress, people were especially reticent to accept that the gifts of non-optical technologies had to be balanced with some potentially lethal repercussions. We now know that damage is possible and this knowledge has changed our view of science-and our consciousness as well.

Finally, the discovery of non-optical imaging tools has articulated the difficulty we have in accounting for the kind of innovation that stretches our capacity to see. Often the various theories concur that good models for parallel processing are similar to Gestalt images such as the one where we see two heads if we look one way and a vase from the other perspective. Yet, what we do not find in this kind of image is a description of how we add to our knowledge base. The Gestalt image offers a means to conceptualise that the whole can be more than the sum of its parts but the image is not versatile enough to speak about how we perceive what is not 'present' in any part of what we know and see. As the X-ray shows and as consciousness theorists have discovered, the kind of revolutionary discovery I am speaking about is not defined when we map the brain either. These maps likewise fail to include a means for introducing information we are not capable of representing. Moreover, as Roentgen's discovery graphically demonstrated, this kind of invisible and unknown information cannot be adequately accounted for simply by adopting traditional philosophy or spiritual ideas about transcendence. In Roentgen's case, there is no reason to assume his insight was related to logic or any kind of *higher* dimension.

In sum, perhaps it is because we cannot define the wonder of creativity that both artists and scientists often see their work with images as mapmaking and the maps as a record of their explorations. As the Japanese physicist Hideki Yukawa said:

> Those who explore an unknown world are travellers without a map; the map is the result of the exploration. The position of their destination is not known to them, and the direct path that leads to it is not yet made (in Hall 1992, p.159).

Note

1   'What appears to be the only recorded prediction came in a talk delivered to a small group of physicians in Philadelphia in 1872 when the physician James Da Costa satirised the state of medicine by describing a future in which 'Dr. Magent, who is a very accomplished physicist, steps forward: 'If you would permit me, I will make you transparent'. And with lenses dexterously passed into the stomach, the fair patient is really rendered transparent.' (Kevles 1997, p.14).

References

Barry, A.M.S. 1997. *Visual Intelligence*. Albany: State University of New York Press.
Beck, R.N. 1994. 'The future of imaging science', in T.S. a. J.Umiker-Seboek Ed., *Advances in Visual Semiotics-the Semiotic Web*, pp.609-42. Berline:Walter de Gruyter, Mouton Publications.
Dondis, D.A. 1973. *A Primer in Visual Literacy*. Cambridge: MIT Press.
Fischbach, G.D. September 1992. 'Mind and brain', in *Scientific American*, 2673, 48-57.
Hall, S.S. 1992. *Mapping the Next Millenium*. New York: Vintage Books.
Henderson, L. 1983. *The Fourth Dimension and Non-Euclidean Geometry in Modern Art*. Princeton: Princeton.
Kevles, B.H. 1997. *Naked to the Bone*. New Brunswick, New Jersey: Rutgers University Press.

Posner, M.I. and Raichle, M.E. 1997. *Images of Mind*. New York: Scientific American Library

Amy Ione writes on discovery, creativity, innovation and historical challenges in art and science. She also speaks about her research in these areas internationally. In addition, Ione works as a consultant on art and science projects, recently bringing her expertise to the San Francisco Exploratorium museum planning process for the 'Our Bodies, Our Facsimilies' exhibition, scheduled to open in the year 2000. She was a panelist at the ArtSci 98 symposium 'Seeding Collaboration' held at the Cooper Union in new York City, and a speaker at The International Congress for Discovery and Creativity at the University of Gent in Belgium. Ione's background includes a commission by the City of San Francisco to create the poster/publication illustration for San Francisco's 40th annual arts festival, teaching the History of Science at John F.Kennedy University, and gallery, museum and conference exhibitions of her paintings in the United States and Europe. More information about her art and academic work is available at <http://users.1mi.net/~ione> or contact her at ione@1mi.net

The power of digital seduction in biomedicine
Nina Czegledy

The common core to recent developments in imaging has been the availability of computers providing rapid mathematical modelling of diagnosed images. The force and speed of these changing technologies have been overwhelming, leaving little room for reflection. Few current developments offer ideas as seductive and captivating as the computer for analysing ourselves. It is, however, a commercially generated instrument and as such, it does not promote extensive philosophical investigation.There are many possible approaches to explore the nature of consciousness and the potential of 'digital' seduction (e.g. cultural, socio-psychological, etc.). This paper examines visual awareness from a neurobiological viewpoint.

The acceptance and use of digital imaging systems have developed as a result of the maturation of key technologies which underlie image processing as well as the increased demand for image and graphics applications in the commercial and scientific sectors. At the heart of the investigations is the digital image itself, which is manipulated, processed, communicated, stored, compressed and archived and which-on account of its intangible nature-is difficult to define. As a consequence, we deal with images whose truth we cannot test and that may represent the projection of an idea rather than a bona fide observation from the 'real' world.

The sophisticated skills of the advertising industry have undoubtedly influenced our visual perceptions and heightened our expectations. This has already led to the 'glossy' standards of commercial journals being extended to scientific publications. Digitisation permits extensive prevarication or even fabrication. The range of manipulation varies from subtle changes in illustrations to markedly altered advertisements. Dramatic diagrams and expertly manipulated images have contributed to a dazzlingly colourful array of scientific illustrations aimed at luring the reader towards acceptance and away from scrutiny. The area of biomedical image interpretation provides an excellent model in which to consider the enigma of this seduction by digital imagery.

Practically limitless data preservation, immediate retrieval and reproducibility are some of the key components contributing to the pragmatic explanation of the power of image enhancement in investigative medicine. The transformation of ideas into visible reality requires the construction of specific models. Reproducible elements-an inherent quality of digitisation-are eminently conducive to this work. Beyond the strictly practical applications however, these technologies acquire an often unexpected life of their own. By offering a seemingly rich and compelling 'life-like' representation of reality, digitally mediated images are extremely captivating. Set beside this embellished vista, real life becomes pale and the distinction between real and virtual becomes more and more ambiguous.

A few decades ago computing and, by extension, digital technologies, appeared to be mechanistic, whereas medicine was considered a humanistic subject. The explosive development of communication and information theories has since reshaped our perceptions of such considerations. Digital manipulations created a dramatic change, a new affinity in the relationship between humans and machines. Medicine, from genetics to telesurgery-through the escalated use of advanced digital technologies-has become increasingly mediated in character. In addition to image adjustments, the new visualisation techniques contributed to a change in scaling, a minimisation of distance, an alteration of space, a loss of physical and conceptual reference points. Kim Sawchuk, communication theorist, coined the term 'biotourism' to describe investigations of the interior space of the body. Fascinated by the mediated relationships and the transformation of scale involved in currently employed diagnostic tools, Sawchuk asked 'What happens to the viewer when you start to play with scale, what does it do to your consciousness?' (17)

The mechanisms of perception and visual awareness related to consciousness and neuroscience have been extensively investigated by Francis Crick and Christof Koch (2,3,4). Arguing from a reductionist point of view, they postulated that to be aware of an object or event, the brain has to construct a multilevel, explicit, symbolic interpretation of the visual scene. According to Crick and Koch, it is not possible to convey with words the exact nature of a subjective experience, although it might be possible to convey a difference between various subjective experiences (4). Despite intensive investigations their search has failed to find the location and nature of the neural correlates of visual awareness. Crick and Koch are not alone in trying to find a key to explain the nature of visual awareness. Neuroscientists have began to use neuropsychological data to investigate the structure of the processing system and to characterise the nature of representation itself (6,11). While some current findings began to clarify cognitive procedures, it is apparent that visual imagery, like perception and visual representation, is not a unitary concept (1). Yet the complexity and far-reaching consequences of digital seduction require scrutiny, especially as the extended use of computer-assisted image processing increasingly blurs the line between the real and the seemingly real-the virtual.

How is our spatial sense altered by new technologies? Advances in computer visualisation and user interface programs have enabled the development of virtual reality (VR) programmes that allow users to perceive and interact with objects in artificial three dimensional environments. Computer generated and digitally enhanced images have also

Post-biological Body

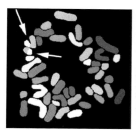

Spectral chromosome analysis using SKYVIEW software

Giemsa stained chromosome metaphase

permitted physical representation of formerly abstract or less defined scientific concepts. An underlying perceptual reason might be that visual information is much easier to perceive and less time consuming to process than text based information. The development of image enhancement has thus reduced the degree of abstraction by the achievement of seemingly 'perfect' practically tactile pictures. It is now suggested that we operate a range of satisfaction levels, determined by how removed abstractions are from direct and hardwired recognition (12).

In specific areas of molecular genetics the development of virtual reality environments has even replaced mathematical descriptions. A long list of molecular modelling software programs-from *'Babel'* to *'Xmol'*-are advertised on the Internet. The advertisement for the molecular visualisation freeware *'Rasmol'* entices prospective customers with 'a little basic biochemistry' knowledge 'to get lots of molecules', because 'learning how to use (*RasMol*) is easy and fun'. The lure of this advertisement is not dissimilar to expectations generated by the ever-growing number of devoted computer game players, sharing a passion for the possibilities opened by new exciting worlds .

In addition to VR techniques, a long list of various imaging technologies, such as FISH (Fluorescence in Situ Hybridisation) for gene mapping or confocal microscopy for

3D molecular modelling have recently been used for diagnostic and research purposes. In contrast to these spectacular techniques, the use of black and white images of traditional electron micrographs has declined. A possible explanation for the preference for 'virtual' representation might be that it has provided a new, interactive factor in scientific investigations. It is intriguing to hypothesise that the digital image created by the researcher has now become an important motivational tool as well as a postulated entity. Mental imagery is used frequently by athletes or musicians to deal with critical events. In this case the researcher 'sees ' the target for his/her creative ideas and actualisation becomes a means of energising towards the critical research 'event'.

Further than the obvious importance of improved diagnostics, what is the true nature of the technological enticement in biomedicine? Is it due to humans being highly visual ? Or is this a definable siren-like lure of the neural network, perhaps a narcissistic process? Is this enhancement due to curiosity or the aesthetic pleasure of discovering something of a fascinating quality? What is mirrored in the inner minds of those who are involved, those who are seduced? Is there an underlaying desire to be seduced? If we postulate that everything stems from within (the investigator), where does this seductive power originate? What leads to a specific visual experience?

Despite intensive research, it has become increasingly difficult to decipher and define the exact nature of mental imagery. The unresolved but well documented phenomenon of *phantom limb* sensation might serve as an intriguing example of unusual subjective experiences. The feelings and perceptions of the patients who suffered loss of a limb are reported to be extremely vivid and real. Mental events and processes may therefore seem as much part of our biological natural history as those affecting any other organism. According to McGinn any theoretical concept which serves to explain a property of the brain or other physical object must have its roots in perception. No such concepts exist, however, for explaining the brain's production of consciousness, since the property to be explained-consciousness-is itself paradigmatically unobservable (13).

In order to find some answers it is important to ascertain which activity in the brain is directly involved with visual awareness and to explore the connections between visual awareness and neuronal activity in the nervous system. For a long time the localisation of mental image generation was considered to be a function of the right hemisphere. Recent reviews postulated left hemisphere specialisation for mental image generation and Sergent argues (16) that both hemispheres contribute to the process. In her review Farrah concluded that image generation-depending on the nature of the image-is either allocated to the left hemisphere alone or to both hemispheres(7).

It is assumed that the brain requires a considerable amount of neural activity to construct a representation, most of which is probably done unconsciously. Research on subliminal perception and implicit memory suggests that events can affect mental functions, although they cannot be consciously perceived or remembered (10). Milner and Goodale suggest (14) two systems in primates: an 'on-line' and a 'seeing system'. The latter is considered conscious while the former, acting more rapidly, is not. Further explanations of multiple cortical streams with numerous anatomical connections have been postulated by Distler (5) and Fuster (8) among others. To determine the validity of these systems further research is essential .

Neural networks seem to provide a framework for considering connections between perception, memory and imagery. Crick and Koch suggest that the biological usefulness of visual consciousness is to produce the best current interpretation of the visual scene in the light of past experience, either of ourselves or of our ancestors (embodied in our genes) (3). In their opinion the brain must use past (even ancestral) experience to assist visual interpretation. Yet it remains debatable how much of our perception is due to early visual experiences or even reactivated embryonic imprinting such as observed in genetics.

In the last decade conceptions of cognition have shifted from theoretical modelling to more contextualised, neurobiological views. The seemingly close relation between brain and computer have produced new definitions such as 'neural networks', 'neurocomputation' 'neurodynamics' and 'computational neuroscience'. Are we being redefined by computer technology?

Traditionally the development of a mechanistic view of the human body/brain in Western medicine has been consistent with the ideas of calculability, predictability and control, according to Max Weber's rationalisation theory (9). But, how is this special type of inner reality created from a 'flat-monitor-picture' environment? How are brains calculating the meaning and sense from this limited perceiving process? Are there any specific correlations between mechanistic and ontological viewpoints? Various assumptions posited ontologically based, experimentally verifiable theories prompting claims that consciousness is an electromagnetic component of ionic currents generated by the human brain in the course of its functioning. (15) Such a hypothesis-as well as other emerging metaphors related to consciousness-have opened the door to the possibilities of humanising electronic technologies. Research into potential systems of artificial intelligence now looks to the brain for models rather than looking to technology for ideas from which to model the brain.

Current theories on brain processes do not seem to provide a model for explaining subjective awareness. Despite detailed investigation of the mind/body question by neuroscientists and neuropsychologists, as yet no comprehensive theory of the brain's activities has been offered. As Krellenstein concludes in his aptly titled 'Unsolvable problems, visual imagery and explanatory satisfaction' review (12), an explanatory gap remains between physiological processes and subjective experience. While the enigma of individual subjective awareness has not been resolved, the relentless development of imaging technologies continues at. A duality is emerging in society where the experts in communication, media and even in genetics are increasingly influencing the fate and consciousness of the general population. Consequently, whatever additional dynamics may arise from resolving the connection between physical mechanism and consciousnes is not only a matter of speculation but also of overriding ethical importance. Who decides this unresolved ethical framework remains a question yet to be answered just as much as determining the future applications of digital imaging and the new technologies.

Acknowledgements
Grateful thanks to Dr Andre Czegledy, Dr Jeremy Squire and Drs M eredith and M alcolm Silver for their invaluable help.

References
1. Cooper, L.A. 1995. 'Varieties of visual representation'. *Neuropsychologia*. 33: 1575-582.
2. Crick, F & Koch, C. 1992. 'The Problem of Consciousness'. *Sci Am* 267:153-59.
3. Crick, F. & Koch, C. 1996. 'Towards the neuronal substrate of visual consciousness'. in: *Toward a science of consciousness: The First Tucson Discussions and Debates*. eds. SR Hameroff, AW Kaszniak and AC Scott. Cambridge MIT Press.
4. Crick, F. & Koch, C. 1995. 'Are we aware of neural activity in primary visual cortex?' *Nature* 375(6527): 121-23.
5. Distler, C. et al 1993. 'Cortical connections of inferior temporal area in macaque monkeys'. *J Comp Neurol* 334:125-50.
6. Farah, M.J. 1995. 'The neural bases of mental imagery'. in: *The Cognitive Neurosciences*, ed. M.S. Gazzaniga. Cambridge: MIT. Press, pp 963-75.
7. Farah, M.J. 1995. 'Current issues in the neuropsychology of image generation'. *Neuropsychologia*. 33:1445-1471
8. Fuster, M.J. 1997. *The prefrontal cortex anatomy, physiology, and neuropsychology of the frontal lobe*, 3rd ed. Philadelphia: Lippincott-Raven.
9. Hewa, S. & Hetherington, R.W. 1995. 'Specialists without spirit: limitations of the mechanistic biomedical model'. *Theoretical Medicine*. 16:129-39.
10. Kihlstrom, J.F. 1987. 'The cognitive unconscious'. *Science* 237:1445-452.
11. Kosslyn, S.M. & Behrmann, M. & Jeannerod, M. 1995. 'The cognitive neuroscience of mental imagery'. *Neuropsychologia*. 33:1335-44
12. Krellenstein, M. 1995. 'Unsolvable problems, visual imagery and explanatory satisfaction'. *The Journal of Mind and Behaviour* 16: 235-253.

13. McGinn, C. 1989. 'Can we solve the mind and body problem?' *Mind* 391: 349-66
14. Milner, A.D.& Goodale, M.A. 1995. *The Visual Brain in Action*, Oxford: Oxford University Press.
15. Rakovic, D. 1996. 'Brainwaves, Neural Networks and Ionic Structures', in *Consciousness, Scientific Challenge of the 21st Century*. ECPD. Belgrade.
16. Sergent, J.1990. *The neuropsychology of visual image generation: Data, method, theory*. Brain Cognit. 13:98-129.
17. Sawchuk, K. 1995. 'Enlightened Visions, Somatic Spaces'. in: *RX Taking Our Medicine* Kingston: Queen's University. pp 31-42

Nina Czegledy, independent media artist, writer and curator. <czegledy@interlog.com>
http://www.interaccess.org/aurora http://www.ostranenie97/cyberknitting http://www.ostranenie97_crossingover

Stasis: The Creation and Exploration of Subjective Experiential Realities Through Hypnosis, Psychosynthesis and Interactive Digital Technology

R.D.Brown

Stasis: The Project

'The combination of feeling and thought of high-tension leads to a higher form of psychic life. Thus in art we have already the first experiments in a language of the future. Art anticipates a psychic evolution and divines its future forms'.[2]

This paper describes the background to the Stasis project with reference to virtual reality, the fourth dimension, hypnosis and pyschosynthesis.

Virtual Reality and the Fourth Dimension

Over the last seven years I have explored the inter relationships of time, space and energy by creating interactive and time-based art works. The most recent of these has been Alembic, an interactive 'Virtual Reality' installation. The creation of this work, using the immaterial medium of VR, stemmed from the desire to create an abstract 3D experience of four dimensional forms.

Many of you will be familiar with the work of Duchamp[3] who was similarly inspired by ideas of non-Euclidean space and an invisible higher dimension. A number of 20th century artists, including the Cubists, tried to represent alternative perceptions of form from multiple view points or the notion of time, either via painting or sculpture. Duchamp explored a variety of techniques, including painting on glass, the use of kinetics and installations as means of evoking higher spaces. I found myself struggling with physical media and, inspired by Boccioni's idea of dynamic form[4] and Kandinsky's idea of an immaterial medium of light[5], I turned to virtual reality as a means of expression. A previous background in programming enabled me to use this technology as an expressive medium, creating the illusion of form with the ability to transform such representations both in time and in response to the viewer.

Alembic[6] is the product of this work, conveying the idea of illusionary and responsive dynamic form, but also including references to alchemy and magic, placing the viewer in the position of creating his own unique reality challenging ideas of objectivity and thus scientific empiricism.

The idea of a fourth dimension, non-Euclidean space and Einstein's theory of relativity had a great impact on 20th century art and science[7]. However, even before the mathematic and philosophical speculations on space/time came into existence, various branches of 'non-science' such as theosophy and the Kabala discussed ideas of invisible spaces, labelling them as etheric, veiled or astral. Duchamp not only practised as an artist but was also familiar with the mathematics of the fourth dimension (Jouffret[8]), whilst Kandinsky and the scientist Sir William Crookes practised theosophy and were familiar with the writings of Madame Blavatsky[9]. It is this synthesis of ideologies represented by the 20th century notion of a fourth dimension which brought together the culturally opposed and disparate ideologies of art, science and mysticism.

The realisation and exhibition of Alembic with its questioning of our accepted notions of the representation of 3D space, encompassed in the term virtual reality, led me to focus on the perception of both real and virtual form, more recently experimenting with representations that convey qualities of life. We use the changing and responsiveness of form to determine whether something is alive or not. It appears that the appropriate use of illusion can bring out an empathy towards something that is not alive, even though it produces a resonance within us that evokes the appropriate emotional response.

The fourth dimension can be seen as a means of encapsulating ideas about space and time, including organic growth and decay found in life. It is difficult to mentally picture a fourth dimensionwhere do we look and how can we perceive or conceive of a higher space?

Charles Hinton[10] tried to produce a technique for this using coloured cubes, and Ouspensky refers to the idea of a 'super-consciousness' enabling a higher form of perception, beyond causality and dualism. In the two-dimensional world of Flatland, Edwin Abbott Abott[11] used analogy as a means of relating to higher dimensions.

Perhaps we must look inward and outward, physically and mentally, spatially and temporally to the micro and macrocosm, abolishing conventional ideas of causality, duality, time past, present and future. It is these ideas of consciousness and awareness, a continuing desire to explore further interpretations of the idea of a fourth dimension, that led to the hypnosis experiments I dabbled with in my youth.

Hypnosis

When I was aged around 14, I came across a pamphlet entitled 'How to Hypnotise People' at a joke shop. The pamphlet was a little thin but did lead to reading serious medical books on the practice of hypnosis and before long I was practising the techniques on school friends. Three of us began experimenting amongst ourselves to find out what hypnosis could actually do. These included: making someone disappear by telling the hypnotised person that when he woke up, a person who was actually still in the room had left; causing a heat blister to appear by touching the subject with a cold pen top which he thought was a hot soldering iron; and posthypnotic suggestions such as falling asleep on command or becoming thirsty or itchy. Other experiments included regression and controlled astral projection. I left off hypnosis

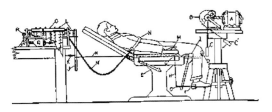

A setup for experimental hypnosis to determine the differences between hypnosis and sleep, 1931.[1]

experiments for some time, feeling that I perhaps was not quite ready or mature enough to investigate the subject as seriously as it deserved.

More recently hypnosis has been brought to the attention of the television viewing public by Paul McKenna, who featured on Radio 4 programmme *Altered States*[12]. Hypnosis tends to be viewed with suspicion by most people; they fear the loss of control associated with the practice, a fear perhaps confirmed by the use of hypnosis as a form of comical entertainment.

The fictional work *Svengali* by George Du Maurier in 1894 also set out the classical fear of the misuse of hypnosis for the control of one human by another for his own wicked ends.

Historically, hypnosis has been rejected by both the scientific and medical communities. Mesmer, in 1847, was frowned upon by the establishment as a fraudulent prankster for his practice of animal magnetism. When other magnetisers used this technique as a means of achieving pain-free surgery, the medical community was relieved when ether was discovered as a more scientific and reliable method of achieving anaesthesia.

I believe hypnosis to be a powerful, underrated and often misunderstood technique that may enable us to explore the inner mind, perhaps opening up avenues of understanding time and space and conceptualising of higher dimensions.

The practice of meditation shares much with hypnosis and is associated with transcendental states of mind. Both techniques utilise body relaxation and controlled breathing leading to a focused state of mind where the general flow of thought activity is quietened. The perception of time is lost in these states, where 20 minutes of 'real time' may be subjectively experienced as a couple of minutes, parallel to the loss of time awareness when we sleep.

Hypnosis enables powerful physiological changes to be induced, as if there are control mechanisms that we consciously do not have access to but which through hypnosis may be affected. Examples include altering visual perception, the senses of taste, smell and touch as well as being able to influence autonomous parts of the body such as the heart and stomach. In some ways hypnosis could be said to provide the perfect tool for creating the ultimate virtual reality.

Hypnosis is also used to help people with psychological problems. It is this area that I wish to ground my explorations. There is a danger that the use of hypnosis as an art form could simply be associated with the current populist trend of art-media sensationalism. I hope that Stasis will manage to tread a fine line between art, science and the mystical, resonating with the philosophical views as outlined by Ouspensky and more recently in the work of Roberto Assagioli[13] in the form of Psychosynthesis. I wish to use some of these principles as a foundation for the Stasis project.

Psychosynthesis

'From the eternal
Out of the past
In the present
For the future"
    (Roberto Assagioli)

Psychosynthesis was conceived by the Italian psychiatrist Roberto Assagioli (1888/1974) and represents an amalgam of psychotherapy techniques. A fundamental and distinguishing principle is the idea of an essential 'I' that is somehow beyond the idea of a single personality. Psychosynthesis recognises that we may have a set of personalities that are often in conflict and that a sense of inner harmony may be achieved by recognising these sub-personalities and getting them to work together.

> 'We are not unified. We often have the illusion of being so, because we do not have many bodies or many limbs, and because one hand does not fight with the other, but in our inner world this is actually the case - various personalities and sub-personalities struggle continuously with each other; impulses, desires, principles and aspirations are in continual tumult'
>
> R. Assagioli.

> 'One of the problems is identification with a sub personality as if this were the true 'I' or self. To undo this a process of disidentification is followed coupled with the training of will to undo any persistent habitual repetitive behaviours. A key aim is the realisation of the super consciousness, that level of the unconscious which generates all that is highest and most meaningful for a human being'.
>
> (What is Psychosynthesi? The Psychosynthesis and Education Trust, London)

One of the techniques used in psychosynthesis is a set of visualisation exercises based on Initiated Symbolic Projection (ISP). It is this process that I wish to use as a basis for Stasis, which has the following form.

The subject is induced into a deep state of relaxation by suggestions of bodily weight and warmth whilst focusing on deep and regular breathing. This has the effect of producing a diminished consciousness of the outside world. The subject is then given what is called a symbol motive which acts like a centre for crystallisation for a specific type of vision which differs from mere images, daydreams or most phantasy, in that it possesses a clarity in form and content, a life of its own, and permits exact description. This vision is determined largely subjectively and consciously and is termed as being autochthonous.

Each motive will evoke a subjective experience which may be influenced by deep and unconscious thoughts. For instance, difficulties in imagining a meadow and experiencing a dusty and arid space a psychotherapist could then use to explore the unconscious forces at work in influencing what should be a positive and calming place to envision.

Below is a set of example symbol motives (taken from *Psychosynthesis: A Manual of Principles and Techniques*),which the subject is asked to visualise, perhaps with suggestions of additional sensations such as sound, touch and smell.

1. A meadow-with the connotations of a safe happy childhood place.
2. Climbing a mountain,overcoming obstacles to reach a plateau, a place where one can see things in perspective.
3. Following the course of a stream, indicative of energy and passage through life.
4. A house/home containing:
   4.1 A friend-sense of well-being or significant aspects of friendship relationships- or uncovering problematic relationships.
   4.2 A picture book-significant imagery-childhood.
5. Ideal Personality-same sex-all the characteristics that one would hope to have/be like/admire-part of a will exercise.
6. Interpersonal relationships via animals: Mother-cow, Father bull or elephant-method of seeing relationships via animal characteristics - humour.
7. Sexuality: male-rose unfolding; female-getting a lift from a stranger in a car on a road.
8. Pool of water in a swamp-expressing darker suppressed elements of the psyche.
9. Waiting for a figure to emerge from a cave-wise person, powerful mystical figure-sage, witch, soothsayer etc.
10 Eruption of a volcano-life force-sexual energy, libido.
11 The lion-figure of strength-how strong the self is viewed.

Stasis would be realised by the playing of digitised audio hypnotic commands and visualisation exercises under computer control. The computer monitors the subject using a bio-feedback interface and is programmed to deliver the appropriate suggestive techniques in response to the physical and mental state of the subject. Suitable interfacing would also allow the subject to be able to navigate through various hypnogogic worlds.

The benefits of digital hypnosis includes the generation of Ericksonian-like[14] techniques of confusion by multiple and rapid suggestions, the continual, automated and monitored assessment of the patient's hypnotic state and the ability to deliver commands that are both repeatable and are as prolonged as necessary to achieve a deep state of hypnosis..

I am researching the possible dangers of automated hyposis and the use of the above motives for the Stasis project by consulting with qualified counsellors and psychotherapists.

I hope to demonstrate a prototype of Stasis at the conference.

References

1 Image taken from <http://www.tranceworks.com/history.htm>
2 Assagioli Roberto. 1975. Psychosynthesis: A manual of principles and techniques. London. Turnstone.
3 Ouspensky P.D. 1923, 1982. Tertium Organum. UK. Kegan Paul, Random House.
4 Henderson, L.H. 1983. The Fourth Dimension and Non-Euclidean Geometry in Modern Art. UK. Princeton University Press. pp. 117-163.
5 [4] p. 111.

6   Ringbom, Sixten. 1970. The Sounding Cosmos, Finland, Abo Akademi.
7   Alembic website <http://www.crd.rca.ac.uk/~richardb>
8   [4] pp. 3-43.
9   [4] p. 72.
10 Blavatsky, Helena Petronava, 1888. Isis Unveiled: A Master-key to the Mysteries of Ancient adn Modern Science and Technology. New York. J.W. Bouton.
11 Hinton, Charles Howard. 1888. A New Era of Thought. London. Swan Sonnenschein & Co.
12 Abbott, Edwin Abbott. 1884. Flatland: A Romance of Many Dimensions by a Square. London. Seely & Co.
13 Altered States. 7.30 pm, 28/2/98. BBC Radio 4.
14 Waxman, David. 1989. Hartland's Medical and Dental Hypnosis. London. Bailliere Tindall, pp. 259-265.

R.D.Brown, Research Fellow, Royal College of Art, Computer Related Design Research. BSc Computers and Cybernetics 1977. University of Kent. MA Fine Art, 1995. Nottingham Trent University.R.D Brown practises as a creator of experiential works that explore the self and our relationship to and understanding of time, space and energy. As part of the creative process I seek to combine the conventionally disparate disciplines of art, science and mysticism. < r.brown@rca.ac.uk>   http://www.crd.rca.ac.uk/~richardb

Emotion, Interactivity, and Human Measurement
Christopher P. Csikszentmihályi

Media theorists are being presented with a dizzying host of new technologies to analyse. Interactivity, for instance, has been considered extensively. One quality of interactive media is that they contain implicit models of the interacting subject. Existing models include navigational or linguistic interfaces but new forms of interaction are being developed called 'affective computing.' These techniques rely on machines which read and make decisions based on the participant's facial emotion. They may be fully interactive but require no conscious manipulation by the participant. While the implications of these interactions may be analysed, it is also useful to know the primary research upon which these models of emotion are based. Thirty years of active research in emotional modelling have generated the algorithms upon which affective computing runs. To understand the technology, the analyst must first understand the underlying theory.

This paper offers a brief introduction to the notion of emotional modelling by looking at the research of one of its founding authors. It then argues that technologies created from a scientific model will often outlive that model and have a separate impact. Finally, it shows that technologies often lose the experimental nature of their parent models, assuming new authority.

Affective computing was introduced to the American public by a recent spot on CNN. This segment depicted a group of researchers using a technology which could 'read' the emotions on the face of a subject and then interactively tailor video programmes for them. In the news report, scientists sit a glum-looking teenage male before a special computer monitor which, in turn, watches the teen through a video camera. The computer displays a few introductory video clips. As the teen watches the clips, the computer digitises his face. As the teen reacts, either voluntarily or involuntarily, his emotional responses are recorded and fed into a model of emotional reactions. So, for a male teen, the system might try to elicit first fear, a laugh, more laughs, anxiety, more laughs, fear, disgust lots of disgust, and then a smile. The system presented on CNN was editing an AIDS awareness video. Looking at the sullen teenager, you couldn't help side with the computer: "Tell him how serious HIV is . . . Crack through his studied disinterest".

This example, affective computing is framed as a way to circumvent a viewer's personal defenses to media and a way to bypass his idiosyncrasies. A traditional, linear film editor chooses shots they hope will convey a particular emotion or forward a particular point. Eisenstein and many others have written extensive systems for evoking particular emotions through specific visual and temporal combinations. Yet contemporaries watching an American ephemeral from the 1950s, or a Stalinist promotional from the same period will not have the emotional reactions that the editor expected. Similarly, many people find themselves laughing at the most serious scenes of *Titanic*. Presumably, a computer which could learn a viewer's reactions, build a specific model for them and then edit based on that model would be a much more effective, and probably a more irresistible, form of communication. The AIDS awareness message presented on CNN would undoubtedly be a great application of this new media technology. On the other hand, it is hard to imagine the scientists demonstrating their system with an ad for a politician or a children's toy, or a cheap sitcom

It is, in fact, difficult to analyse a technology simply by weighing its applications. For every socially problematic use of affective computing there exists one equally beneficial. For a 20th century liberal, regardless of applications, affective computing will seem suspicious on several levels. First, it presumes to 'read' a person below the level of cognition and as such resembles lie detectors, subliminal suggestion or other technologies which bypass a person's will. On another level, the technology's mechanism presumes a certain universality in emotions, a hereditary factor that a liberal would have trouble admitting. Finally, metrics of the face has happened before, several times, and the (invariably tagged) 'pseudo' or 'so-called' sciences of phrenology, cranioscopy and criminal anthropology carry dark legacies.

One of the chief inventors of affective computing, a University of California researcher named Paul Ekman, is not particularly fazed by these concerns. He situates his work more in the 19th than the 20th century and has spent considerable energy linking his agenda to Darwin's bestseller *The Expression of the Emotions in Man and Animals*. Darwin postulated that emotions were universal across humans, and even animals, because we share common genetic history. Until recently, observers of the 20th century have almost universally determined emotions to be as specific and arbitrary as language. In contrast to Darwin's work and his own, Ekman compares this environment-based cultural relativism to a kind of Lysenkoist metaphysics (Ekman 1973).

Ekman believes that emotions are very much innate and universal. His research rests on two trips he took to New Guinea in 1967 and 1968, the first a year after he started studying emotions. Much of his research centers on the notion that the people he studied in New Guinea were 'untouched' by Westerners and their media. Ekman, like many ethnographers of his time, felt these 'stone age' people represented a naturalised society, one with, perhaps, less culture and personal mediation than ours. "Their society entailed a permissive style of child rearing and generally egalitarian social relations without chiefs, medicine men, or patriarchs." Much has been written about these types of presumptions and their importance in ethnographic practice (see Clifford 1986). Ekman is perhaps a little guilty of romanticism, as when, in 'Faces of Man', his photographic monograph of New Guinea, he includes a section of photos which "were not selected to illustrate facial expression but show the beauty and appeal of these people." Ekman stresses that these villagers resemble our neighbors that their expressions of emotions are the same ones we might show. Indeed, many of the photos he selected are easily read as such. Many, however, look perplexingly ambivalent. Ekman never mentions how many photos were discarded as inferior representations of emotional universality.

While Ekman seeks to naturalise his human subjects, he also constructs and maintains boundaries between emotion (which he says is universal) and things which appear not to be universal. So, in a typical study of emotion in literate cultures, he tests for six emotions: happiness, fear, surprise, anger, disgust/contempt and sadness. (Ekman 1980). In describing this case, and in others, Ekman seems to reduce the number of human emotions to those expressed and readily measured in the face. He calls these emotions the 'basic set of universal facial expressions of emotion.'

In contrast to Ekman, other scholars have chosen to focus on culturally specific emotions, or the ways that people describe their emotions, rather than on the universals. Such researchers might focus on the Japanese emotion *amaeru*, roughly 'to presume upon another's love,' for which no English translation exists, yet which Japanese theorists observe even in puppies (a proof technique favored by Darwin, and used (though with primates) in Ekman's own edited books) (Harré 1986). Others choose to discuss loneliness, an emotion which is described in many cultures but difficult for anthropologists to observe. Such analyses raise questions about the notion of a 'basic set' of emotion and also about the issues of translation in Ekman's cross-cultural methodologies.

Ekman repeatedly allows for some emotions to be expressed in culturally-specific ways but his language soon favors the ones which are easily tested:

> Regardless of the language, of whether the culture is Western or Eastern, industrialised or preliterate, these facial expressions are labeled with the same emotional terms: happiness, sadness, anger, fear, disgust and surprise. And it is not simply the recognition of emotion that is universal, but the expression of emotion as well.
>
> Our theory holds that the elicitors, the particular events which activate the affect programme, are in the largest part socially learned and culturally variable, and that many of the consequences of an aroused emotion are also culturally variable, but that the facial-muscular movement which will occur for a particular emotion (if not interfered with by display rules) is dictated by the affect programme and is universal (Ekman 1980), pp 138-9)

While Ekman is primarily attempting to understand emotion, he is part of a school of psychology which seeks to find easily automated processes in human cognition, and part of the proof of discovering such a process is to embody it within computer code. It is not coincidental that he calls his emotional model 'the affect programme.' His 1993 application for a continuing NSF grant, 'Automating Facial Expression,' explains that his project will 'develop a neural network system to dynamically code facial expression, automating a process which, though widely used, is now highly time, labour, and cost intensive.' As is standard in NSF grants, he mentions potential applications for the resulting technology:

> From a practical standpoint, different aspects of expression elucidate whether a listener is empathetic or hostile (important in politics and business), distinguish abusive from nonabusive caretakers (social work), predict divorce in dysfunctional married couples, and may incriminate dissembling witnesses (forensics) (Ekman, 1993).

Ekman is one of the leading researchers in this field and researchers who follow his path tend to marginalise the other emotions (and consequently those that read them) even further. For instance, in attempting to isolate each of these emotions, Ekman trained actors to flex only certain muscles which he had identified as representing a particular emotion. The resulting photographs can look oddly unnatural. One computer science researcher involved in affective computing explained to me that he had trained neural networks to read Ekman's photographs of actors false emotions. The computer scientist then compared his machine's identification scores to those of college students at his University and announced that his machine could 'read emotions 10% better than the average college student.' This story shows that by the time Ekman's model is adopted, it carries an autonomy and finality which it does not have in Ekman's texts. This is perhaps especially true for researchers using it as a tool to read emotions, rather than as a basis for further investigation. Consequently, the computer scientist declared the model running on his machines to be the victor, without pausing to question what was meant by 'read' or 'emotion' in the context of his study.

Ekman may have proved that there are five emotions which are, indeed, innate and universal. Let us assume that he is correct. Nonetheless, it should be clear is that in the process of naturalising human subjects, of creating models of those subjects, of ignoring data which does not fit the models and of finally creating technologies which act upon those models, several steps of abstraction and reduction occur. What is finally meant as emotion in an affective computing program is now a highly stylized, specific and contested subset of what emotion means, contested even by primary researchers in the field.

Affective computing follows a string of technologies which rely on models of human intelligence and behaviour. As mentioned before, these tools often outlive the models on which they are based. For instance, expert systems, based on symbolic logic schemes of human cognition, are busy working in corporate and government organisations throughout the world. Their popularity, especially as a means of assisting corporate downsizing, continues despite the declining popularity of symbolic logic in the artificial intelligence community. Similarly, while most of the science behind theories of hereditarian IQ has been disproved, not just once, but in several different epochs, they are still used in many critical situations. Tools, like the IQ test, the expert system or the commonly used Myers-Briggs personality test (based on Jungian psychoanalysis), are able to generate their own logic and continue to forward the assumptions of models used to build them. Affective computing, even when used for entertainment, art or AIDS awareness messages, contain presumptions about emotion which may be disseminated regardless of 'content.'

Understanding the assumptions and conventions underlying a media technology may be difficult, perhaps because while tools indisputably have strong effects on culture, nonetheless scientists maintain that culture has relatively little to do with their work. Scientists involved in primary research typically refuse to associate themselves with the eventual uses of their technologies. As Western culture is more inextricably connected with technological production, the links between the sciences and their technological progeny are increasingly ignored. In what has become the masterful quick-switch of Western culture, scientists purport to discover a timeless and valueless world outside of society but in doing so may nonetheless dramatically shift and shape society. Scientists are carefully isolated from notions of ethics or from confronting the eventual implications of the technological culture they create. Bruno Latour observes that 'If Westerners had been content with trading and conquering, looting and dominating, they would not distinguish themselves radically from other tradespeople and conquerors. But no, they invented science, an activity totally distinct from conquest and trade, politics and morality.' (Latour 1994 ,1991). Latour believes this to be a dissimulation: Science is, in fact, involved in all of these things. One way to show science's involvement in all of these ethical realms is to simply reconnect it to the technologies it midwifes rather than to ignore these connections.

Chase, Allan. 1980. *The Legacy of Malthus: The social costs of the new scientific racism*. Urbana: University of Illinois Press.

Clifford, James, ed. 1985. *Writing culture : the poetics and politics of ethnography*. Berkeley: University of California Press.

Ekman, Paul. 1973. *Darwin and Facial Expression*. New York: Academic Press.

Ekman, Paul. 1980. *The Face of Man*. New York: Garland Publishing, p. 4.

Ekman, Paul. 1993. 'Automating Facial Expression.' NSF Award #9129868.

Gould, Steven J. 1979. *The Mismeasure of Man*.

Harré, Rom. 1986. *The Social Construction of Emotions*. Oxford: Basil Blackwell Ltd.

Latour, Bruno. 1994. *We Have Never Been Modern*. Cambridge, MA: Harvard University Press, p. 97.

Christopher P. Csikszentmihályi <ccsiksze@ucsd.edu> is an artist who ruminates on new technologies and the history of computing. Documentation and writings may be found at: http://jupiter.ucsd.edu/~csiklet/

Writing the Post-Biological Body

Steve Tomasula

It's getting harder to avoid the sense that our bodies are slipping away. As they drift off into what some call a post-biological future, increasing numbers of artists compose visual poems of longing (or recrimination) to the flesh we inhabit.

This 'body art,' art that predicates the body as a text about the body, is set apart from art that uses the body as metaphor, e.g., Christ nailed to a cross. That is, these works of body art are of and about flesh and sperm and blood and saliva. They are about gender and reproduction (or do you say sex?) and the way we perceive them.

Of course, 'body as text' is nothing new. From tribal scarification practices to the Human Genome Diversity Project, people have always conceived of the body as something that could be written and read. Usually these assumptions are simply absorbed with the culture. For example, artists in the Greco-Roman tradition commonly depicted sexual intercourse as an act performed by people who were standing upright - the 'natural' way to embody man's bestial nature. Today it is natural for us to represent the body (and beasts) as folders of digitised information, as in the on-line Visible Man project (http://www.nlm.nih.gov). Indeed, the post-structuralist mindset we live within has pushed into consciousness a conception of the body as a page inscribed by class, by gender, by the colonisation manifest in cosmetic surgery as, for example, performed on women by men, sometimes to the degree of Cindy Jackson: the woman who tried to turn herself into a living Barbie doll by absorbing over 20 operations.

Artists writing the semiotic body into existence have also done so by foregrounding what had previously been unspoken. Like self-reflexive paintings that work against representation as window-on-the-world, body art calls attention to the window itself, often by distressing its surface. Thus one photo by Miguel Rio Branco is a picture-window-sized print of an open gash across a person's back. Sculptor Marc Quinn casts his head out of his own blood, while performance artist Gina Pane gazes into a mirror as she slices her body. Jake and Dinos Chapman's sculptures of prepubescent girls with vaginas for mouths; Rudolf Schwarzkogler's photos of crabs pinching the head of a penis . . . We're definitely in the tradition of the abject. But we are at the same time in the tradition of the sublime, of works that use beauty as a subversive means to call attention to the body and its fluids: Andres Serrano's own piss glowing like the diffuse light of a church, Robert Mapplethorpe's nudes posed as exquisite vases on pedestals and the work of so many other lesser known artists that they beg the question always put to art: Why? - and why now? Specifically, what are we to make of a genre that emphasises the body as meat, as a container for fluids, mostly vile?

From a certain remove the answer is always the same: At different points in history, different heuristics have become a kind of *lingua franca* by which to transact ideas, artistic and otherwise. For example, the works of Freud, Heisenberg, Stein, Joyce, Picasso, et.al. although radically different, nevertheless share so many assumptions that if you squint you can see an over-arching consciousness - consciousness itself coming to function as an ordering principle, much as geometry did for Renaissance painters.

It's not difficult to get a sense of what occupies us by looking at our popular culture and its plethora of fashion tattooing and scarification, disaster films based on genetics, and so on . . . At the State Museum for Technology and Labour in Mannheim, Germany, over 200,000 people recently lined up to see an exhibit of skinless corpses that make Damien Hirst's sliced animals look quaint. If you want to see the future, though, look to the lab.

'I divide my life into B.D. and A.D. [Before Dolly and After Dolly]'
R. Alta Charo, US Presidential Commission on Human Cloning

— *Post-biological Body* —

After Dolly the Sheep, it became clear that we had all been gathering at an intersection and were waiting for the light to change. Then it did - and Dolly signalled our passage as powerfully as the first pictures of earth from space had signaled an earlier shift. Anyone old enough to remember a world too big to be held in a single glimpse can remember how those images of a cloud-swirled globe suspended in utter darkness changed our world. Like those looking through Galileo's telescope, we felt ourselves shift a bit further from the centre of the universe. The sidestep contained not a little anxiety, for as Thomas Kuhn points out, those critics of Galileo who refused to believe that the earth moves were not entirely wrong; up until then, earth was synonymous with 'fixed, immovable position.' Galileo was not asking people to simply employ a neutral technology. He was asking them to step off into the unknown - to re-imagine Earth as just another planet. How unique and fragile that blue planet became again, 350 years later, when we were able to look back at it from space and see ourselves as travellers on a fragile space ship. Surely, Dolly requires us to do something similar in terms of another autonomous world, the individual. Specifically, Dolly forces us to consider the individual person - especially the body - in much more flexible terms than we are accustomed. Dolly forces us to viscerally confront a place we had already arrived at intellectually: a reconsideration of the body, the last stronghold of the individual's claim to uniqueness in a world where the self was already seen to be constructed.

Kant is often called the watershed between (natural) earth and (constructed) worlds. For Foucault and others, Velázquez's *Las Meninas* is an icon of this shift: a painting of a cycle of gazes that incorporates the gaze of the viewer as an integral component. Four hundred years over that watershed, the epistemology we live with is not directed at revealing God's hand, nor mechanical cause and effect, but to reconstructing that which is hidden, that which is said between the lines - a project made conscious by Freud. In fact, Freud's most lasting achievement seems to be that he taught the 20th century how to read.

Perhaps the widespread dissemination of his method, a method of literary criticism performed on a living text, is what ultimately opened the breach between our selves and our bodies. Freud himself found he could dispense with the body on the couch; the structuralist reading of language he performed worked just as well on works of art. Likewise, anthropologists stopped measuring cranium capacity or classifying genitalia and began doing semiotic readings of funeral customs, for example. Or garbage. Or as Foucault puts it in *The Order of Things*, the same turn of mind that cast man as a linguistic being waxes until the unique 'I'

Not Even Philosophers Can Jump over Their Own Shadows

morphs into a locus for its representational doubles: the psychological self, the sociological self, the anthropological self, the economic self. It was only a matter of time before the method of reading that generated these selves was turned on the practices that created them in the first place.

Which is to say, the deeper we look into ourselves, the more clearly we see Descartes: If I am in the representations that I make of myself and if the centre of my representations will not hold, where does that leave me?

Up until now, the body has served as a firewall against complete subjectivity. Even if the self was given over wholly to an infinite regress of linguistic mirrors facing mirrors, even if we kept faith with a mind/body split, body and self were inextricably linked. But body technologies have pried even this relationship loose. There is a fundamental difference between the new body my cells continually generate out of themselves and a cloned duplicate. The natural body and my self begin and end more or less as one. The cloned body continues without my self. In fact, multiple selves can inhabit multiple extensions of my singular body and this very real possibility has translated onto the literal body a vocabulary of instabilities generated by the proliferation of body texts: imitation, pastiche, influence, quotation, irony, the pun, or significantly, plagiarism, given the fact that a bio-tech company can clone my cells without my knowledge and an original meaning of plagiarism referred to the kidnapping of a child.

Writing about mechanically reproduced multiples such as posters, Walter Benjamin pointed out that to ask which is the original is a meaningless question. When cloning is a routine farm technique, will the same be said of cows? Or will the debate simply continue in altered terms, the way the meaning of the word 'earth' shifted to fit a new understanding of the world's diminished position in the cosmos? As we've already seen in the world of the text, in the absence of the author and other origins, ownership is decided by legal force. For example, according to the trademark on the box of Glad-lock sandwich bags in my kitchen, the phrase 'yellow and blue make green' is owned by the First Brands Corporation. And there's an odd echo here with courts called upon to determine the legal 'owner' of an embryo who (that?) is the product of sperm from a bank and an egg from a donor, nurtured in the womb of a surrogate rented by a fourth party who contributed none of the other 'components.'

Each purchase of a cheese-food product instead of real cheese, each facsimile that is given legal authority, represents a shift, however small, towards an acceptance of the double, not as the original's equivalent but as a substitution that blurs one with the other. In the ubiquity of doubles lies one source for the rise of connoisseurship for originals: an Olympic record unaided by steroids, for example. But outside the world of the connoisseur, fetishisation of the original bows to practicality and we find that we cannot only get along quite well without originals, but that we don't even want them. GenVec, Inc., for example, is working to offer BioByPass, a genetic treatment that allows a heart to replace itself with a stronger version of itself. Designed for medical purposes, the treatment is already being examined by others for its ability to genetically make athletes (or soldiers?) faster, stronger. To put it another way, how much respect will we have for the unmediated Olympic record once kids in the street are running two-minute miles? The question pales against the growth industry predicted by the US president's Recombinant DNA Advisory Committee for other techniques of 'self-directed evolution': e.g genetic enhancements to "treat" skin color, mental concentration, memory, emotional stability, resistance to AIDs, prostrate cancer, etc., etc., etc. To get a sense of how strong market desire might be, consider that two out of three of you reading these words will die because of a genetic time bomb that is ticking away within you. One that may be able to be engineered away.

Against this backdrop we conduct our affairs: we marry, we make love (or do you say fuck upright?), we procreate and fight in court over whether or not genes can be patented and by whom - the person they were taken from or the pharmaceutical company that duplicates them? And, of course, we make art.

Historian Simon Schama has shown how landscape painters provided the public with an idiom by which they could formulate their nation's identity. And there seems to be a parallel here with the way body artists work with themes of personal identity. Like landscape painters, body artists create an idiom out of nature - but this nature is one of body fluids, of bone and punctures and operatic flesh. They mould metaphors that argue out the nature of the human. And they seem to be compelled to do so for much the same reason that many early American painters, for example, adopted the land as their subject: the encroachment of technologies (railroads then, artificial hearts now) and the breach between people and nature exposed by their spread. That is, at a time when the humanist project has changed so radically that it begs for a different name, artists ask, In what image will we recreate this post-human human? The rise of the body as a forum for this debate seems almost predictable; in hindsight, inevitable. This seems especially true when one considers how pointedly body-art bridges the world of real bodies and cultural, or aesthetic, artifice - a zone opened up most consciously by feminist thought and explored by artists like Orlan.

Using her own flesh as her medium, Orlan makes literal the metaphor 'surgery theater' by staging performances which consist of her body being surgically sculpted. The extremity of Orlan's body distortion calls attention to her body's ontological status for the reason Heidegger claims that a hammer is most itself when broken: its broken condition makes it impossible for everyday users to take it for granted. But, of course, a rhinestone-encrusted hammer would have the same effect.

Indeed, beauty can be an even more compelling means to subvert invisibility brought on by familiarity, especially when the ordinary is socially constructed as grotesque: piss, for example, or a finger in a penis, as Dave Hickey has shown in *The Invisible Dragon*. In either case, the artist puns *heimlich* (the familiar) with *unheimlich* (the uncanny), as Freud once dreamt it. And expecting him or her to do so, we bring to the work something we normally don't bring to surgery - the difference in reading that separates Duchamp's *Fountain* from the ordinary urinal or one of Jeffery Silverthorne's corpses from any John Doe.

Like Renaissance viewers coming off a long diet of iconography, we are struggling to imagine what the self could be, developing a new idiom of the body for thinking about the self in relation to the body. We are pulled towards affirmations of the body's physicality the way our Renaissance counterparts must have found a realistic crucifixion both startlingly grotesque and seductive. How disembodied the art of the medieval must have suddenly appeared against this reinvestment in the aura of the body, those ravishing atrocities!

Even more to the point, we find it so hard to look away from body-works because their flesh-and-blood materials reduce to insignificance any fashion in art. Bound up as the body is with the humanity it bears, to posit the body as construct, even in denial, is to navigate the deconstruction of the individual, the collapse of the self and this collapse's own rebuttal - with all of its ethical consequences and ramifications for what we will see when we look at one another. Whether history shows the self to be essential or a heuristic whose time has passed, there will still be the body (whatever we decide that is), constructed or not.

Steve Tomasula's fictions and essays have appeared in numerous publications, including most recently Emigre, Fiction International *and the issue on narrative theory and image of* The Electronic Book Review *(www.altx.com/ebr), which he guest edited. He teaches creative writing at the University of Notre Dame and can be reached at Tomasula.4@nd.edu.*

5   Space and Time

For a Spatiotemporal Sensitivity
Ginette Daigneault

Acoustic Territory
In 1996, with cellist and visual artist Andrée Préfontaine, I began an exploration of the videoconference as a medium for creation. *Acoustic Territory* was a telepresence event that enabled two cellists to improvise long-distance, between Hull and Montreal, for 20 minutes. This improvisation was the pretext for an exploration of the telematic space and the elements of formal language that allow us to grasp the specific nature of this medium. Our intention was not to 'make' television or present a concert, but to explore a *spatiotemporal aesthetics*. The extremely poor image definition and the visual and auditory interval were used as characteristics specific to the medium.

Variable Geometry

Physical Organization of the Event
At the front of the room, Andrée Préfontaine was seated before a giant screen onto which was projected the enlarged spectral image of the body of Ian, the second cellist. In Hull, in a corner of the room, Ian was seated with his cello in front of the camera. A miniature train circled a track next to Ian, pulling a miniature cello. At the front of the room were two monitors side by side. The first displayed the framing chosen with the camera prior to its transmission to Montreal. The second monitor showed the image of the two cellists sent by Montreal. The exaggerated pixillation of the spectral image of Ian's body in comparison to that of Andrée made the spectator aware of the distance travelled. Thanks to the presence of the two monitors in the Bell Room in Hull, we experienced the overlapping of three spaces:
1. Ian, in the flesh, playing his cello;
2. the selected framing of Ian's body as he played (image selected by Hull);
3. the spectral bodies of the two musicians facing each other and improvising (image transmitted by Montreal).

Rather than creating an effect of delocalisation, the presence of the three spaces created the impression of a territory with *variable or shifting geometry*. We sensed a shift, a transformation of place through what could be seen and touched at a distance. Paradoxical though it may seem, the sense of tactility was heightened. An awareness of here, now and elsewhere coincided with a perception of the visual and auditory interval that seemed to want to determine the rhythm for this space-time continuum. There was a time lag between the movement captured by our camera and its spectral image retransmitted from Montreal, and this lag emphasised the distance.

The event began when the two cellists *touched* hands, virtually. The gesture awoke the memory of our tactile sensitivity. A spectator in Hull remembers this sequence as the most moving moment of the event: a moment that connected perception and action, that allowed for the communication of energy to physical objects, for an encounter between two persons.

Spatiotemporal Sensitivity

Normally, touch is inseparable from the body; it defines us as material beings. Touch implies gesture, action and perception. In fact, Annie Luciani avoids the use of the term 'touch,' too closely identified with the purely tactile, and replaces it with *sensorimotor gestural channel*. For Luciani, 'the sensorimotor gestural channel is simultaneously and indissociably a motor and perceptual pathway' (Luciani 1996, p. 80). This channel emits and receives information, but above all, it serves as a link, a bridge, a passage for the exchange of energy. In the telepresence events, the sight of the act of touching awakens in both the actor and spectator the memory of the sensation of an exchange of energy. We touch with our eyes. The gaze appropriates the tactile abilities of touch, of skin contact. This is proprioception, 'the direct and immediate perception we have of the world in our bodies.' (De Kerkovc 1996, p. 135). The telepresence events made us aware of human responses that a few years back were unknown to us, or perhaps forgotten through lack of stimulus. We experienced the sensitivity given to muscles, to articulations, to the skin, to movement in space, and our sensitivity responded to new, more subtle solicitations.

The integrating axis of *Acoustic Territory* was the idea of a connection, a bridge between two spaces. Andrée Préfontaine and I wanted to visually mark this idea of linkage. In replacing the transmission of the image of Ian's body by that of the little locomotive pulling a miniature cello, we opened up the space. The image of the train in movement entered the visual field by the upper left part of the monitor and exited on the lower right, tracing a circle in front of the body of Andrée Préfontaine. The locomotive seemed to travel between Hull and Montreal. The train's displacement created a visual bridge. A visual, mental and sensorial loop was activated. Also, at a given moment we replaced the image of Ian's body by a close-up of his instrument. On the image arriving from Montreal, we could then see Andrée Préfontaine improvising, nose to the strings, the bow and giant hand of Ian.

Despite immateriality, despite the distance and the spectral work, I do not think that we are in the process of becoming pure spirit, detaching ourselves from our bodies. On the contrary, we are rediscovering our bodies, discovering or rediscovering a sensitivity and responsiveness that lay dormant because it was not solicited. Distance touching distances itself from the simple transmission and reception of information, placing the accent instead on the exchange of energy. The caesura that over the years has existed between our bodies and spirits, prioritising one or the other throughout the ages, is in the process of disappearing. I believe we are entering an age of exploring what we are becoming as human beings.

The Art/Communication Connection

Through telepresence, it is not only the tool that is questioned, but the idea and the very essence of communication, i.e. of being a human being in contact with other human beings. For a long time, the art world refused to acknowledge any link between art and communication, and some still persist in refuting this connection. Artists who use technology are accused of subordinating the imagination to technocracy, of collaborating with those in power. But the question here is not one of the *communicability* of the message, but of the components essential to the constitution of an art event, of what can be seen and felt, what participates in the constitution of meaning.

I think this controversy is to a large extent due to a problem of the definition of the term. Currently, for the majority, the term 'communication' relates to the direct and effective transmission of information. But we have to remember that the definition of the term has been constantly modified over the years. In fact, it was only in the 18th century with the appearance of *communicant tubes* that the meaning of the term 'communication' shifted from 'communion' and 'sharing' to 'transmission.' It was also during the eighteenth century that means of transportation were developed which crystallised the meaning of the term around the notions of transport (roads, canals). The meaning of the word would take on a new acceptance with the development of the mass media during the first part of the 20th century. Researchers at the Chicago School sought to develop a communication theory through interaction, through something that is done by people, and communication came to be considered a *sine qua non* condition of human life. Ray Birdwhistell declared "An individual does not communicate, he takes part in a communication, becoming an element of the communication. He can move, make noise . . . but he does not communicate. In other words, he is not the author of the communication, but a participant in it." (quoted by Winkin 1984, p. 75). The focus is on the act, on participation, on the shared experience.

Thus the art/communication connection takes place through interaction and can also be based on the definition given by members of the Invisible College and Birdwhistell, who say that 'it never happens that nothing happens,' or that of Bateson and Watzlawick, for whom 'even if an individual can cease speaking, he cannot prevent himself from communicating through body language,' or 'we cannot not communicate.' (Winkin 1984, p. 74).

Conclusion

Through telepresence, I have used telecommunication networks as both the medium and surface of creation, in order to explore the new frontiers of our spatiotemporal sensitivity. We now seem to be entering into a new aesthetic age (Poissant 1994) or aesthetic ambiance (Maffesoli 1988) of feeling and touching combined. Here I mean aesthetic in the sense of 'a common faculty of feeling' (Maffesoli 1988), in the sense of the effect, of emotion, of a receptivity toward others, of individuals together in a common space, of what we experience with others: the first meaning of the term 'aesthetic.' And this aesthetics that we would qualify as affective in keeping with Maffesoli cannot be dissociated from the characteristics, challenges, effects and impacts of communication and telecommunication; it cannot be dissociated from the relationships between human beings. New telepresence art practices seek to establish a connection, to prove the existence of a distance relationship and participate in the definition of our new human identity.

References

De Kerkhove, D. 1996. 'Propriodéception et autonomation,' *Les cinq sens de la création art, technologie et sensorialité*. Mario Borillon and Anne Sauvageot. p. 130-43. Mayenne: Éditions Champ Vallon. (eds.)

Luciani, A. 1996. . 'Ordinateur, geste réel et matière simulée,' *Les cinq sens de la création art, technologie et sensorialité*. Mario Borillon and Anne Sauvageotp. 80-90. Mayenne: Éditions Champ Vallon. (eds.).

Maffesoli, Michel. 1998. *Le temps des tribus, le déclin de l'individualisme dans les sociétés de masse*. Paris: Méridiens Klincksieck.

Poissant, Louise. 1994. *Pragmatique esthétique*. Québec: Hurtubise HMH.

Winkin, Y. 1984. *La nouvelle communication*. Tours: Éditions du Seuil.

A media artist and professor in the Arts and Design programme at the Université du Québec à Hull, Ginette Daigneault earned her bachelor's and master's degrees in the visual arts from the Université du Québec à Montréal. She has exhibited her work in Quebec, the United States and France. She is currently enrolled in the joint doctoral programme in communications at the Université du Québec à Montréal, the Université de Montréal and Concordia University, where her field of research is artistic practices in network art.

Virtual Geographies, Borders and Territories:
GPS Drawings and Visual Spaces
Andrea Wollensak

Recent investigations on the meanings and significance of space and ideas related to the spatiality of human life articulated by Edward Soja in *Thirdspace* (1996) provides an important framework in considering the impact of digital technologies in our lives. The trialectics of spatiality - perceived, conceived and lived - defined by Henri Lefebvre and mapped by Soja categorise dynamic social conditions.[1] Conditions inherent to this triad, satellite navigation, data sharing and teleconferencing, have shifted our understanding of space, connectedness and place. In this age of digital self-awareness, we have the ability to see more information and ourselves shifting in a virtual geography—a matrix of mutability that is dynamic, elusive and unpredictable. Satellite navigation technologies (such as the Global Positioning System, GPS) act as a beacon in this new geography by pinpointing place, articulating our connectedness and reconstructing our notions of navigation in space.

The art work explained in this paper conceptually and visually defines spaces as lived, perceived and conceived through digital map drawings and timebased computer animations. *Drifting: Position Drawings*, *The Global Positioning Series* and *Expression with Global Technologies* are a collection of recent artworks that investigate ways of apprehending

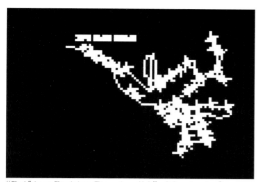

"Drifting: Position Drawing #1". 20 minute drawing sent remotely to gallery and projected on sculptural frame

information spaces using new instruments that articulate visuality of movement, time and location in space.

The Geographic Gaze

Edward Soja's central argument is that spatial thinking has been bipartite - limited to two approaches. Firstspace is seen as concrete material forms that can be mapped, analysed and explained (empirical) and Secondspace is seen as mental constructs, ideas about and representations of space and its planned social structure. Thirdspace questions this duality, joins and transcends both the material and mental dimensions to new and different modes of spatial thinking.[2]

Firstspace and Secondspace have limitations both in representation and utopian ideals of place often found in maps as a form of power and authority. J.B. Harley states that 'the practices of visual representation of the map serve to disguise the power that operates in and through cartography. Maps are not empty mirrors, they at once hide and reveal the hand of the cartographer. Maps are fleshy: of the body and of the mind of the individuals that produce them, they draw the eye of the map-reader. The map does not simply itemise the world: it fixes it within a discursive and visual practice of power and meaning; and, because it naturalises power and meaning against an impassive and neutral space, it serves to legitimate not only the exercise of that power but also the meaningfulness of the meaning.'[3] Perceived space, as both Harley and Soja note, are material and materialised physical power systems in measurable configurations. These days, GIS (geomatic information satellites) performs much of the empirical data collection of the earth and atmosphere. More than merely recording landmass, GIS expands the notion of perceived space to information that was never recorded with earlier technologies. Satellite imagery is able to visualise air, wind and temperature. Satellite photos create a virtual photo from a virtual image/surface which is actually a three-dimensional environment in constant flux. In contrast to overall image collection of spaces, GPS visualises movement of particular bodies, not as image but as coordinates. GPS uses 24 Navstar satellites that allow one, by triangulation, to determine longitude, latitude, altitude and time. A navigator looks not out to the

"Global Positioning #4", 1996. Video still extracted from computer animation

landscape for location recognition but instead to the steady stream of data coming from the satellites.

The following section describes of artworks exploring real and imagined places through the use of GPS satellite connectivity. The works are a collection of movements and trackings - in, through and above different sites. Varying site conditions include borders, territories, climates, population and cultures. The work is collective in that the movement is determined not only by myself, but collectively and also by people inherent to the site.

Drifting: Position Drawings

Drifting: Position Drawings is a project using GPS satellite connectivity to track[4] my movement and also other person movements throughout the streets of New York City. The data is sent in real-time to ExitArt, a gallery downtown on Broadway. In the gallery, there is a computer running mapping software and modem that receives the stream of coordinate data. The drawings document continuous and broken paths created by travelling throughout the streets of the city (lived spaces). As abstract linear notations, the manifestation of the satellite connection renders both the availability and inavailability of connectivity to the satellites as well as the drift of inaccuracy in coordinate positioning. The visual tracking information sent remotely to the gallery overlaps on top of itself and develops patterns of individual activities from experiences in a complex urban landscape. The visual work thickens as different paths overlap one another and begin to build up on the surface of the computer screen. *Drifting: Position Drawings* makes visible a re-mapping of the subject in an evolving cartographic form notating a Thirdspace condition. Not only is this work empirical and subjective but it is constantly changing and transforming from representations of space to lived spaces of representation.

In the gallery space, a map is constructed from the remote feed of GPS coordinates from the tracking of someone's movements in the city. This map is a virtual geography (conceived spaces) a terrain of media and networks crisscrossing the globe ". . . these vectors produce in us a new kind of experience, the experience of telesthesia - perception at a distance. This virtual geography is no more or less "real". It is a

"Chihuahua, Mexico Desert Drawing". 200 mile drawing set with GPS receivers in a Cesna airplane for 7 hours during the full moon on July 18, 1997

different kind of perception, of things not bounded by rules of proximity, of "being there". If virtual reality is about technologies which increase the "bandwidth" of our sensory experience of mediated and constructed images, then virtual geography is the dialectically opposite pole of the process. It is about the expanded terrain from which experience may be instantly drawn.'[5]

Global Positioning Series

This work was created with two differential GPS (DGPS) receivers capable of determining my position within a few feet.[6] Government control of signal accuracy (selective availability) does not interfere with DGPS because the signal measures and transmits the error between the known position and the measured position to users of the same radio navigation, establishing a very accurate reading.[7]

Global Positioning Series is a group of three-dimensional animations that and document the human condition. As purely subjective interpretations of place and time (conceived space), the hikes are wanderings, pacing back and forth and standing still, a pure data recording of idle moments and pauses. *Global Positioning #4*, a three-dimensional computer animated environment, represents an alternative application and aesthetic for GPS use - one that invites calm and weightlessness, as well as a sense of place. *Global Positioning #4* represents place by way of transparency, undulating folds and layers in space, aerial photography and subtle movement. The visual linear path that the viewer follows in the work is the original data collected from a hike in the Canadian Rockies.

The first phase of the work *Global Positioning #4* is based on field data that I translated into two visual components: the path, the space around the path seen as aerial photographs and images of skin. The path (hike) data is created with a DGPS that is programmed to determine my position every five seconds within an accuracy range of a few feet in four dimensions: longitude, latitude, altitude and time. The hike is approximately six hours long and throughout that time the receiver collects a string of coordinates digitally defining the path. The images in the work are aerial government photographs as recognisable associations of place as well as black/white close-up photographs of skin which are clear enough to discern its texture (micro) yet similar to landscape (macro). Both images have associations of power and surveillance.

Through a series of translations that included importing the data to three-dimensional animation software, I altered the original position coordinates representing the walk with a series of parameter setups that include time, space, speed, the viewer position to the path, the visual form of the path and the environment. I altered the linear path by extruding it to represent a tunnel, assigned it a colour (pink) with transparent attributes. The data coordinates are translated into individual animation frames. Each frame represents a position and is notated by the exact coordinates at the bottom of each frame in a black bar (from left to right is time, longitude, latitude and altitude).

Expression with Global Technologies

This work is a collection of recent collaborative digital drawings organised by Andrea diCastro[8] at the Multi Media Centre in the National Centre for the Arts, Mexico City as part of an ongoing project initiated in 1997. The goal for the year 2000 is to create a collective work of national and international artists around the globe using GPS and the internet. This work is a series of large scale virtual drawings, measuring several miles in size, and they are made following the path of the creators via satellite. The concepts of the project involve the global character of the new millennium, the use of global technologies, tracking of movement becomes image and the environment/site as canvas. The plots are the result of the GPS data that allows a very precise register of the paths into visible shapes. The images are viewed on the internet where the users can watch the development and the final graphics of the plots. The project will be completed at 24:00 hrs. Central Time on 31 December 1999.

Technologies available today have permanently changed the spaces we live in. Virtual geographies of networked information blanket the entire physicality of the earth. GPS satellites are part of this new topographic layer allowing interconnectivity of place and individuals in real-time. As we approach the *fin de siècle*, we experience and think of space more critically and actively. The trialectics of spatiality: perceived, conceived and lived are dynamically interconnected to historicality and sociality and cogently presented by Soja within the magical realism of Borges writing as a tangible example of Thirdspace:

> The Aleph: What eternity is to time, the Aleph is to space. In eternity, all time - past, present, and future - coexists simultaneously. In the Aleph, the sum total of the spatial universe is to be found in a tiny shining sphere barely over an inch across.
>
> (Borges, 1971:189; Soja 1996:54).

References

1. Soja, E. 1996. *Thirdspace Journeys to Los Angeles and Other Real-and-Imagined Places*. Oxford, UK and Cambridge, MA: Blackwell, p.5
2. Soja, E. 1996. Ibid. p74.
3. Harley, J.B. 1988. 'Maps, Knowledge, and Power', In Cosgrove. D., ed. *The Iconography of Landscape*. Cambridge: Cambridge University Press, pp.277.
4. The equipment used for this project was a Garmin handheld receiver, Nokia cell phone, interface box and Precision Mapping from GPS TRAC.
5. Wark, M. 1994. *Virtual Geography*. Bloomington: Indiana University Press, p vii.
6. The equipment Smart Base, by Premiere GPS, Inc. is typically used by surveyors and allows for real-time and post processing data generation. Smart Base is a CPU 386-SX-40 MHz with a hard disk storage of 250 MB; it has a real-time clock/calendar and battery backup, one mutli mode parallel port, 8 channel continuous parallel tracking, 1 pulse per second output and differential input and output corrections.
7. GPS was used during the Gulf War by the United States Department of Defence to track missiles and locate targets. Regularly, but especially during times of perceived threat to national security, the Pentagon intentionally scrambles the GPS signal through a maneuver called selective availability. This procedure involves the degradation of GPS signal accuracy by clock dithering and the incorporation of calculated errors into satellite orbits. Paul Virilio describes this deceitful combination of self-sabotage and self-protection in *The Art of the Motor:* 'In the case of a declaration of war against the United States, the Pentagon automatically reserves the right to tamper with what amounts to a "Public Service" by falsifying indications of proximity in order to guarantee the operational superiority of its armed forces!' Paul Virilio, *The Art of the Motor*, Minneapolis: University of Minnesota Press, 1995, p156.

8. The crew and collaborators for the northern Mexico project 'Expression with Global Technologies' included Jose Firez Kuri, Humberto R. Jardsn, Andrea Wollensak, Edmundo Dmaz, Sebastian, Alberto Gutiirrez Chong, Armando Lopez, Hictor Moreno, Carlos Salom, Ignacio Del Rmo and Alejandro Nava.

Bibliography

Baudrillard, J. 1983. *Simulations*. New York: Semiotext(e).
Borges, J. 1971. *The Aleph and Other Stories*: 1933-1969. New York: Bantam Books.
Gregory, D. 1994. *Geographical Imaginations*. Oxford and Cambridge: Blackwell.
Hayden, D. 1995. *The Power of Place*. Cambridge and London: MIT Press.
Keith, M. and Pile, S., ed. 1993. *Place and the Politics of Identity*. London and New York: Routledge.
Lefebvre, H. 1991. *The Production of Space*. Oxford and Cambridge: Blackwell.
Virilio, P. 1995. *The Art of the Motor*. Minneapolis: University of Minnesota Press.

Andrea Wollensak, Assistant Professor of Studio Art and Associate Director at the Centre for Arts and Technology at Connecticut College (ajwol@conncoll.edu)
Born in the USA in 1961. Received Bachelor of Fine Arts from the University of Michigan in 1983 and Master of Fine Art from Yale University in 1990. Wollensak's professional work (national and international) includes graphic design, exhibition design including interactive kiosks, site research with Global Positioning System (GPS), sculpture and holographic installations. The institutions involved with the mentioned work include the Museum of Modern Art, the New Museum of Contemporary Art, Total Design and UNA (Amsterdam), Stichting Kunst and Complex (Rotterdam), the Banff Centre for the Arts and the Nationale Centre for the Arts in Mexico City.
Funded research has included site projects in Banff, New York City, Montreal, Rotterdam and Mexico City, interactive work at Alias Research and Fringe Research in Toronto and a Group Fulbright Grant to the Czech Republic for research on regional design.
Prior to teaching at Connecticut College, Wollensak taught at Concordia University in Montreal. Other institutions where she has lectured include SUNY Purchase (adjunct lecturer), The University of Texas at Austin (visiting artist), The Cooper Union in New York City (visiting critic), Nova Scotia College of Art and Design (visiting artist).

An Exploration of Spatial/Temporal Slippage

Chris Speed

Introduction

The aim of the work reported here is to promote the exploration of spatial understanding and identity in a time when our abilities for spatial perception are under constant duress from the influx of new media technologies. These technologies demand a perceptual shift between different ideas and representations of spaces local, global, networked, actual and virtual. Given the complexity and subtleties of these shifts, it is unsurprising that the Cartesian model of space proves unable to provide a satisfactory framework for understanding this new, non-linear geography. In order to illustrate the nature of these shifts and 'slips', presented below is an example of our perception of time and space under the stress of new media as it breaks conventional forms and structures. These shifts are then discussed in relation to Deleuze and Guattari's rhizome principle, which it is felt offers an appropriate means of explaining our uncanny ability to move between such diverse manifestations of space. The author then presents some of his own fax work that actively attempts to reconfigure conventions of space and questions their stability when bound together.

New Media Spaces

Establishing 'where we are' is becoming an increasingly common task in our everyday existence. We are required to work out where we want to go and how we wish to get there via path, road, rail, air, phone line, radio frequency, television channel. Our expectations for travel, movement and speed, demand we orientate ourselves in anticipation of the next move. It is not so much the new technologies of immersive virtual reality environments or the stunning sensory effects now routinely found in films or theme parks that concern us. Rather, what is focused on here, is a seemingly insignificant event which might infiltrate our daily life.

BBC Television Breakfast News often attempts to carry out debates around current affairs in the 20 minute slots between hourly and half hourly news bulletins. Traditionally, the group is made up of three people in 'the studio' and one other contributor who is located in some 'far off' place, joining the debate via telematics, appearing framed on a screen behind the breakfast presenter. Once the debate begins, the programme controllers are able to cut between cameras to construct a coherent 'in-studio' discussion chaired by the presenter using a well-established cinema technique, the 180× rule. Confusion sets in when the virtual participant is introduced, switching from studio camera to the live feed coming from another location. The introduction of the new member to the group gives the viewers just time to observe the mystification on the faces of the other participants as to where to look for their adversary. Since the new member of the debate is superimposed onto the stage via 'blue screen' technology, the 180× camera rule is rendered useless since he cannot be seen to exist in the studio. Unfortunately, on this occasion some of the participants recognise the new guest on an off-set monitor and can be seen to look off screen, to where the virtual participant can be seen.

Further to the complex mixing of spatial models, the broadcast team appear to make things worse by attempting to locate some people and not others. The backdrop behind the virtual guest is clearly superimposed and goes some way to help locate him by way of a digitally blurred and colour saturated still of the former BBC Television headquarters with the words 'Central London' supposedly clarifying the point across the top of the screen. During one of the more lengthy comments, another framing device is introduced to enable us to see the presenter and his virtual guest at the same time. Composed within a saturated blue background hover the two moving images of presenter and guest, clearly in different frames and labelled accordingly 'Studio' and 'Central London'.

If the presenter is in the 'Studio'and our virtual guest in 'Central London', where is the frame located? Where is home to ourselves, if home is where we locate our own point of view on the proceedings? The naming of the 'Studio' and 'Central London' seem to suggest a breakdown in spatial conventions. By introducing Central London as a means of locating the virtual guest, the production team are forced to locate the studio but without a geographical context for it, it must remain as the 'Studio', an ambiguous term, but one that the audience have taken on board because of the introduction of such new media.

Rhizome

In their book *A Thousand Plateaus*, Deleuze and Guattari extend their thoughts upon what they mean by the rhizome. The name is lifted from biology and refers to a plant underground storage system that grows horizontally through the ground and also partly above it. This system that upon inspection, does not follow the hierarchical structure that is explicit in the tree, which for so long has provided the model for linear and deductive thinking. It is a concept that can be applied to our apparent skill in moving between spaces and places, which Bourdieu in his 'Habitus' and Foucault's 'dispositif' for space set out to do. Bourdieu's work looks to a structural approach to establishing key methods for making sense of space and time by constructing a habitus, a system using different class patterns and activities to represent a vocabulary for our actions in time and space. The vocabulary of habitus consists of aspects of routine and as Bourdieu would have us believe, habits, the characteristics by which we identify many jobs and tasks.

Foucault represents a move away from Bourdieu's causal proposals, the result of a social situation, and instead looks to identify where the motivation comes from. Through his dispositif, Foucault outlines the complexity that our position or state is founded on by identifying three parts: 1. From formative acts that are the result of a response to a new situation ('emergency') comes a strategy for action; 2. A 'jurisdiction' is developed as the action becomes stable and ordered; and 3. The actions become heterogeneous as we struggle and succeed to fit new events into the initial strategy. In this way, Foucault finds we interpret space through strategies that form order and in turn, procedures.

Clearly the apparent skill to navigate through the confusion of spaces that viewers are subjected to in much contemporary media does not fit into Bourdieu's or Foucault's systematic proposals of spatial interaction primarily because we now travel so much from the couch as well as by foot, car or plane. Nevertheless, as spaces are so often bound and presented by new media, the rhizome becomes more and more appropriate as the structure of the tree appears increasingly too linear and procedural.

The moment at which the rhizome is most exciting is when it changes direction abruptly and yet still fulfils its objective, taking a route that otherwise appears inefficient or nonsensical. It is this moment that I would refer to as 'rhizoming', and I would suggest that it is similar to the moment in which we search for a new model of space to explain the slippages

Figure 1: London to Sofia Fax

between alternative models for space to explain an unusual situation, be it on screen or in actuality.

Fax works

In an attempt to explore the moment of 'rhizoming' as we search outside of one framework for space to find another in order to make sense of a slip, I will turn to two pieces of fax work that were carried out on flights to different European cities. Both attempt to embrace different languages for space and thereby force the audience or reader to flex a given spatial convention in order to make sense of the work. On 18th January 1997 a fax (Fig. 1) was sent from a computer in London to three companies that were directly below the flight path of a Boeing 757 that was travelling from Gatwick, England, to Sofia in Bulgaria. Three passengers including myself each had a digital watch on his wrist. The digital watches were alarmed and set to go off in turn when the aeroplane flew over the three companies in Roeselare, Belgium; Augsburg, Germany; and Graz in Austria. The objective was to set up consecutive situations that demanded the senders and the recipients to reflect upon an event that referred to multiple spaces at one time.

The fax itself consisted of different representations of space and identity: a map of Europe, a photograph of the planet, logos and addresses for the organisations and photographs of my colleagues and myself. Whilst I have received replies back on previous occasions from many recipients of my faxes, unfortunately no replies were received from Roeselare, Augsburg or Graz. Nevertheless, the experience for my colleagues and myself in the aeroplane was particularly interesting as we were forced to conceive of a number of spaces simultaneously:

- Space One: is the actual space, that the interior of the aeroplane that we were within from where we observed each watch denote the sending of a fax from London to each respective location.
- Space Two: is the memory space, that the interior of the North London flat where the computer programmed to send the faxes to the locations was located.
- Space Three: is the imagined spaces, that the offices where the faxes arrived, places we had never visited and were never likely to visit.

Indeed as we grappled with the three spaces as we attempted to map and correlate them, the most interesting slip in consciousness occurred as we watched each other and shared the concept that our target recipients were also having to construct an equally

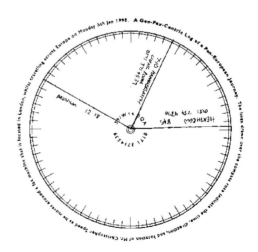

Figure 2. London to Munich.

complex reading of the situation, one that might have been made up of an alternative three spaces:
- Space One: the actual space, within which the fax was received.
- Space Two: the imagined space, that of a room or place somewhere in London where a fax was being sent.
- Space Three: the projected space, that was cast at an aeroplane flying above or nearby within a few minutes of receiving the fax, a space that also encapsulated an image of the three passengers pictured in the fax.

The complexity of the situation certainly caused me to move between models of space that were of very different orders, and whilst it could be said that I merely had to contextualise the event on a European flight map, the process of juggling with them all at once forced slippages that would fit into the ambiguity of the rhizome principle.

Fax two (Fig.2.) was sent out on 28 December and was a personal investigation into the mapping of a journey from a flat in Hammersmith in West London, to Heathrow Airport and arrival at Munich Airport in Germany. Primarily meant as an exploration in the documenting of a journey from the alternative perspective of that of the European map, the relative time and orientation differences from the flat itself. The experiment instead project revealed similar slips in consciousness to those experienced in the previous work.

Very simply, three faxes were sent to a fax machine located in the Hammersmith flat, each one sent from alternative times further and further away from the start of the journey. Fax one was sent from the flat itself and marked the beginning of the journey, the second was sent one hour and 35 minutes away from Heathrow and the last was sent two hours and 33 minutes away from Munich airport. No distances were used in the 'mapping', only their time differences, whilst the direction of each location from the fax machine was marked around a compass, providing a means of locating the journey through space.

Upon returning to England after the event, the three faxes were then superimposed on top of each other to provide a single document that describes the journey away from the Hammersmith flat in time and space from the perspective of the fax machine that effectively became an alternative North Pole.

In doing the work, I found myself again juggling with spatial conventions that do not easily sit together and in fact leap or slip between models as though I might be 'rhizoming'.

The process of sending the airport faxes consisted of recording my time away from the fax machine and measuring the orientation of my position with a compass as I looked towards Hammersmith with the help of a map. In these situations I was forced again to juggle with different spaces:
- Space One: the actual space, that of the airport from where the fax was sent back to Hammersmith.
- Space Two: the memory space, that was an image of the flat and the fax machine as I left it at 7:10 am.
- Space Three: the map space which marked out the projected direction of the flat across England, from where I stood in the airport.

Certainly I dealt with the situation at the time and the paper evidence of the experience reminds me of the mental conundrum that I was placed in but at no point did the juggling cease, only perhaps did the proposed 'rhizoming' become quicker and more fluid in order to ease the comprehension.

Conclusions

Whilst the appropriation of Deleuze and Guattari's rhizome appears to be effective in describing the moment of 'slipperiness' as one searches for spatial security upon receipt of the faxes, it could be said that such situations are orchestrated to generate such an experience. However, our apparent ability to deal with the increasing number of occurrences that put the traditional frameworks for space under stress, such as the example at the beginning of the paper, does happen to correspond to the nature of the rhizome extremely well. Whether this may help us observe and describe the human condition is perhaps at best hopeful as so much shifts under the duress of new technologies; but what it certainly may aid us in is the design of systems that describe or represent spaces that do not rely upon the residue of Renaissance systems.

References

Bourdieu 1977. *Outline of a Theory of Practice*. Cambridge University Press.

Deleuze,G and Guattari, F. 1987. *A Thousand Plateaus: Capitalism and Schizophrenia*. London:Athlone Press.

Foucault, M. 1980. *Power/Knowledge*. London: Harvester Wheatsheaf.

Patton, P. 1996. *Deleuze. A Critical Reader*. London: Blackwell.

Shields, R. 1991. *Places on the Margin*. London: Routledge.

Chris Speed is currently Senior Lecturer in Interactive Media on the BSc MediaLab Arts programme at the University of Plymouth. Following a BA (Hons) in Fine Art Alternative Practice at Brighton University, where he exploited multimedia and telecommunications technology to develop interactive installations, he moved between the roles of Telematic artist and multimedia freelance designer.

He joined the Research group in September 1996 after working in the New Media Department of Marshall Cavendish in the capacity as Art Editor and was involved in the direction and production of multimedia titles such as Images of War, Science Lab and Murder in Mind.

In a research capacity that stretches between London and Plymouth, Chris Speed is active in the production of practical and written work, he is a member of CAiiA-STAR Research Centre headed by Roy Ascott, and he is also presently studying MA Design Futures at Goldsmiths College in London. Preoccupied with how we identify and make sense of space and time, his research concentrates on the manipulation of the frameworks that enable us to orientate and navigate ourselves through the world. Adopting many forms of new media technology, the work aims to reconstruct and present information through events and experiences, to explore alternative ways of reading geography and time, and in turn question our relationship with the global village.
<chriss@soc.plym.ac.uk> http://CAiiA-STAR.Newport.Plymouth.ac.uk

– Space and Time –

Thinking Through Asynchronous Space
Mike Phillips

Background
The intent of "MEDIASPACE", whether in its "dead" paper-based form or the "live" digital forms of satellite and internet, is to explore the implications of new media forms and emergent fields of digital practice in art and design. "MEDIASPACE" is an experimental publishing project that explores the integration of print 1. WWW 2. And - interactive satellite transmissions 3 (incorporating live studio broadcasts, ISDN based video conferencing, and asynchronous email/ISDN tutorials). The convergence of these technologies generates a distributed digital "space" (satellite footprint, studio space, screen space, WWW space, location/reception space, and the printed page). This paper focuses on the relationship of this space to time and its 'inhabitants'.

Space and Time
Today, we take on space and time.

> The space-time continuum is being challenged. The notion of communication is changed for ever. All the information in the universe will soon be accessible to everyone at every moment."
>
> (Tracey 1998).

The *Futurist* inspired rhetoric that surrounds so much digital technology ignores the subtler aspects of the human condition. However, the shifts in our perception of time and space caused by these technologies can be compared to the impact the 'machine' had at the turn of the 20th century. It is as if the "poop-poop" mentality of *Mr Toad* was just a rehearsal for the rapid acceleration into cyberspace. Attempts to map cyberspace, as an alien terrain or an extension of our consciousness, challenge our understanding of the dimensions that engulf us and initiates the search for a 'cyber-cartography' in order that we may understand the complexity of this spatial-temporal flux.

The evolution of our spatial-temporal perception is acutely linked to technology. Toffler describes the shift from the cyclic idea of time, enjoyed by pre-industrial, pre-clock societies, as the 'linearisation of time', which resulted in fragmentation of 'space' into many 'spaces'. The need to control and measure space and time inevitably led to the standardisation of time.

> It was in effect synchronisation in space. For both time and space had to be more carefully structured if industrial societies were to function
>
> (Toffler 1983).

The 'global embrace' of McLuhan's (1973) extended nervous system, which contracted the globe into 'no more than a village', did not actually liquidate space and time. It gave the 'linearisation of time' nother 'dimension', the ability to pass through many streams of 'geographical' time, but time was still essentially linear. Pallasmaa articulates this notion of the second 'dimension' of time:

> The experience of space and time have imploded and become fused by speed. As a consequence of this implosion we are witnessing a distinct reversal of the two dimensions, a temporalisation of space and a spatialisation of time. We live increasingly in a perpetual present, flattened by speed and simultaneity, and grasped by instantaneous perceptions of the eye. The only sense that is fast enough to keep pace with the astounding increase in the technological world is sight. But the world of the eye is threatening to turn into the flat world of the present.
>
> (Pallasmaa 1996)

Figure 1, 'The Humming of Strings', illustrates a proposal for a system which plays with the notion of a 'flat world of the present' and explores the potential for asynchronous activity. The piece suggests the opportunity to jump in and out of the space-time fusion, by adding another 'dimension' to time. It uses the time delay in a satellite signal to create a 46,000 mile echo chamber/feedback loop. The reception of the signal returning from the satellite to its point of origin is delayed by a quarter of a second (23,000 miles x 2 at 186,000 mps). This delay places the received signal permanently in the 'past' and allows interventions to be made to the structure of the signal. The signal (video and audio), past and present, can be manipulated in 'real time' in multiple locations.

'MEDIASPACE' explores this other 'dimension' by offering a combination of synchronous and asynchronous interaction. The construction of this system is designed to allow individuals to communicate synchronously or asynchronously from a variety of locations through the new space, 'MEDIASPACE'. Consequently time and space are not limited by the flat present or by the fast-forward and rewind button. The system can shift the space-time that its participants inhabit. A recursive version of Palasmaa's flat present is envisioned by Vonnegut:.

> Yes, and when the timequake of 2001 zapped us back to 1991, it made ten years of our pasts ten years of our futures, so we could remember everything we had to say and do again when the time came.
> Keep this in mind at the start of the next rerun after the next timequake: *The show must go on!*
>
> (Vonnegut 1997).

The environment facilitated by 'MEDIASPACE' provides another vista, one of multiple spaces slipping in and out of synch, sometimes simultaneously sometimes asynchronously.

(I/You/We) (Am/Are/Was/Were) (Here/There)

> If the concept of space is not a space, is the materialisation of the concept of space a space?
>
> (Tschumi 1996).

– Space and Time –

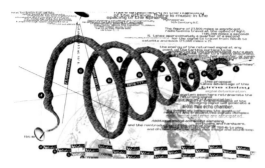

Figure 1: 'The Humming of Strings"

The screen grabs shown in Figure 2 are taken from a video conference between Paris and Plymouth 4. Are the two locations contracted to one or expanded to three? Or duplicated at each end to four or five? Is the shortest distance between two points a video-phone? Paik (1984) compares the art of the *ninja* to function of a satellite.

> The first step for a *ninja* is to learn how to shorten distances by shrinking the earth, that is, how to transcend the law of gravity. For the satellite, this is a piece of cake.
>
> (Paik 1984).

The shrinking of space through digital communications generates new spaces. The flat 'screen-space' of a video conference presents rectangular 'portals' to the other participants and a feedback 'portal' showing, in real-time, the place of origin. The 'portal' opening on the other screen shows similar views but at the point of reception the delay and strobe indicates that a space between has been crossed. However, this space may not be geographical; it may be generated by technological inadequacies. This 'space between' is a conceptual and temporal space.

'MEDIASPACE' attempts to harness these temporal spaces within and beyond the transmission times. Figure 3 shows a 'map' of the 'MEDIASPACE' system: the wire-frame circles are the transmission times (60 minutes), the dark spheres are the ISDN conferences and the central wire-frame cube the WWW site. The floating figures/ boxes are addressed in the next section. Multiple 'pathways of communication' connect each point on a plane (time of a transmission) and between transmissions through email or ISDN tutorials. The WWW site acts as an central focus which evolves as the transmissions progress.

The system only functions when it is populated. The interaction of individuals within this system generates a 'social' space which, according to Harré (1985), is the 'space' where understanding and knowledge are exchanged and learning takes place. It is possible that this space allows individuals to externalise their "inner worlds" to generate a 'spatial' consciousness.

> I look to my left, and I am in one city; I look to my right, and I am in another. My friends in one can wave to my friends in the other, through my having brought them together.
>
> (Novak 1991).

Novak's 'Liquid Architecture' momentarily solidifies to form a 'city', with waving

Figure 2: 'Visio' Conference

– Reframing Consciousness –

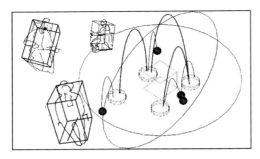

Figure 3: Map of 'MEDIASPACE' (scale 1 : × ÷ n)

inhabitants. This next section explores the relationships that may develop between the participants of 'MEDIASPACE' through the process of 'waving' to others and/or watching others wave.

Watching People Watching...

The walls of Saint Jerome's 5. study are punctured with an array of rectangular 'portals', each one giving access to a different space: internal, information and external sky/landscapes. Jerome ignores these spatial frames, focusing instead on the symbols on the printed page. He also ignores the gaze of the viewer. This information space is pierced by windows which, while allowing him to see out, allow the viewer (and viewers at the other windows?) to see in. Is this the perfect model of a web-cam?

Whether or not Adolf Eichman, on trial in his glass box 6. in Jerusalem, was the inspiration for the wire-frame structures which hang around Bacon's figures, their 'spatial tension is significant. This spatial anomaly impacts on the confined and entropic condition of 'Organisation Man' and parodies the structure of 'Saint Jerome in his Study'.

Like Saint Jerome, Eichman and 'Organisation Man', the architect of the Panopticon 7. sits in a 3-D frame, some 250 years after his death. Jeremy Bentham's AutoIcon, still sits in the corridor at UCL, his body preserved in wax, his head mummified and his vital organs conserved. The public display of the mortified body of the creator *of "Inspection House"*, anticipated the global voyeurism with this ultimate act of exhibitionism.

Figure 4 is a diagram of these seated figures. The exaggerated structure which surrounds them separates them from each other and the viewer but also functions as a window on their activities. It is possible that the relationship of the passive occupants of 'MEDIASPACE', to each other and the space, fall somewhere between the matrix created by these structures that house the likes of Saint Jerome, Bentham and Organisation Man. Figure 3 places these 'lurkers' in the vacant space, just off the main 'pathways of communication' within the 'map' of 'MEDIASPACE'. It is also possible for 'active' individuals within the space to drift in and out of these 'pathways of communication'.

These people in this space and time require a new nomenclature. They are no longer an 'audience'; such a definition is too passive. Telematic activity, as Sermon (1997) says,'is nothing without the presence and interactions of the participants who create their own television programme by becoming the voyeurs of their own spectacle' Turley (1997) identifies the term 'spect-actor' to define

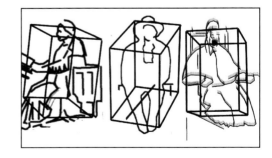

Figure 4: Boxed Philosophers

"forms in which members of the audience cross, ultimately *consciously* cross the boundary between watching and taking part in the action.

Terminus

The 'MEDIASPACE' project attempts to provide a framework for 'cyber-time', 'cyber-space' and the social interaction that occurs within them. To cross boundaries within this framework we must experience the spatial-temporal shifts that occur, as they are too elusive to be articulated. Epps (1997) terms the reality of cyberspace 'subjective', and that 'it is this non-local space, as Novak terms it, which transcends language and reinforces the spaces which we already know'. In order to understand the 'cyber'-spaces that we do not already 'know', the human mind requires 'time' and 'space' to evolve through a social dialogue within the complexity of this 'subjective' spatial-temporal flux. And in cyberspace there are plenty of 'times' and 'spaces'.

Notes

1. MEDIASPACE WWW can be found at: http://CAiiA-STAR.Newport.Plymouth.ac.uk
2. 'MEDIASPACE' is published as part of the CADE (Computers in Art and Design Education) journal Digital Creativity by SWETS AND ZEITLINGER.
3. 'MEDIASPACE' interactive satellite transmissions were funded by the European Space Agency (ESA), the British National Space Centre (BNSC) and WIRE (Why ISDN Resources in Education) and use Olympus, EUTELSAT and INTELSAT satellites via a TDS-4b satellite uplink.
4. Visio Conference: 13e Expolongues Salon International des langues. 28 janvier 1er fevrier 95, Grande Halle de la Villete Paris.
5. Saint Jerome in His Study, Antonello da Messina, active 1456; died 1479, Oil on lime, 45.7 x 36.2 cm, The National Gallery, London.
6. Examples of Bacons 'glass box' device: 1949 Head VI (93 x 77 cm, The Arts Council of Great Britain), through to 1991 Study from the Human Body (oil and pastel on canvas, 198 x 147.5) the Estate of the Artist.
7. *The Panopticon; or the Inspection House.* was proposed by Jeremy Bentham in 1791.

References

Epps, C. Objective and Subjective Virtual Realities, *Digital Creativity* Vol 8 No. 3 and 4 1997 Mediaspace 4, pp.7-12.
Grahame, K. 1908, *The Wind in the Willows*. Methuen & Co. Ltd. P.41.
Harré, R. et al. 1985. *Motives and Mechanisms: An Introduction to the Psychology of Action*. London: Methuen & Co, Ltd,. P.72.
McLuhan 1973. *Understanding Media*, Abacus. Pp. 11-13.
Novak, M. 1991. Liquid Architecture in Cyberspace. *Cyberspace First Steps*. MIT Press. p225.
Paik NJ. 1984. 'Art and Satellite', in Stiles K. et al (eds.), *Theories and Documents of Contemporary Art*. University of California Press. P. 435.
Pallasmaa J, 1996, The Eyes of the Skin, in Polemics. Academy Group Ltd, p.12.
Sermon P. 1997, From Telematic Man to Heaven 194.94.211.200, Conference Proceedings Consciousness Reframed 1997, CAiiA, University of Wales College, Newport.
Toffler 1983. *The Third Wave*, Pan Books. P.119.

Tracey M. 1998. Advertisement in the New York Times on 5 January 1994, for MCI.
The Decline and Fall of Public Service Broadcasting Oxford University Press, p.192.
Tschumi, B. 1996. 'Questions of Space', *Architecture and Disjunction*, The MIT Press, p.53.
Turley, S., 1997. 'Designing for Audience Response' *Intelligent Tutoring Media*, Vol 7 No. 3 and 4, Mediaspace 3, p.28.
Vonnegut, K. 1997 *Timequake*, London:Jonathan Cape, p. 20.

Mike Phillips is currently leading the Interactive Media Subject Group at STAR (part of the CAiiA-STAR integrated research programme), in the School of Computing, University of Plymouth. After a BA (Hons) in Fine Art (1984), Phillips received a scholarship to the University of Massachusetts, USA, and a Higher Diploma in Fine Art (Experimental Media) at the Slade School of Fine Art. Experimental work included interactive electronic experiences through kinetics, collaborative computer networks, installation and audio-visual work, making a steady shift from the analogue to the digital realm. Collaboration through and with technology has been a key theme, leading to interactive kinetic work and telematic performances in Europe with EAT88-UK and a regular collaboration with Donald Rodney (ICA, TSWA and South London Gallery).
As Lecturer in New Technology at Exeter Faculty of Arts and Design, Phillips worked in Fine Art (4D), Media and Publishing, and ran digital media courses for industry. In 1992 he moved to Plymouth to co-ordinate the BSc MediaLab Arts Programme in the School of Computing, an innovative interactive multimedia degree which has gained significant industrial sponsorship (Macromedia, Sony, SGI). During this time he has been involved with grant funded research projects (ESF, EC, EPSRC, HEFCE, Henry Moore Foundation) in the field of interactive media. Phillips is on the Board of Plymouth Arts Centre and the Editorial Advisory Board of Digital Creativity and regularly designs and produces Mediaspace through WWW, Print and Interactive Satellite Transmission.
<mikep@soc.plym.ac.uk>. http://CAiiA-STAR.Newport.Plymouth.ac.uk

The Space between the Assumed Real and the Digital Virtual

Dan Livingstone

In writing this paper my first task is to engage the reader in the conceptual framework proposed in order to identify the space I envisage. Some clarification of the terms I use will be necessary in order to focus our attention within this framework, as owing to its interdisciplinary nature, more avenues for exploration and debate arise than are resolved.

I use the *assumed real* to identify the physical space our bodies inhabit; reality is too definitive a term and presupposes that we all are in agreement as to the space that this term encompasses. Our individual realities are informed by our experience of the world, more frequently mediated through technology. An obvious but nevertheless compelling example of this is the exponential increase in mainstream broadcast television framing the output of a network of live CCTV and surveillance systems as infotainment where the recursive spectacle of us consuming our various levels of antisocial behaviour is as much a part of the content as the digital reprocessing which allows us to view it.

> Although we know the imagery from video cameras in banks and supermarkets is relayed to a central control room, although we can guess the presence of security officers, eyes glued to control monitors with computer aided perceptions – visionics - it is actually impossible to imagine the pattern, to guess the interpretation produced by sightless vision.
>
> (Virilio 1994).

By *digital virtual* I encompass screen-based representation and simulation systems that provide interfaces for either individual or group experiences that are manifested within digitally generated spaces. This would include gaming environments, collaborative virtual workspaces and other forms of networked virtual reality.

If we were to attempt to reach agreement about how one of the sensorimotor systems that informs our perception of physical space operates, we would become sidetracked, as these biological and psychological mechanisms do not operate independently. Francisco Varela et al. provide an excellent illustration of this potential debate, discussing visual perception by asking what came first, the world or the image and suggesting that researchers from both cognitivist and connectionist persuasions would support the "chicken position".

> The world out there has pregiven properties. These exist prior to the image that is cast on the cognitive system, whose task is to recover them appropriately (whether through symbols or global subsymbolic states
>
> (Varela et al. 1991).

The alternative to this very convincing statement is given as the egg position:

> The cognitive system projects its own world, and the apparent reality of this world is merely a reflection of internal laws of the system."
>
> (Varela et al. 1991).

These inner and outer extremes focus on the visual as the central issue. This is also a recurrent theme in the discussion of the perception of physical space, whether from a philosophical or scientific perspective.

> Inside and outside form a dialectic of division, the obvious geometry of which blinds us as soon as we bring it into metaphorical domains
>
> (Gaston Bachelard *Poetics of Space*).

The space I envisage not only falls between these divisions but in a sense is manifested through the friction that is generated between them. It can only exist because we are prepared to engage with it, simultaneously authoring and documenting our own experience through interaction with both real and virtual space in the same moment. This space could be seen as augmented reality but this term is misleading. Conventional manifestations of augmented reality are usually dependent on the subject having additional information processing power integrated with his existing sensorimotor capacities; a pilot with digital overlays of objective navigational and strategic data has an increased view of the physical world through the system. The emphasis is on augmenting what is already there. The space between the assumed real and the digital virtual is not 'already there'. We are complicit in generating it through interacting with a system that enables us to occupy this space both on a physical and conceptual level.

Sorcerers Apprentice (Livingstone 1997) is a current project that enables us to occupy this space. It is a tri-part system that traverses perceived space, conceived space and lived space (Lefebvre 1991) simultaneously. This space could be considered in terms of Soja's 'Thirdspace' although in its current form the social and collective implications of the mechanism have not been resolved.

> Everything comes together in Thirdspace: subjectivity and objectivity, the abstract and the concrete, the real and the imagined, the knowable and the unimaginable, the repetitive and the differential, structure and agency, mind and body, consciousness and the unconscious, the disciplined and the transdisciplinary, everyday life and unending history.
>
> (Soya 1996) .

The system is described in phases in order to identify the various processes through which this integrated space is realised. This system depends initially on collaboration between a subjective human agent and an objective artificial agent, although as it evolves, the boundaries between agents blur.

Phase One: Perceptual Shifts

The first goal is to move the human agent from a position of spectator/user to that of performer/instigator. It is necessary to strip away the critical and physical limitations of a screenspace and to relocate the human agent's perspective from a position of viewer of graphical virtual representation to conceiver of spatial relationships through interaction with a limited but focused digital mechanism. The mechanism is not intended to augment our perception of physical reality but to refocus our cognitive processes on the physical space we occupy, thereby enabling us to overlay our own individual subjective responses to that space. An environment is suggested and extended through our interaction with it; sound is manipulated to describe our perception of this space. As we drag objects from the screenspace into the actual space our bodies inhabit, the transient perceptual shifts that occur allow us to operate in the grey area between representation and interpretation.

> Tight linkage between visual, kinesthetic, and auditory modalities is the key to the sense of immersion that is created by many computer games, simulations and virtual- reality systems"
>
> (Laurel 1993).

In order to break our preconceptions of what interacting with a computer mediated experience might be and anchor this experience in the *assumed real,* a graphical metaphor on a screen is used. Objects on a screen can be manipulated via infra red. We are both part of and separate from this interaction, emphasising our participatory complicitness in the generation of subjective spatial referencing. The design of each object references the compass, our tool for navigating physical space. As we drag it across the screen its readout changes, numerical coordinates update. Even as we drag the objects from this screenspace, the readout continues to relay its position in space. As the visual representation is limited to the screenspace, three dimensional spatialised sound is used to sustain our perception of the object that we have moved from the digital virtual into the physical space we currently occupy. We are engaged, we perceive the object to be there and if we doubt it's presence, a glance back to the screen will affirm its location in degrees. To further reinforce our suspension of disbelief we can continue to manipulate the object, drawing it closer to us or moving it away. In harnessing the physical movement of the human agent as a means to manipulate perceptually generated objects within physical space, we are using a form of embodied interaction as discussed by Lakoff and Johnson's 'experientialist approach to cognition'.

> Meaningful conceptual structures arise from two sources: (1) from the structured nature of bodily and social experience and (2) from our innate capacity to imaginatively project from certain well structured aspects of bodily and interactual experience to abstract conceptual structures. Rational thought is the application of very general cognitive processes -focusing, scanning, superimposition, figure-ground reversal"
>
> (Lakoff 1988).

As we continue to explore this mechanism by dragging additional objects into being, the conceptual relationships we make between the *assumed real* and the *digital virtual* form the foundations of a new space. Through perceiving the objects we have co-created as coordinates within physical space, we have produced a perceptual construct, that is, a series of relationships that can be documented and explored.

> The business of interaction between an individual and a machine (as opposed to interaction with an installation and in public) is to reinvent passion, which I would understand as the flash of fire that sparks from the crunching of familiar gears in
> unfamiliar combinations"
>
> (Cubitt 1995).

Phase Two: Translation

As we engage with the perceptual construct, the artificial agent translates our activity and broadcasts it in a remote location and our interactions are recreated. This 'remote location' can either be a physical space such as a gallery, using physical projection such as a laser system where transition between points is reminiscent of a child drawing in space using a sparkler, or within a virtual space created dynamically using VRML. In either case, the perceptual construct, the broadcast of this visual mapping of our auditory interactions has been disembodied from its physical context. But as the artificial agent learns through participation in the recreation of the perceptual construct, it's own ability to collaborate in the realisation of further constructs emerges. This process of translation is significant as it provides documentation of the perceptual constructs generated by different subject's, thereby enabling us to build comparative data in order to explore the nature of each subjects response to this speculative mechanism.

Overlaying a series of perceptual constructs within one virtual space would allow a far deeper exploration of the conceptual relationships that each individual brings to the system. For example, if we provide constraints to our subject's interaction within the system, the perceptual differences between constructs will become more acute. We will use a very simple example: Visualise, if you will, eight points in space; now compose them within the physical space you currently occupy. We will impose constraints: the points must be positioned so that a cube can be created by joining them; no point can fall outside the room you occupy. Each manifestation of the cube will be different; each individual's construct will be different. The more we engage with the concept and react to the constraints, the more creative the solutions will become. The key here is to underline the fact that however you compose the cube mentally, it has a relationship with the physical world, or more accurately, our experience of this world. The constraints have sharpened our awareness of the perceptual constructs context. If we could now retrieve all the diverse instances of the cube that had been previously generated within this space, by overlaying our perceptually constructed cube with diagrammatic references to previous cubes, how would this level of experience affect our strategies for co-forming more sophisticated perceptual constructs. I use the example of the cube to illustrate how the system could interpret and reinforce this perceptual process; but within this critical space there are no such constraints. How an individual interprets these points is not defined:

> A point in space seems perfectly objective. But how are we to define the points of our everyday world? Points can be taken either as primitive elements, as intersecting lines, as certain triples of intersecting planes, or as certain classes of nesting volumes. These definitions are equally adequate, and yet they are incompatible: what a point is will vary with each form of description"
>
> (Varela et al. 1991).

Phase Three: Integration.

A human agent within this world could attempt to describe the physical space occupied to the artificial agent within the system; this process of communication would not be a one-way dialogue, as the system would have the capacity to 'question' the perceptual construct being generated by integrating it with the physical environment, either by real-time projected diagrammatic overlays through a robotics or augmented vision system, or more simply; by modelling the space through VRML. It would be equally valid for the artificial agent to generate space by storing, reconstructing and extending previous perceptual constructs in order to question our methods of communication.

The third layer of this mechanism for spatial generation and critical exploration is the integration of the two previous phases, providing a critical space that is formed through the collaboration of both human and artificial agents. Interacting with this system has the potential for refocusing the way we perceive and engage with both physical and digitally generated spaces. We have the potential to bring compositional and spatial aesthetics into these relationships. Will the system also develop this potential? And by interacting with this tri-part system, can we evolve a more objective or critical approach to the way we interact with technologically mediated experiences?

References

Bachelard, G. 1994 *The Poetics of Space*. Boston: Beacon Press.

Cubitt, S. 1995. From an essay on interaction for VideoPositive.

Lakoff, G. 1988 "Cognitive Semantics". In *Meaning and Mental Representations*, Umberto Eco et al. Bloomington: Indiana University Press.

Laurel, B. 1993. *Computers as Theatre*. Addison-Wesley. p.161.

Lefebvre, H. 1991. *The Production of Space*. Oxford: Blackwell. p38-50.

Livingstone, D. 1997. 'The Sorcerers Apprentice'. In 'Mediaspace', contained within. *Digital Creativity*. Intellect Ltd.

Varela, Francisco, J. 1991. The Embodied Mind. Cambridge, Massachusetts: MIT Press. pp172,.232.

Virilio, P. 1994. *The Vision Machine*. London: BFI Publishing, p62.

Dan Livingstone is a Senior Lecturer in Interactive Media with the School of Computing at the University of Plymouth. His current role within the School is that of programme manager for the BSc MediaLab Arts course. He is a member of STAR (Centre for Science Technology and Arts Research) His current research activity deals with the design of interactive environments and augmented realities that investigate new forms of perception and consciousness. In 1996 he was an active contributor to the 'Mediaspace' satellite broadcasts, a series of interactive programmes subscribed to by a number of European universities. Other collaborative research activity includes a project with contemporary artist John Skinner and professional musician and composer Chas Dickie (of 'Startled Insects' fame). In 1995 a series of collaborative experiments were held at the Abbotsbury Studio in West Dorset, investigating the critical aspects of interactive performance and composition. Livingstone moved to Plymouth in 1994 after completing a Masters Degree in Sequential Design, specialising in Interactive Media, at the University of Brighton. During his MA studies he won the Rediffusion Award for Multimedia and was sponsored by the Rediffusion Simulation Research Centre, to continue his research into human facial expression. 'Darwins Expressions' was exhibited in September 1993.

Part III – Art

6   *Strategies*

7   *Projects*

8   *Architecture*

9   *Creative Process*

6   Strategies

The Work of Art in the Age of Digital Historiography

Brett Terry

Introduction

Walter Benjamin's famous essay 'The Work of Art in the Age of Mechanical Reproduction' is deeply concerned with the sociological, philosophical and aesthetic implications of film and photography and their effect on the art object. Benjamin's critique has been widely adopted to analyse the relations surrounding the aura of the art object and the socio-political implications of authenticity and legitimacy within the dominance/marginisalation spectrum. Were Benjamin alive to experience the digital information age, he might well have comported his critique from concerns of authenticity, filmic discourse and linear narratives, to a critique of the inherent biases and constraints of the way in which *digital* thinking influences the writing and making of history.

Edward Soja (Soja 1996), arguing against the marginalization of spatial and diachronic methods of historiography, states Benjamin's views thus:

> Benjamin was particularly concerned with the alluring and illusive 'narcotic of the narrative', the discursive and non-discursive constraints imposed by narrativity on the writing as well as the making of history. He saw this addiction to the narrative form, with its compulsion toward linear, sequential, progressive, homogenising conceptions of history, as requiring a withdrawal strategy involved 'blasting' the embedded historical subject out of its temporal matrix and into a more visual, imagistic, and spatial contextualisation.

Soja's concern regarding the prioritisation of spatiality over temporality as a historical *modus operandi* refers to both geographical theorist Derek Gregory and historical theorist Hayden White.

Digital historiography refers both to the construction of historical 'writing' in a virtual informational landscape as well as to the user experience of traversing such a landscape in one's experience of History *vís-a-vís* historical writing and simulacra. My use of the *landscape* terminology is couched in the domain of cognitive flexibility theory, which, in turn, borrows from Wittgenstein's notion of the 'criss-crossed web' as a epistemological and pedagogical metaphor.

Defining Digital Historiography
The use of the term digital in this context is necessarily multivariate. I will put forward some relevant polarities that, collectively, should help to form a cohesive yet pluralistic methodological and empirical definition of 'digital historiography.'

Atoms vs. Bits
Thinking along the lines of Nicholas Negroponte, one might contrast information as *bits* as opposed to *atoms*. This distinction implies the adaptive, reconfigurable nature of bits as well as their innate differentiation from atomic paradigms of publishing, temporal and geographical fixity and, ultimately, ownership and history.

Analogue vs. Digital
As opposed to the continuity of analogue (e.g. the waveforms of signal processing), being digital is typically understood as fundamentally discrete and quantified (e.g. sampling and switching). The associative and navigational infrastructure through which digital information is imparted necessarily relies on naming discrete localities (however transient) and the quantification of relevance (however fuzzy).

Linear vs. Non-linear
There is nothing constituitively non-linear about digital information (in fact, sequential access predated random access). The typical user experience of digital information involves making choices that influence what is viewed by the user. Since the granularity of a computer screen (or for that matter, a VR headset) is much finer than a 'chapter,' retaining linear/sequential channelling for any significant length of time is ultimately antithetical to the user-driven experience. This distinction is actually far more complex as well: the non-linear domain includes many forms of adaptive and enframing modalities, which might make use of a universe of past user choices and preferences.

Technology
Information can, of course, be non-linear without being digital. *Digital* is meant here as the progressive step beyond Benjamin's use of the term *mechanical*. As Benjamin uncovered specific relations in the domains of art, history and technology, I intend to do similarly.

History and Time
History's temporal nature looks out at the otherness of the past and articulates connective metaphors that narrate, differentiate and evaluate distinct historico-temporal strata. The writing of history makes use of these metaphors - progress, evolution, accumulation, causality, influence - within the contexts of the predominant art-historical constructs: the artist's vita (the Vasarian the-man-and/as-a-work), the intrinsic domain of the art object, the tropologies of style and genre and the socio-political influences on artists and their works.

    Consider the relation of digital historiography's methods to time: it is not a passive temporal object; rather it is an adaptive, malleable nexus of information whose only real claim to fixity is its being named as an entity. In the same way in which the King's Singers can remain a diachronically coherent entity (for marketing concerns) even as its entire constituency is replaceable, a locus of any digital information diminishes the legitimacy of authentic versions of its content, enshrined by authorial intentionality. Essentially, a version of the landscape as it stands at any one particular moment in time can be preserved but the notion of a *truer* version is ultimately deconstructed by its innate predilection for revisionism, its lack of control over its own dissemination and reception paradigm and the inherent difficulty of forcing atomic conceptions of ownership and possession onto slippery digital bits.

The Artifacts of Discourse

Every medium enframes its discourse through its modalities of production and reception. Historical intrepretations and narratives are understood concommitly through the windows of the paramount paradigm of written historiography - the *document*. The document, enunciatively positioned on its own definite chronological point, is a participant in a wealth of relations: its authorial presence and related diachronic strains of influence, its positing of a futural reference point and the assertion of its legitimacy *vís-a-vís* disciplinary (or cross-disciplinary) spectra of dominance and marginalisation.

Digital historiography ultimately stands to override the ubiquitous and influential document model with its own informational landscape. At present, the World Wide Web (the modern poster child for informational landscapes) continues to use the document model (i.e. the web 'page') even as experts in information architecture, personalisation, and information retrieval attempt to mount a conceptual paradigm shift. Inevitably, as users experience and learn about history through traversal and relevance evaluation (rather than serial narratives), the dominance of the document/page metaphor will be weakened.

Most significant will be the transition from atemporal chronological point to dialectic temporal flow. The mutability of digital information will challenge notions of definitive chronological reference points (which are, in a certain sense, by-products of the economic dictates of the modern publishing model and the accountability of scholarship). Authorship, while valiantly attempting to enshrine legitimacy and protect intellectual 'property,' remains likely to be effaceable unless (or until) there occurs a legitimised standardisation of meta-bits (Negropontes's 'bits about bits'). Adorno's characterisation about the historical trajectory of the critical function of art objects is one possible example of what is meant here by 'dialectic temporal flow'. The temporal evolution of the informational landscape would directly participate as a ontical epistemic framework evolving in relation to the plurality of interpretations and the critical power of the art object(s) to which it refers.

Reception and Visualization

The post-document historiographic landscape is enframed by paradigms of information design and technologies of viewership, among them: interface, quantifications of relevance and (ultimately subjective) informational hierarchies. How does this compare with what Hayden White might characterise, in 'traditional' historigraphic practice, as the employment of privileged historical narratives? In White's pro-narrative view, a sequence of historical events could be represented:

a,b,C,d,e, ... , n

Certain events (e.g. C) are endowed with explanatory force 'either as causes explaining the structure of the whole series or as symbols of the plot structure of the series considered as a story of a specific kind.'

(White 1978).

In the traversal of the historiographic landscape there is a multiplicity of privileging relations, including: relevance, correlation, dominance, linkage, personalisation and authenticity (i.e. in addition to explanatory focus). Given the lack of an *a priori* narrative totality, any attempts to privilege and influence the interpretation of the traversed landscape are not holistic. In contrast to the static intentionality of narrative employment, these influences may, however, be adaptive, melding a history of the user's traversal and preferences to their specific (if singular) programme.

Edward Soja (Soja 1996) makes the point that geography and history were intertwined prior to the long 19th century (1789 - 1914) and that there emerged, in socialist, idealist and empiricist thought a reconstitution of the temporal-spatial configuration that came to privilege idiographic and diachronic temporality at the expense of nomothetic and synchronic spatiality. Compare the spatial-organisational complexities of the history textbook and exhibition space to the nodal informational landscape:

A typology of the landscape nodes might include nodes for retrieval, adaptive information or interpretation, channelling, relevance differentiation and stratagems for influencing the traversal path such as gateways, forced impositions, spaces of participation and resources for reference. In any event, it is clear that the level of visualisation involved in the construction and design of such spaces reconstitutes the temporal-spatial spectrum in its own way, while also creating a relatively uncharted realm for the visualization and design of information and its relations.

Post-Historical Art and Post-Narrative Histories

Arthur Danto (Danto 1997) describes what he means by the 'End of Art' as when there is no longer any defining style, when "nothing is historically mandated: one thing is, so to say, as good as another". The notion of a crisis, prevalent in 20th century

Comparative Visualisations of User Experiences

cultural history, is partially predicated on the deconstruction of its historiographic methodological paradigms of periodicisation and progressive development, and their lack of applicability to the pluralistic tropology of artistic praxis in the latter part of the century. Leonard Meyer suggests, in a similar fashion, that in contrast to the relentless emphasis on diachronic style and syntax identification and, above all, stylistic and vernacular *progress*, that temporal *stasis* must be understood, in all likelihood, as a central defining characteristic of the contemporary cultural epoch.

Digital historiography, itself temporal, is independent of the sequential narrative form so inexorably linked to the form/content of progressive reification. The landscape of digital historiography holds up a mirror to the idiographic panorama of works and artists, intimating that nomothetic narratives have formed a dialectic with the interactive behaviour of the landscape's explorer. Temporality's primacy is, in fact, challenged by its mandatory participation in the virtual geographic informational space.

Interpretation and the Objects of Analysis

The following points, detailed aphoristically, engage several points raised in Donald Preziosi's *Rethinking Art History - Meditations on a Coy Science*.

History and Ideology.

Donald Preziosi notes: '... art history in its formative phases in the latter half of the nineteenth century was one of the important sites for the manufacture, validation, and maintenance of ideologies of idealist nationalism and ethnicity, serving to sharpen and to define the underlying cultural unity of a people as distinct from others.' (Preziosi 1989: 41) This ideological/hegemonic function was critically deconstructed in the cultural transition from historically redemptive Habermasian modernity to post-modern micro-, marginal-, and cultural history. Digital historiography posits the illusory neutrality of information (recombinant bits) but it operates distinctly in the realm of the technology enfranchised and serves to further bifurcate the technological have and have-nots.

Historical Objects Lie Orderly.

The disciplinary apparatus of the disembodied historical 'voice,' written in the third person and possessive of its distinct enunciative modality, tacitly assumes that historical objects will *lie orderly* within its purview of panoptic distinctions. In the transition from document to landscape, the potential de-privileging of authoritative versions seems likely to invite a pluralism of correlative associatives that illuminate the relativity of the interpretative function. If, indeed, 'the power of the museum lies precisely in its ability to elide alternative signifying practices,' it can be further observed that the informational landscape's entelechy is the homogenisation of alternative ideologies *sine qua non*. Instead of *lying orderly*, art-historical entities must be reconstructed to participate in the relevance-oriented parameterization of informational topologies.

Interpretation as Cryptographic Method.

If interpretation is to contend with a new geographic discursive logic, it seems likely that distinct interpretations could be made to participate in a coplanar informational arena. If the possessive totality of the interpretative modality is deconstructed by the informational apparati of digital technology, then there will occur a general pluralistic fragmentation of the decoding practices:

The artwork is invariably taken as autonomous and unique - a bafflingly intricate public text to be read. Interpretation, within this regime, is established as the supreme critical-historical activity. The network of new disciplinary technologies operates to suspend the analyzad in a discursive space as an essentially communicative token, as a medium through which an author (or, in a broader sense, a society) speaks: its mode of existence becomes primarily that of saying." (Preziosi 1989: 83).

In the fragmented competition of coplanar interpretations, the decoding of the analyzad becomes subject to relevance correlation and adaptive personalisation. It is likely that artistic works would actively participate in this dialectic by comporting themselves to this new interpretative arena.

References

Adorno, Theodor. 1984. *Aesthetic Theory*. London, Boston: Routledge.
Belting, Hans. 1987. *The End of the History of Art*, trans. Christoper S. Wood Chicago: University of Chicago Press.
Compagnon, Antoine. 1994. *The Five Paradoxes of Modernity*, trans. Franklin Philip New York: Columbia University. Press.
Danto, Arthur. 1997. *After the End of Art*. Princeton: Princeton University Press.
Iggers, Georg G. 1997. *Historiography in the Twentieth Century - From Scientific Objectivity to the Postmodern Challenge*. Hanover, London: Wesleyan University Press.
Jenkins, Keith. 1991. *Re-thinking History*. London, New York: Routledge.
Lévi-Strauss, Claude. 1966. *The Savage Mind*. New York: Harper & Row, p. 258.
Meyer, Leonard. 1994. *Music, the Arts, and Ideas: Patterns and Predictions in Twentieth-Century Culture*. Chicago: University of Chicago Press.
Negroponte, Nicholas. 1995. *Being Digital*. New York: Knopf.
Preziosi, Donald. 1989. *Rethinking Art History - Meditations on a Coy Science*. New Haven: Yale University Press.
Soja, Edward W. 1996. *Thirdspace Malden*, MA: Blackwell Publishers Inc., p. 167.
White, Hayden . 1978. *Tropics of Discourse - Essays in Cultural Criticism*. Baltimore, London: Johns Hopkins University Press.
White, Hayden. 1987. *The Content of the Form - Narrative Discourse and Historical Representation*. Baltimore, London: Johns Hopkins University Press.

Brett Terry is currently involved in design, information architecture, and programming at the Burnett Group in New York City USA. He has published papers on Hypertextuality, Object-Oriented Software Engineering; and Chaos Theory. In addition, his choral and electro-acoustic musical works are performed at many venues including the International Computer Music Conference, the Society for Electro-Acoustic Music in the United States, the San Francisco Bay Area New Music Marathon, Sound Culture and the Connecticut College Arts & Technology Symposium (New London, CT, USA). His research interests include Information Theory, Software Enginnering and Interface Design, interfaces for computer music synthesis, composing music and innovations in website design. A graduate of Wesleyan University, he holds graduate degrees in Computer Science and Music Composition from the University of Illinois, Urbana-Champaign, where was affiliated with the Experimental Music Studios and the Computer Music Project.
<bterry@burnettgroup.com>   www.burnettgroup.com/bterry/

Mind Memory Mapping Metaphor: Is Hypermedia Cognitive Art?

Dew Harrison

The practice within contemporary art which retains a conceptual base and involves some form of cultural narrative results in art works of great complexity, many requiring computer management for their multi-media environments. Hypermedia would seem particularly suitable for such complex databases in that it has an affinity with concept-based art where both are concerned with connecting associated ideas into meanings or concepts. Hypertext has expanded beyond text into hypermedia to include multi-media items. Conceptualism has expanded beyond language into a concept-based art practice which embraces a variety of media including electronic multi-media.

Art practitioners using hypermedia confront the problems apparent in hypermedia system design of data construction and interface. Hypermedia researchers have sought ways of addressing these problems through navigation, mapping mechanisms and metaphor. The limitations of the geographical metaphor have stimulated a search for a richer metaphor in the field of AI which has provided the semantic network metaphor. AI researchers are continuing to reduce the complexity of semantic networks which may improve search mechanisms for hypertexts; however, hypermedia when used for creative activities such as multimedia fiction does not require precision searching. Creative hypermedia systems require interaction with the viewer to generate new meanings and interpretations from the information in the database.

Hypermedia?

The initial vision of hypermedia was for a mechanised method of augmenting human memory. The work of Vannevar Bush was committed to mechanical analogue technologies and concerned the design of machines bound neither to, nor by, mathematical logical operations. He first conceived of the *Memex* machine in the early 1930s, searching for a mechanical analogy to the human mind for locating and representing knowledge in order to support human intellectual work. Bush finally published his ideas about the *Memex* in 1945 in the article '*As We May Think*' a text now generally acknowledged as the first vision of hypermedia.[1] The purpose of the *Memex* machine was to expand our intellect by emulating the process of memory for the speedy access of related information. *Memex* was the inspiration for the work of Douglas Engelbart, Ted Nelson and others resulting in Hypermedia. Hypermedia is the non-sequential software applied to the sequential processing of the von Neumann machine architecture, which is the digital computer. Hypermedia offers an analogically human way of accessing information from knowledge stored in electronic memory for the promotion of more human deep thought upon one concept of interrelated items.

Creating in Hypermedia
Hypermedia allows for the contained, discrete system of a unique art installation or the mass-produced work of art in CD-ROM form. Hypermedia is also the Internet system, our cultural information conduit that can accommodate concept-based art by acting as a hyperlinked platform for the creation and presentation of global art works. Artists using hypermedia encounter the difficulties of navigation and interface that confront all hypermedia system designers. The methods and techniques apparent in hypermedia research which attempt to overcome these problems fall into two camps:
- The stringent models of cogent linkage and material organisation from the fields of Cognitive Science, AI and Information Systems. These models are derived from linguistic models and therefore address hypertext.
- General design strategies with emphasis on viewer interaction from hypermedia system designers. These guidelines are offered mainly from the authors of multi-media fictions and pertain to hypermedia.

There are a number of understandings of hypermedia within hypermedia research but all agree that a hypermedia system consists of a database and an interface which provides the viewer with a mechanism for accessing stored data. The data is interlinked, cross-referenced and non-sequential and requires high-level information organisation if it is to be retrieved in a meaningful way by the viewer. An efficient interface design should enable the viewer to engage with the database as comfortably and intuitively as possible. Accessing the database is the process of moving through the multimedia items stored in the computer's electronic memory and in hypermedia research this process is termed "navigation" through its similarities with geographical navigation. Constructing a complex, multi-linked hypermedia system is understood to be problematic with an overiding concern for "navigation". Navigating through hypertext is also known as the 'Travel Metaphor' and is one of the most common metaphors in the field.

Metaphor as Design Aid
Hypermedia designers use metaphor in order to establish conventions which enable viewers to overlook the interface through which they find the information they are searching for. A metaphor provides familiarity which gives the viewer confidence and speed of access through prior experience. Metaphors commonly used are the "book" with its bookmarks, cover, pages, index and "contents page" e.g 'ToolBook', the "card" or "file box" with its card stacks and index, e.g. 'HyperCard' and the "film" with it's cast and stage, e.g. "Director". Nielsen (1990)[2] asserts that we must rely on metaphors such as 'navigation' to describe the active processes we engage with when accessing a hypermedia system. 'An appropriate visual metaphor can serve to compartmentalise a large hyperbase comprised of motion video and various stills and help to focus reader's attention on the subject matter. . . To structure their experience and plan their moves within the hyperdocument's information space.' The problem arises from metaphors based on the familiar non-computer world. Nielsen later states that "Although the use of metaphor may ease learning for the computer novice, it can also cripple the interface with irrelevant limitations and blind the designer to

new paradigms more appropriate for a computer-based application'"[3] Our perception of the information space behind the computer screen is informed by the non-hierarchical net-like quality of interconnected data items indicating a three-dimensional domain. According to Carlson (1989)[4] a hypertext system attempts to model the deep structure of human idea processing by creating a network of nodes and links, allowing for three-dimensional navigation through a body of information.

The nodes and their links within a hypermedia system form a web-like structure identified by Trigg (1983)[5] as analogous with a Semantic Network. Semantic Networks are an AI concept for knowledge representation consisting of a directed graph in which concepts are represented as nodes and the relationships between concepts are represented as the links between them. These networks are "semantic" in that the concepts in the representation are indexed by their semantic content rather than by some arbitrary (e.g. alphabetical) ordering. Trigg's analogy with hypertext holds where the nodes are hypertext nodes representing single ideas with links between them representing the semantic interdependencies among those ideas. In the Semantic Network of a hypermedia web, associations are not bound to strict rules as for the other information structures and can incorporate hierarchical grid and tree sequence structures together with the linear-sequential. Hypertext can then capture an interwoven collection of ideas without regard to the ability of a machine to interpret them. Quillian's original work on Semantic Networks[6] concerned the understanding of how meanings may be stored to enable computers to 'reason' as humans do, his Semantic Networks were a method of storing meanings and opened the way for AI tools, such as the Expert System, to explore computed 'reasoning'. Littleford[7] proposes three AI programming techniques which he views as particularly relevant to the future of hypermedia - Rule-based Expert Systems, Neural Networks and Blackboard Systems.

Systems such as Expert Systems are limited by set rules which address discreet chunks of memory and result in a rigorous mapping method which is non-adaptive and non-flexible. The greatest downside for using an Expert System from the artist's point of view is that it cannot incorporate her common-sense, playfulness or intuition. The highly developed, formal methods of AI and the models and mechanisms of information organisation can inform the hypermedia system author in structuring items and defining link relationships. However, Kolb (1997)[8] in his "wish list" of software capabilities to enhance and support scholarly enquiries, dismisses AI methods as "computing on the fly", unable to accommodate arguments over the worth of a new metaphor, alternative translations and where what is in question is the criteria of relevance and connection presumed by an automatically linked search mechanism. These methods are useful where the hypermedia system is understood as an information database for selective searching by the viewer but are detrimental to the designer, such as the multimedia fiction writer, who intends the hypermedia system to be accessed through the viewer's curiosity and intrigue.

The Semantic Network metaphor appears particularly apt according to Landow (1992)[9] who understands hypertext as sharing a commonality with contemporary critical theory in that they both emphasise the model or paradigm of the network. 'The analogy, model, or paradigm of the network so central to hypertext appears throughout structuralist and poststructuralist theoretical writings.' Landow sees the network as an example of the antilinear thought apparent in our post-modern culture and hypertext as the non-linear narrative form through which we may articulate our cultural thoughts. He suggests that the nature of narrative invites reflection on the nature of culture and asks, 'What kind of culture would have or could have hypertextual narration, which so emphasizes non- or multilinearity?' In answer he refers to the work of Lee (1973) [10] on the Trobriand Islanders whose narratives have no plot, no lineal development or climax and whose culture of non-linear thought is based on the idea of clustering. Landow sees Lee's description of Trobriand structurisation by cluster as offering means of creating forms of hypertextual order.

Hypermedia Art

Lee considers most, if not all, Western cultures to be predominantly verbal and linear, left-brain hemisphere cultures entrenched with a linear, sequential understanding of the world, in constrast to the right-brain hemisphere culture in the Trobriand people with non-linear elements in their language, daily lives and geographical navigation. Until recently, research undertaken in hypermedia has come from Western computer science, cognitive science, AI and Information systems. When engaging with art works we are employing the right-hemisphere of our brains. Therefore accessing an art work in hypermedia would require right-hemisphere system navigation methods informed by geographical navigation evident in predominantly right-hemisphere cultures and not the left-brain mapping systems offered from the domain of computer science. My own research has concerned the transposition of Marcel Duchamp's conceptually complex *Large Glass* as a Semantic Network into hypermedia and a possible form of non-linear navigation has emerged as a means of addressing both Duchamp's work and the implications of the new medium involved. Searching for semantic associations within the *Large Glass* in keeping with its overall content has directed me to consider an analysis of this piece as an example of the bipolar dichotomy of mind according to the left and right hemispheres of the brain.

Duchamp's *Large Glass* can be investigated as two separate domains forming one whole, one male theother female, as in the dual mode concept. Duchamp clearly intended this, calling his piece *The BrideStripped Bare by her Bachelors, Even* and stating it to have '2 principal elements: 1. Bride 2.Bachelors. . . Bride above-bachelors below.' [11] His interest in the male/female aspect to his work isfurther evident in the signing of works under the name of his alter ego Rose Sélavy. The two sections of the *Large Glass* encompassing male and female attributes can bear a direct analogy to the left and right hemispheres of the brain considered responsible for these attributes. The right or female hemisphere being responsible for the non-linear, unconscious free association of thoughts exemplified in dreams, in the Bride's thoughts. The left or male hemisphere is responsible for the precision perspective and sequential mechanisation of the bachelors. This reading of the *Large Glass* implies differing hypermedia system navigational methods for the 'Bride' half and the 'Bachelor' half or, as the 'Bride' is the central character in the work, a unifying feminine/right brain navigation system for non-linear movement may be indicated. The navigation possibilities for such a hypermedia art system could be explored with the aim of informing artistic practice in hypermedia.

The creation of concept-based art works in hypermedia as a method of inter-linking semantically associated thoughts and ideas in multimedia form may become common through the ubiquitous Internet. The idea of an art practice using a medium developed to emulate and therefore augment human thought leads one to question, - Are we now moving away from the conceptual art practice apparent throughout the second half of the century and towards a form of cognitive art practice more appropriate for a new Millenium?

References

1. Bush, V. 1945. 'As We May Think' in *Atlantic Monthly*, 176 (1), July, pp.641-9.
2. Nielsen, J. 1990. 'The Art of Navigating Through Hypertext'. in *Communications of the ACM*, 33 (33), March. p.299.
3. Gentner, D. Nielson J. 1996. ' The Anti-Mac Interface'. In *Communications of the ACM*, 39 (8).
4. Carlson, P.A. 1989. 'Hypertext and Intelligent Interfaces for Text Retrieval'. In : BARRETT, E. (Ed). *The Society of Text: Hypertext, Hypermedia, and the Social Construction of Information*, Cambridge. MA: MIT Press, pp.59-76.
5. Trigg, R. 1983. *A Network Based Approach to Text Handling for the Online Scientific Community*. Ph.D Thesis, University of Maryland.
6. Quillian, M. 1968. 'Semantic Memory'. In MINSKY, M. (ed). *Semantic Information Processing* . Cambridge, MA: MIT Press.
7. Littleford, A. 1991. 'Artificial Intelligence and Hypermedia'. In : BERK, E./DEVLIN J. (Eds). *Hypertext/Hypermedia Handbook* . New York: McGraw-Hill Publishing Company, Inc.
8. Kolb, D. 1997. 'Scholarly Hypertext: Self-Represented Complexity'. in *Proceedings for Hypertext '97*. ACM Press, pp.29-37.
9. Landow, G. 1992. *Hypertext: The Convergence of Contemporary Critical Theory and Technology*. London: John Hopkins University Press.
10. Lee, D. 1973. 'Codifications of Reality: Lineal and Nonlineal'. In : ORNSTEIN, R.E. (Ed.) *The Nature of Human Consciousness: A Book of Readings. Section 2, Two Modes of Consciousness*. London:Freeman. pp.128-142.
11. Duchamp, M. 1934. *In*: SANOUILLET M. PETERSON, E.Eds. 1973. *Salt Seller:and The Writings of Marcel Duchamp*. New York: Oxford University Press.

The affinity between concept-based art and hypermedia constitutes the research undertaken as a doctoral student at CAiiA. The outcome of this research has been a CD-ROM (StarGlass) and an Internet piece (4D Duchamp) concerning the Large Glass of Marcel Duchamp together with a text thesis.
<dharriso@newport.ac.uk> http://caiia-star.newport.plymouth.ac.uk
4D Duchamp can be seen @ http://caiiamind.nsad.newport.ac.uk/lead.html

The House That Jack Built: Jack Burnham's Concept of "Software" as a Metaphor for Art[1]

Edward A. Shanken

Like the famous cumulative story to which my title refers, this paper explores the complex, interrelated convergence of myriad elements in the exhibition *Software. Information Technology: Its New Meaning for Art*. Art historian Jack Burnham curated the show in 1970 at the Jewish Museum, then one of the premier venues for experimental art in New York. My research identifies how this 'house that Jack built' was constructed of, and drew parallels between, computer information technology, conceptual art practice and structuralist art theory and was predicated on the idea of software as a metaphor for art. *Software* was designed to function, moreover, as a testing ground for public interaction with 'information processing systems and their devices.'[2] Many of the displays were indeed interactive and based on two-way communication between the viewer and the exhibit. In this and other respects, I interpret much of the work in *Software* as heralding post-modernist strategies for art-making. Finally, as will be discussed below, the architecture for the physical installation in the museum was based on the two-tiered model of Marcel Duchamp's *Large Glass*, which Burnham interpreted as a signpost announcing the demise of art as 'a separate facet of life.'[3]

Jack Burnham's first book, *Beyond Modern Sculpture: The Effects of Science and Technology on the Sculpture of Our Time*, 1968, established him as the pre-eminent champion of art and technology of his generation. Building on this foundation, his second book, *The Structure of Art*, 1971, developed one of the first systematic methods for applying structural analysis to the interpretation of individual artworks as well as to the canon of Western art history itself. Many of his articles for *Arts* magazine from 1968, where he was Associate Editor (1972/6) and *Artforum* from 1971-3, where he was Contributing Editor (1971/2), were collected in his third book, *The Great Western Salt Works*, 1973. These essays still remain amongst the most insightful commentaries on conceptual art, already suggesting what he now sees in retrospect as the 'great hiatus between standard modernism and postmodernism.'[4]

In 1970, at the invitation of Jewish Museum director, Karl Katz, Burnham curated *Software*, the only major show he has curated to date. In contrast to the numerous art and technology exhibitions which took place between 1966/1972 and which focused on the aesthetic applications of technological apparatus, *Software* was predicated on the ideas of 'software' and 'information technology' as metaphors for art. He conceived of 'software' as parallel to the aesthetic principles, concepts or programmes that underlie the formal embodiment of the actual art objects, which in turn parallel 'hardware'. In this regard, he interpreted 'Post-Formalist Art' (his term referring to experimental art practices including performance, interactive art, and especially conceptual art) as predominantly concerned with the software aspect of aesthetic production.

– *Strategies* –

It is significant that Burnham organised *Software* while writing *The Structure of Art* and conceived of the show, in part, as a concrete realization of his structuralist art theories. Drawing on Claude Levi-Strauss's idea that cultural institutions are mythic structures that emerge differentially from universal principles, Burnham theorised that Western art constituted a mythic structure. And he theorised that the primary project of conceptual art was to question and lay bare the mythic structure of art, demystifying art and revealing it for what it is.[5]

Such ideas were already present in Burnham's 1970 article 'Alice's Head.' True to the title, he began the essay which focused on the work of conceptual artists Joseph Kosuth, Douglas Huebler, Robert Barry, Lawrence Wiener and Les Levine - with the following quotation from Lewis Carroll's *Alice in Wonderland*:

> 'Well! I've often seen a cat without a grin,' thought Alice, 'but a grin without a cat! It's the most curious thing I ever saw in all my life!'[6]

By selecting for his preamble Alice's curiosity over a disembodied presence, Burnham suggested that, like a grin without a cat, a work of conceptual art is all but devoid of the material trappings of paint or marble traditionally associated with art objects. Similarly, he explained *Software* as 'an attempt to produce aesthetic sensations without the intervening "object;" in fact, to exacerbate the conflict or sense of aesthetic tension by placing works in mundane, non-art formats.'[7]

Burnham directly interacted with computer software when he was a Fellow at the Centre for Advanced Visual Studies under Gyorgy Kepes at MIT during the 1968/9 academic year. Having received his MFA from Yale in 1961, he was invited, as an artist, 'to learn to use the time-sharing computer system at MIT's Lincoln Laboratories.' In a paper entitled 'The Aesthetics of Intelligent Systems' delivered at the Guggenheim Museum in 1969, Burnham discussed this experience of working with computers, comparing the brain and the computer as information processing systems, and drawing further parallels between information processing and conceptual art. He stated, moreover, that 'the aesthetic implications of a technology become manifest only when it becomes pervasively, if not subconsciously, present in the life-style of a culture,' and claimed that 'present social circumstances point in that direction.'[8]

As Burnham explained in the paper, given the artistic limits of the computer system at his disposal, he focused on the 'challenge of . . . discovering a program's memory, interactive ability and logic functions,' and on 'gradually . . . conceptualis[ing] an entirely abstract model of the program.' In this regard, he was especially interested in how 'a dialogue *evolves* between the participants - the computer program and the human subject so that both move beyond their original state.' Clearly he recognized how his interaction with software altered his own consciousness, which in turn simultaneously altered the program. Finally, he drew a parallel between this sort of two-way communication, and the 'eventual two-way communication in art.' In 1969, he wrote,

> The computer's most profound aesthetic implication is that we are being forced to dismiss the classical view of art and reality which insists that man stand outside of reality in order to

observe it, and, in art, requires the presence of the picture frame and the sculpture pedestal. The notion that art can be separated from its everyday environment is a cultural fixation [in other words, a mythic structure] as is the ideal of objectivity in science. It may be that the computer will negate the need for such an illusion by fusing both observer and observed, 'inside' and 'outside.' It has already been observed that the everyday world is rapidly assuming identity with the condition of art.[9]

The metaphorical premise of *Software* permitted Burnham to explore convergences between his notion of the mythic structure of art, emerging information technology and the increasing conceptualism characteristic of much experimental art in the late 1960s. These components were conjoined in works that emulated the sort of two-way communication he experienced with computer programs and which he advocated in art. The catalog emphasized the importance of creating a context in which "the public can personally respond to programmatic situations structured by artists," and explicitly stated that the show "makes no distinctions between art and non-art."[10]

Burnham was careful to select works of art that demonstrated his theories. I contend that many of these works anticipated and participated in important trends in subsequent intellectual and cultural history. In this sense they contributed to the transformation of consciousness. Quoting McLuhan, Burnham identified this shift from the 'isolation and domination of society by the visual sense' defined and limited by one-point perspective, to a way of thinking about the world based on the interactive feedback of information amongst systems and their components in global fields, in which there is 'no logical separation between the mind of the perceiver and the environment.'[11]

For example, in the hypertext system *Labyrinth*, a collaboration between Xanadu creator Ted Nelson and programmer Ned Woodman, users could obtain information from an 'interactive catalog' of the exhibition by choosing their own narrative paths through an interlinked database of texts, then receive a print-out of their particular 'user history'. The self-constructed, non-linear unfolding of *Labyrinth* shares affinities with structuralist critiques of authorship, narrative structure, and 'writerly' (as opposed to 'readerly') texts, made by Barthes. Needless to say, with the advent of powerful Internet browsers like Netscape, and the proliferation of CD-ROM technology, the decentred and decentring quality of hypertext has become the subject (and method) of a growing critical post-structuralist literature and arguably a central icon of postmodernity. It should be noted that this first public exhibition of a hypertext system occurred, and this was perhaps not just a coincidence, in the context of experimental art.

Hans Haacke's *Visitor's Profile* encouraged visitors to interact with a computer by inputting personal information, which was then tabulated to output statistical data on the exhibition's audience. Such demographic research as art opened up a critical discourse, following Foucault and others, on the exclusivity of cultural institutions and their patrons, revealing the myth of public service as a thin veneer justifying the hierarchical values that reify extant social relations. Similarly, *Interactive Paper Systems* by Sonia Sheridan engaged museum-goers in a creative exchange with the artist and 3M's first commercially available color photocopying machine, dissolving conventional artist viewer object relations. In *The Seventh Investigation (Art as Idea as Idea)* Joseph Kosuth utilized multiple forms of mass media and distribution (a billboard, a newspaper advertisement, a banner and a museum installation) to question the conceptual and contextual boundaries between art, philosophy, commerce, pictures and texts.

In works such as these the relationship Burnham intuited between experimental art practices and 'art and technology' problematised conventional distinctions between them and offered important insights into the complementarity of conventional, experimental, and electronic media in the emerging cultural paradigm later theorised as post-modernity. In this regard, Levi-Strauss's models from structural anthropology, along with Thomas Kuhn's critique of the history of science, led Burnham to question what he saw as the structural foundations of art history's narrative of progressive and discrete movements, a critique he elaborated in *The Structure of Art*.

As a final example, Nicholas Negroponte and the Architecture Machine Group (precursor to the MIT Media Lab, which Negroponte now directs) submitted *Seek*, a computer-controlled robotic environment that, at least in theory, cybernetically reconfigured itself in response to the behaviour of the gerbils that inhabited it. I interpret *Seek* as an early example of 'intelligent architecture,' a growing concern of the design community internationally.[12] By synthesizing cybernetics, aesthetics, phenomenology and semiotics, *Software* emphasised the process of audience interaction with 'control and communication techniques', encouraging the 'public' to 'personally respond' and ascribe meaning to experience. In so doing, *Software* questioned the intrinsic significance of objects and implied that meaning emerges from perception in what Burnham (quoting Barthes) later identified as 'syntagmatic' and 'systematic' contexts.[13]

A further abiding metaphor in Burnham's concept for *Software* was Marcel Duchamp's *Large Glass*, 1915/22, which served as an architectural model for the actual installation. Burnham described the relationship of *Software* to Duchamp's magnum opus in a 1970 interview with Willoughby Sharp. Iconographically, he explained, the *Large Glass*,

> has a lot of machines in the lower section scissors, grinders, gliders, etc . . . it represents the patriarchal element, the elements of reason, progress, male dominance. The top of [it] is the female component: intuition, love, internal consistency, art, beauty, and myth itself.[14]

Burnham claimed that 'Duchamp was trying to establish that artists, in their lust to produce art, to ravish art, are going to slowly undress her until there's nothing left, and then art is over.'

The curator then went on to reveal *Software's* organisational logic:

> As a kind of personal joke. . . I tried to recreate the same relationships in *Software*. I've produced two floors of computers and experiments. Then upstairs on the third floor, conceptual art with Burgy, Huebler, Kosuth, and others, which to my mind represents the last intelligent gasp of the art impulse.

Burnham's point, following his interpretation of Duchamp, was not that art was dead, or dying, or about to dissolve into nothingness. Rather, he believed that art was 'dissolving into comprehension.' He claimed that conceptual art was playing an important role in that process, by 'feeding off the logical structure of art itself . . . , taking a piece of information and reproducing it as both a signified and a signifier'. In other words, such work explicitly identified the signifying codes which define the mythic structure of art. Instead of simply obeying or transgressing those codes, it appropriated them as motifs, as signifiers, thereby demystifying the protocols by which meaning and value have conventionally been produced in art.

In this regard, Burnham became very critical of the role of emerging technology in art. Having lost faith in its ability to contribute in a meaningful way to the signifying system that he believed to mediate the mythic structure of Western art, in *Software* he purposely joined the nearly absent forms of conceptual art with the mechanical forms of technological non-art to 'exacerbate the conflict or sense of aesthetic tension' between them.[15] Given his interpretation of Duchamp, such a gesture also can be seen as an attempt to deconstruct the categorical oppositions of art and non-art by revealing their semiotic similarity as information processing systems.

How relevant today is Burnham's position *c.*. 1970 on art and technology? Beginning in the 1970s, how has the convergence of art, technology and theory influenced the work of contemporary artists? In what ways, if at all, have advances in technology opened up new signifying potentials for artists? As my research progresses, questions such as these will be further discussed and problematised.

References

1 Burnham, J. 1970. 'Notes on Art and Information Processing', in *Software Information   Technology: Its New Meaning for Art*. New York: Jewish Museum, p. 10.
2 Burnham, J. 1972. 'Duchamp's Bride Stripped Bare: The Meaning of the *Large Glass*'. In Burnham J. 1973. *Great Western Salt Works: Essays on the Meaning of Post-Formalist Art*. New York: George Braziller, p. 116.
3 Burnham, J. 1998. Personal correspondence with the author, April 23.
4 Burnham, J. 1971. *The Structure of Art*. New York: George Braziller.
5 Burnham, J. 1970. 'Alice's Head', in *Great Western Salt Works*, p. 47.
6 Burnham, J. 1998. Personal correspondence with the author, April 23.
7 Burnham, J. 1969. 'The Aesthetics of Intelligent Systems', in Fry, E. F. (ed.) 1970. *On the Future of Art*. New York: The Viking Press, p. 119. The three subsequent quotations come from same page. The 'present social circumstances' to which Burnham refers here can only be the increasing pervasiveness of computer information-processing systems, which he described in the Software catalogue as 'the fastest growing area in this culture.'
8 Ibid, p. 103.
9 Burnham, J. 1970. Notes on Art and Information Processing, p. 10.
10 Burnham, J. 1970. 'Alice's Head', p. 47.
11 See for example, Gerbel K. and Weibel P. (eds.) 1994. *Intelligente Ambiente* (Ars Electronica 1994 catalogue). Vienna: PVS Verleger.
12 Burnham, J. 1971. *The Structure of Art*, pp. 19-27.
13 Sharp, W. 1970. 'Willoughby Sharp Interviews Jack Burnham', in *Arts* 45:2 (November) p. 23. All subsequent quotations are from this page.

Edward A. Shanken is a doctoral candidate in Art History at Duke University and the recipient of Luce/Amercian Council of Learned Societies Dissertation Fellowship in American Art. His online publications include: 'Life as We Know It and/or Life as It Could Be: Epistemology and the Ontology/Ontogeny of Artificial Life', 1998, [http:// mitpress.mit.edu/e-journals/LEA/ARTICLES/zeddie.html]; ' Technology and Intuition: A Love Story? Roy Ascott's Telematic Embrace', 1996 (http://www-mitpress.mit.edu/Leonardo/isast/articles/shanken.html#2]; and ' Jeffrey Shaw's Golden Calf: Art Meets Technology and Religion', 1994 [http://www-mitpress.mit.edu/Leonardo/reviews/shankencalf2.html]. Forthcoming articles include 'From Drips to ZOOBS: The Cosmology of Michael Grey', in Artbyte 1:3 (1998), 'Gemini Rising Moon in Apollo: Attitudes Towards Art and Technology in the U.S., 1966/71', in ISEA97 Proceedings; and Le Coq, c'est moi!' Brancusi's Pasarea Maiastra: Nationalistic Self-Portait?' in Art Criticism (1998). His dissertation analyses art and technology in the US in the 1960s, focusing on Jack Burnham's 1970 exhibition, Software. He is editing a book of essays by British artist/theorist Roy Ascott, tentatively titled Is There Love in the Telematic Embrace? Roy Ascott's Visionary Writings on Art, Technology and Consciousness, forthcoming from University of California Press, 1999.

Streams of Consciousness: Info-Narratives in Networked Art
Christiane Paul

'Stream of consciousness' became the literary technique of choice in the literature of classical modernism. Authors such as James Joyce and Virginia Woolf invented and perfected the technique by striving to locate their narratives completely within the thoughts and perceptions of their novels' protagonists.

'Modernism' as the international artistic movement of the early 20th century responded to the sense of a loss of centres the breakdown of established structures, whether social, political, religious or artistic. The modernists considered the traditional ideas of order, sequence and unity in works of art as expressions of a desire for coherence which was artificially imposed on the flux and fragmentation of experience. The formal characteristics of the modernist work are its construction out of fragments drawn from diverse areas of experience and shifts in tone, voice and perspective. The coherence of most modernist works lies beneath the surface: their dynamic pattern often is the quest for the coherence they seem to lack on the surface.

The 'modern' belief that forces eluding the consciousness of the author/artist constitute inspiration in writing and that the creative process is a dictation from the unconscious and tradition resulted in the idea of the depersonalisation of the author and the ideal of impersonality. In fact, the theory of impersonality has a long genealogy. It derives from many sources philosophical, poetic and political and has always been eclectic and inconsistent. Its meanings range from the destruction to the glorification of the self, and it often tends to fall back into the very ethics of personality it aims to transcend. The use of masks, personae and polylogue in modern literature became manifestations of this ideal of impersonality. The artist, as James Joyce famously put it, 'remains within or behind or beyond or above his handiwork, invisible, refined out of existence, paring his fingernails.'

In the age of the Internet, the notions of fragmentation, multiplicity and 'streams of consciousness' are revived in a different context and carried to further levels. The networked society has profoundly affected our concepts of self and identity: the on-line self is now commonly understood as a multiple, distributed, time-sharing system. On-line identity allows a simultaneous presence in various spaces and contexts, a constant 'reproduction' of the self sans body. Identity in the age of the Internet is characterised by multiplicity, heterogeneity and fragmentation. What was experienced as a loss of centers in classical modernism has now become the basis for a self-conscious role play in which identity is constantly reinvented. Avatars are the personae and masks of today's networked multi-user environments.

The networked, hypertextual environment is by nature polyvocal, favours a plurality of discourses and frees the reader from 'domination' by the author. Readers and writers collaborate in the process of remapping textual, visual, kinetic and aural components, not all of which are provided by what used to be called the author. The author does not have to be refined out of existence; the very notion of authorship itself becomes questionable.

The computer age promises to open up new dimensions for consciousness: the ultimate (utopian and dystopian) dream is to download consciousness into the machine and stream it 'live' over the network. Computers are now commonly seen as extensions of the human mind 'tools to think with.' Hypermedia, in particular, support this perception: hypermedia applications are intended to mimic the brain's ability to make associative references and use these references in order to access information. The hyperlinking in electronic writing/imaging systems is supposed to be close to the complex electrodynamics of consciousness. The computer may be a medium in the sense that it is the environment in which a user interacts with a structure, a mode of communication; but what the user actually interacts with is the structure existing in the virtual space of the software. The machine becomes invisible, while the structure becomes the medium and the user may even be understood as a part of this medium, an extension of the machine. Standard interfaces require us to pound a keyboard and use a mouse or joystick while staring at a screen; the human body is forced to conform to the shape of the computer setup (such as monitor and CPU). Total immersion in virtual environments and the downloading of consciousness into the machine may not be a (virtual) *reality* yet but the hybrid man-machine may soon become obsolete since its components become more and more indistinguishable. On some levels, consciousness is already undergoing a process of disassociation from the body.

Digital technologies and interactive media have not only significantly challenged notions of self and identity but also of the art object. Even if art objects have been continually reinvented over centuries through the eyes of their viewers, they have

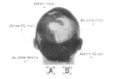

mondrate
From Reynald Drouhin's 'Alter Ego'
http://panoramix.univ-paris1.fr/UFR04/rhizome/ateliers/reynet/files/reynald.htm

been characterised by a certain amount of closure. For the creator of a 'traditional' work of art, the creative process, at some point, used to end in a 'finished' object be it a sculpture, a painting or a book. Even less physical media such as film and video provide a certain amount of closure.

Interactive art, on the other hand, involves reciprocity and collaboration between the creator or creators, a project and the audience. Many of the art works on the Internet are collaborative and time-based; the art object has been transformed into a sculpture in process and info-narrative that relies on a constant flux of information and engages the viewer/collaborator in the way a performance might. In a media-enabled and information-based society, content is information in flux, unbound from the physical object. Data streams form informational exchange corridors which turn into the strands of a narrative that is in constant reorganisation.

face
The interface of 'Last Entry: Bombay, 1st of July'
http://www.aec.at/residence/lastentry/index.html

In our postmodern world, it has become a common expression to read 'the text of the world' to perform this text within referential frames, fulfil its requirements and respond to it in a self-discovering way and make multiple connections while reading. A reader's experience with any work is always a negotiation with text, author, society and self. But it still seems to be difficult to 'read' interactive, net-based works of art, where the control of pace is not only a mental event but shifts further to the side of the respective reader through the possibility of interaction.

In interactive net art, the boundaries between self and virtual art object often appear to collapse, and the art work is not necessarily perceived as an 'otherness,' because it is the user/viewer who assembles or even creates it. Virtual (art) objects are open-ended they are info-narratives with a fluctuating structure, logic and closure. In this scenario, both self and object may ultimately be understood as data streams. The process of 'reading/viewing' interactive, networked art is less "other-directed" than the experience of a traditional work of art, such as a painting or sculpture. While the computer cannot perceive and experience anything as 'other,' computer users may perceive the machine as an extension of their consciousness. According to Baudrillard (1993) the relationship between man and machine is characterised by a peculiar kind of indifference; neither of them perceives the other as otherness: 'The computer has no other.'

In *The Transparency of Evil* (1993) Baudrillard establishes a connection between interactivity and otherness. Although his theory does not refer to interactive media art and relies on a more radical understanding of interactivity than common usage of the term, it might be interesting to apply his concept to the process of experiencing networked art.

According to Baudrillard, it is the absence of otherness (which has been erased through discovery and comprehension) that leads to interactivity. Since consciousness and intelligence rely on the awareness of both internal processes and external objects, states or facts an 'other' that is and distinguish man from machine, we have to resurrect the other or at least dramatise the absence of the other where it has been erased: 'where there is no longer anything, there the Other must come to be.' Baudrillard argues that we start living the psychodrama of otherness: 'All we do in psychodrama—the psychodrama of contacts, of psychological tests, of interfacing—is acrobatically simulate and dramatise the absence of the other.'

Otherness is absent in this artificial dramaturgy and the subject grows more and more indifferent towards its own subjectivity, since there is nothing but subjectivity: 'the subject becomes transparent, spectral . . . and hence interactive.' In interactivity the subject is nobody's other and becomes a candidate for all possible connections and combinations without defining itself in terms of an other. Baudrillard claims that 'The only interaction involved, in reality, belongs to the medium alone: to the machine become invisible.'

The experience of net art thus becomes something similar to Baudrillard's psychodrama, a process of simulating and dramatising the absence of the other in an artificial dramaturgy without an other. Potentially, the virtual art work does not represent an otherness both the work and its user/viewer are ideally open for all connections and combinations.

One of the most prominent net narratives today is the attempt to dramatise the heterogeneity of identity in the digital realm. A growing number of web-specific projects explore the possibilities and difficulties brought about by virtual existence. A (quite traditional) example of this kind of narrative would be 'Last Entry: Bombay, 1st of July' a project that establishes the connection to modernist experiments with fragments and perspective.

Starting from Virginia Woolf's novel *Orlando* (written in 1928), 'Last Entry: Bombay, 1st of July' creates a 'virtual profile' of identity in the networked community. Woolf's *Orlando* the biography of a figure who travels through time and space, centuries and cultures, switching gender and identities now reads like a visionary metaphor of the Internet. The website takes the form of a travel log/diary/narrative montage: visitors are invited to contribute their stories, thus involving Orlando into their own (virtual) background and bringing him/her into contact with others. Users can also go to a page where they can download images of Orlando's latest appearance in order to change and use in their own episode. The interface organises episodes in the form of the profile of a head consisting of micro-profiles. The web of episodes, interweaving the realities and fictions of identity, opens up possibilities for an elaborate game with casual encounters and destiny (even if contributions don't necessarily live up to the potential of the project).

The notions and concepts underlying the discussion on identity in the networked age fragmentation, multiplicity, heterogeneity, polyvocality have been prominent issues at least since the beginning of the century. Previous artistic experiments with these concepts have predominantly been 'mind games' where interaction relied on internal, mental events. Computer networks and communication technologies make it possible for the first time to externalise the experiments and to make various levels of multiplicity become alive in an environment of connectivity. If the 'total immersion' in virtual environments becomes a reality, the boundaries between internal and external objects or states collapse and consciousness is reframed by the machine.

Reference

Baudrillard, J. 1993. *The Transparency of Evil, Essays on Extreme Phenomena*. London: New York: Verso, p. 124 ff.

Christiane Paul is the publisher and editor of Intelligent Agent, a print and on-line magazine on the use of interactive media and technology in arts and education. She is the author of the hypertext Unreal City: A Hypertextual Guide to T. S. Eliot's The Waste Land (published by Eastgate Systems, Watertown, MA, 1995) and has written extensively on hyperfiction and the use of hypermedia in education. Dr Paul was a Visiting Scholar at New York University from 1988 to 1991 and has taught at New York University and Fordham University, NY.

Consciousness and Music
Dr. Mladen Milicevic

Introduction

For a long time the human mind and consciousness fell under the province of metaphysics and were generally considered as topics improper for scientific investigation. Even today there is an enormous reluctance on the part of scientists to deal with such subjective phenomena. Those who dare to wrestle with these slippery issues often address the complexity of the mind via the Darwinian theory of evolution and cognitive psychology; or, as neuroscientists do, focus their interest on the structure and workings of the brain and its functions. The best results will probably come from a marriage of these two strategies.

An important point to make is that the human brain developed over a period of more than 500 million years under the pressures of natural selection. We have reptilian, paleomammalian, and neomammalian brains layered on top of each other, all running at the same time.

> If we remove the cerebral cortex—that part of our brain that has evolved over the past two million years or so—we essentially eliminate our humanity. Beneath the cortex is a brain that is not far different from that of a Bengal tiger, a French poodle, or an Arctic fox. We could, if we wish, remove even more to approximate the brain of a salamander or a rattlesnake (Restak 1984)[1].

Our brain has an archaic multilevel design comprising of several brains each developed in different periods of evolution for different purposes and priorities. Built as a massive complexity of parallel-interacting neurons which lack a hierarchical order and central organiser, the human brain produces an emergent property consciousness, which is mainly serial and subjective in its nature. This is the reason why we have a sense of subjective unity regardless of the multilevel and multi-component make-up of the neurons involved.

Thinking and Logic
Biological objects under evolutionary time have proper functions that depend on their evolutionary history. A human ear has a proper function to transduce a sound into a bio-electrical signal. There is also an evolutionary explanation for the construction of such a component (an ear) in a species and this justifies the correspondence of this organ to "normal" ears in that species. Human ears work well or not; badly functioning ones, like Beethoven's, are abnormal and may need a hearing aid.

Within the evolutionary context, functions that propel the survival of the fittest are labelled as 'normal' functions. Thus, each set of functions may be recognized as 'normal' in relation to how the system redundantly manages to perform that function. Since we know that the neural making of every human brain is different, as well as the environmental and historical context in which each brain develops, we may consider our brain/minds to be quite different systems. Different systems have different contexts that render all kinds of different 'normals.'

> Meaning derives from embodiment and function, understanding arises when concepts are meaningful in this sense, and truth is considered to arise when the understanding of a statement fits one's understanding of a situation closely enough for one's own purposes. (Notice the pragmatism!) Thus, there is no absolute truth or God's-eye view·· (Edelman 1992)[2]

The Matter of the Mind
Humans categorise on a constantly changing culturally adopted set of values. It would be nice if we could talk about mind, as some psychologists do, only in terms of algorithms. Looking at different brains as if they were replicable machines or black boxes which can be understood solely in terms of inputs and outputs is an oversimplification. Neuroscientists are going inside the 'black box' trying to figure out the mechanisms which operate our brains. The tissue organization and composition of the brain in form of groups of interacting neurons may be doing this job.

> These groups compete with each other in an effort to create effective representation, or maps, of the infinite variety of stimuli entering from the world. Groups that form successful maps grow still stronger, while other groups wither (Horgan 1996)[3.]

It is not by chance that the brain of a person musically gifted possesses a thickened area that processes sound the auditory cortex - while a person with photographic memory has more neurons in its visual cortex. This is the feature that we call talent and it is genetically predisposed. The precise structure of a certain neural tissue of the brain exposed to a certain external environment will allow that tissue to develop, for example, a musical talent. For that reason it is wrong to believe that the human brain of the youngster is an empty book that experience will write its story on. If the neural circuits are not receptive to the environmental pressures, the experiences 'written' in the brain will be faint.

Brain's Age

So far I have dealt with two conditions required for the design of another Mozart: a) genetic predisposition, and b) a stimulating environment for that predisposition. There is yet another very important aspect in this picture to be considered and that is the age of the brain. When the brain is young it displays a neural plasticity which is capable of developing more and more neural connections required for a particular talent. This plasticity diminishes as the brain ages and may be one of the reasons why we get no musical talents suddenly erupting at the age of 50. Even if the person had a thickened auditory cortex but the environment did not externally stimulate it at the most appropriate time no Midori will be produced.

> Some circuits are remodelled over and over throughout the life span, according to the changes an organism undergoes. Other circuits remain mostly stable and form the backbone of the notions we have constructed about the world within, and about the world outside. The idea that all circuits are evanescent makes little sense. Wholesale modifiability would have created individuals incapable of recognising one another and lacking a sense of their own biography. That would not be adaptive, and clearly does not happen. (Damasio 1994).[4]

Generally speaking, development of neural mappings for the 'talents' that involve some kind of intensive body coordination, such as playing a violin or a game of basketball, tend to be locked-in very early in life. Other, more 'disembodied' talents retain their neural circuit plasticity for a much longer period of life.

> A famous pianist said to me, about forgetting a familiar piece of music, 'Muscle memory is the last to go,' meaning by that term playing the piece automatically and without thinking about it.
>
> (Crick 1994).[5]

Emotion and Feeling

Let me now try to explain why the embodiment is so important for our understanding of the mind. The human body, as represented in the brain, provides a fundamental frame of reference for the neural processes that we experience as the mind. We use the physical state of our very organism as the ground reference for the mental constructions which we make about the environment we live in. It is extremely important to understand that the human body and the brain constitute an inseparable interconnection that produces our consciousness. Everything we do is derived from the structural and functional ensemble of these two, rather than from brain alone.

> Can one fancy the state of rage and picture no ebullition of the chest, no flushing of the face, no dilatation of the nostrils, no clenching of the teeth, no impulse to vigorous action, but in their stead limp muscles, calm breathing, and placid face (James 1950).[6]

In short, the background state of our body landscape provides a rather neutral 'mood', against which we can judge any changes shaken by emotions. When the brain consciously appraises emotional changes in that equilibrium it releases endorphins, the powerful opiate-like chemicals, and we are having an emotional response - a feeling. Pretty much like being aware of the goose bumps while listening to an effective piece of music. Conscious feeling of these goose bumps creates in your brain a memorised history of your body state under the given circumstances. Feeling depends on the juxtaposition of an image of the body, correlated to an image of something else, such as the auditory image of a piece of music. Thus, later on in life, under the similar listening conditions, your brain may recreate this correlation of the images and you may experience the feeling of goose bumps again.

Our individual identity of selfhood is firmly grounded on this illusion of living sameness, against which we can be aware of the infinite changes in our environment and consequently in our body. For this reason the sheer fact of a change in a room temperature, while listening to piece of music, may dramatically affect your emotional responses.

Attention and Memory

We also have to be aware that not all operations of the brain correspond to consciousness, as I mentioned earlier in the case of the background body feeling. Awareness starts when we consciously focus on a point of interest. Here, I would like to concentrate on consciousness which involves very short term of memory and is closely associated with attention. Attention is, as William James said, 'withdrawal from some things in order to deal effectively with others'. In general, there is a loose agreement that attention and primary consciousness involve some kind of a bottleneck. At first, the brain is processing the vast amounts of incoming information in parallel. Then the selective attention of hearing, for example, concentrates on one or a few objects at a time using the serial processing of the bottleneck - attending to one object after another. This is done by temporarily focusing on the objects of our interest while filtering out unattended information. Pretty much like listening to Bach's fugue while focusing our attention on the thematic workings of the dux and comes.

We may assume that primary consciousness of short-term memory deals with attention which is value-free perceptual categorisation. This takes place *before* perceptual events contribute further to the alteration of neuronally structured and experience shaped value-dominated long-term memory. When short-term memory starts to contribute to the modification of a subjective long-term memory (this could be called a learning process) events are no longer in the remembered present, that is, they are no longer in primary consciousness.

Musical Experiences

When the brain receives sequences of musical tones, it does what it does with other patterns: it attempts to 'interpret' them by using the information stored in its long-term memory about previous, similar experiences. This information may allow some aspects of a future signal to be anticipated - as it happens when we hear the first line of a familiar song. This ability to extrapolate forwards on the basis of past experience is one form of that ability that we call 'intelligence'; it can dramatically enhance an organism's chances of survival. Thus,

> . . . we constantly judge by comparison, and our judgment of any item depends upon what we are comparing it to at that moment.[7]

Ornstein (1986)

Addressing the past experiences we have to also consider the way our brain handles the long-term memory. These memories are not stored in the brain photographically as intact individual events and there are no stores of audio tapes or albums of pictures. This is completely unlike computer-based memory, which deals with exact reproductions. The human brain operates with a reconstructed version of the original - an interpretation. In order to compile a musical tune, the brain has to fire a certain set of neural mappings as a means to paraphrase 'the music.' Your interpretation of music today depends on who you are, what you are doing at that moment and your past experiences stored in the long-term memory. But the next day you are going to be different; what you will be doing is going to be different and your past experiences in long-term memory will change as well.

The 'mass of soothing sound' your mother made while singing lullabies to you in childhood is reduced to *Twinkle, Twinkle Little Star*, later on in life. Our memory of a certain musical piece is influenced not only by previous knowledge but also by events that happen between the time an event is perceived and the time it is recalled. Furthermore, we can only recall memories that are related to our present situation - where we are and what we are doing. If you are composing an orchestral piece your brain most likely focuses on recalling memories related to the instrumental ranges rather than on memories of how to change a flat tyre on your car.

> So our memories, as exact, recorded, fixed images of the past, are an illusion. We believe we are stable, but this is one of the built-in illusions of the mental system. We believe we remember specific events, surely. Yet we don't. We make them up on fly. We change our minds all the time, from our estimate of the odds on a bet, to how we view our future. And we are unaware that the mind is doing this (Ornstein 1991).[8]

All this is pointing out that the human mind deals exclusively with subjective phenomena and what we call objective is nothing but what most people agree to within a given socio-cultural context. Yes, we may listen to Beethoven's symphony in terms of air-pressure waves and on that level probably most people would have similar experiences. The question is, what is the use of doing that?

Notes

1. Restak, Richard M. 1984. *The Brain*. New York: Bantam Books, p. 136.
2. Edelman, Gerald, M. 1992. *Bright Air, Brilliant Fire*. New York: Basic Books, p. 250.
3. Horgan, John. 1996. *The End of Science*. Reading, Massachusetts: Helix Books, p. 168.
4. Damasio, Antonio R. 1994. *Decartes' Error*. New York: A Grosset/Putnam Book, p. 112.
5. Crick, Francis. 1994. *The Astonishing Hypothesis*. New York: Charles Schribner's Sons, p.67.
6. James, William. 1950. *The Principle of Psychology*. Vol. 2, New York: Dover.
7. Ornstein, Robert. 1986. *Multimind Boston*: Houghton Mifflin Company, p. 27.
8. Ornstein, Robert 1991. *The Evolution of Consciousness*. New York: Prentice Hall Press, p. 190.

References

Dawkins, Richard. 1976. *The Selfish Gene*. Oxford: Oxford University Press.
Dennett, Daniel C. 1995. *Darwin's Dangerous Idea*. New York: Simon & Schuster.

Dennett, Daniel C. 1991. *Consciousness Explained.* Boston: Back Bay Books: Little, Brown and Company.
Gazzaniga, Michael S. 1992. *Nature's Mind.* New York: Basic Books.
Ornstein, Robert 1997. *The Right Mind: Making Sense of the Hemispheres.* New York: Harcourt Brace & Company.
Ornstein, Robert and Sobel, David. 1989. *Healthy Pleasures.* New York: Addison-Wesley Publishing Company.
Restak, Richard M. 1994. *The Modular Brain.* New York: Simon & Schuster.
Restak, Richard M. 1991. *The Brain Has a Main of its Own.* New York: Harmony Books.

Dr Mladen Milicevic is Associate Professor of Media Arts at the University of South Carolina, Columbia, SC 29206, USA. To learn more about him you may visit his website at http://www.cla.sc.edu/art/faculty/milicevicm/mladen.html or send him an e-mail at MMladen@SC.edu

The Irreducibility of Literary Consciousness
Philipp Wolf

Ever since the installation of internal representational levels in parallel computers, and ever since the neurological findings of functionally different but linked regions in the brain, a number of psychologists and neuroscientists have been convinced that they can explain and eventually do away with consciousness. The 'hidden level' in so-called 'neural network' computers allows, they point out, for the comparison and matching of previous inputs with new information. This amounts, they further maintain, to an associative and adaptive learning process and, consequently, to the possibility of the full simulation of human consciousness in an artificial system. And since certain 'maps' or 'modules' in the cortex may be taken to represent the perceived sense-data as an image to the subject, they assume the material loci of our conscious realisation of images and emotions are manifest and analysable. For Francis Crick, 'your joys and your sorrows, your memories and your ambitions . . . are in fact no more than the behavior of a vast assembly of nerve cells and their associated molecules' (Crick 1994/ 3). For Daniel Dennett (1992/ 74), as for Steven Pinker (1997), the brain is not only identical with mind or consciousness; the latter's qualitative content, moreover, does not matter and can be eliminated.

If we were to rather innocently accept these eliminative or reductionist propositions, we might come to the naive conclusion that any subjective experience in the field of something like art or poetry would soon be completely transparent as 'nerve cells and their associated molecules'. And we might further deduce that our aesthetic judgements could be consequently considered as merely irrelevant self-attributions, devoid of all hermeneutic opacity. In this paper I not only want to show that poets need not be worried about their craft but that it is precisely within the field of literature (and aesthetics) that these attempts at reducing consciousness to a set of objective data can be countered more cogently than by philosophy. Although the theory of literature will have to remain within phenomenology, its advantage over philosophy is that it is able to describe its phenomena more concretely. Literature enacts consciousness and demonstrates it performatively. Philosophy confines itself to merely appealing to one's common sense, to the 'obviousness' of one's qualitative experience.

For the present I shall use 'consciousness' as an umbrella term for all those subjective phenomena which are still essentially mine, confined to the perspective of the first person. They belong to me in the sense that I find it impossible to causally (or categorically) share them with you. If a person who had always been confined to a white room was given all the physical and physiological data that make up the experience of a particular blue, he or she would nevertheless not know what it feels like to have this particular experience. If you opened my skull, all you would discover, already on a microscopic level, are neurons, dentrites, synapses, transmitters and so on. Each of these could in turn be analysed on a molecular level where all physical means of analysis would finally fail. But for epistemological and perhaps ontological reasons it is highly unlikely that one would ever come across that immediately pleasing 'blue' which I experience when reading Wallace Steven's poetic or seeing Yves Klein's visual representation of the colour.

In literature as well we do not directly meet with consciousness, even though ever since Romanticism it has been considered the genuine medium of subjective experience. First of all, we again meet with a material substratum. This time, however, it is made up of signifiers which are organised to have not only an immediate qualitative or aesthetic effect; the reader is also given the objective or representational equivalent, which enables him or her to reflect on this very effect. In successful works of art, for example in 'Sighs trapped by Liars' by the British Group Art & Language_ or in Yeats's poem 'The Second Coming', there is an inextricable interlocking of sensuousness, materiality and sense. Unlike the brain, art therefore makes both possible: a first-order observation or experience and the second-order observation of how 'raw feels' have been structurally shaped in order to produce this particular qualitative effect of euphoria, anxiety and so on. It is in fact in literature where subcutaneous and vague emotions are transformed into conscious experience. This may then be shared with other persons and reflected upon intersubjectively. Accordingly, the pertinent code of an (aesthetic) evaluation is not true/false but consistent/inconsistent, appropriate/inappropriate. Whereas the neuroscientist's claim of an inner relationship between material basis and phenomenal effect can be considered epistemologically insufficient, the literary critic may well be justified in assuming a necessary correspondence or even identity between signifier and signified, between experience and phenomenon. (For this assumption, though, one must not be part of the waning deconstructive camp.)

Let me focus on four aspects of consciousness which also turn up frequently in philosophical discussions on the subject: 'intentionality', 'qualia', 'perspectivity' and the phenomenon of 'pre-discursive self-awareness'. In the reading process such phenomena not only come to the fore, they are fundamental to the functioning of literature itself.

Contrary to some philosophers, I do not distinguish between intentionality and consciousness as two different aspects of mind.

Although there may be intentional, psychological or linguistic states, that are non-conscious, they can be for 'our' minds only, insofar as they are 'at least potentially conscious' (Searle 1992/ 132, 156). Intentional states, as I understand them, have a phenomenal or aspectual character, whether they consist of images or states of desire. They are in a certain qualitative way for us. Even though conceptually questionable, I will nevertheless speak of the intentional aspect or intentionality of consciousness. It is due to this feature that literature refers, in spite of the absence of a real referent. When reading, our consciousness is *nolens volens* about something represented in the text; it is always directed at something. To exclude intentionality from a literary text is impossible, because it is structurally and performatively built into it by prosopopeia, apostrophy or ecphrasis. As we synchronise fictional objects or events into our presence, say Keats's Grecian urn, we usually cannot help establishing a qualitative and relational connection with the object the corresponding texts picks out or, more precisely, with the content or representation of this fictional object in our mind. We seem to be able to get hold of Keats's Grecian urn directly through the (material) signifier. And while the intentional act occurs, we do not simultaneously distance ourselves from this virtual referent. However rhetorically contrived we may find the poem in retrospect, it actually affects us in the performative process of our reception, because it is consciously and intentionally really existent. For the present it forms the actual content of our mind. We are, however, highly unlikely to detect the vivid urn in the physical world, be it in our brain or beyond.

Intentional contents like Yeats's 'rough beast' in 'The Second Coming' or Stevens' 'color of ice and fire and solitude' in '*Auroras of Autumn*' are concurrent with such conscious modes as anxiety or desire. These states are not merely fictive but actually present. Neurophilosophers call these experiential phenomena 'qualia' and consider them to be the most difficult problem of consciousness (see Guezeldere 1997). To some degree it may be possible to furnish computers with intentional and functional properties similar to the brain but we cannot imagine the computer telling us 'what it is like to' to receive photons of a wavelength of 475 nm, that is, the pure colour blue. Also, we cannot say what it feels like to be a dog sniffing excitedly its way along a dogtrail. There is no causality, and the qualia of this particular experience may well be ontologically intrinsic or confined to the particular experience of this specific dog. While reductionists offer negligable material or even basically Lockean explanations for the particular experience of colour, eliminativists hold qualia to be 'ineffable' (Dennett) with the concept itself confused. Yet it is precisely the complicating fact that colours are perceived immediately that has led numerous poets and painters to draw our attention to the qualitative dimension of this sensuous experience. Unlike syntagmatic scientists, poets transform the quantifiable colour blue into a cultural blue for us by proceeding paradigmatically, synthetically or analogically. This not only enables them to give us a comprehensive, plausible and, indeed, irreducible feedback to an endless variety of colour qualia, they are in addition able to widen our consciousness for further possibilities of subjective qualitative experiences with these colors. The qualitative blueness of the steel-blue sky may itself become the embodiment of the 'black bile' or the 'blues'. (One should not simplify the relation by merely putting it down to linguistic convention or rhetorical transfer.) The arts, according to Nietzsche, effect a continuous refinement of our senses. Thus, texts like Joyce's *Ulysses* or Stevens' *Man with the Blue Guitar* (not to mention other media like Derek Jarman's film *Blue*) not only semantically structure our colour experience into something essentially meaningful, they make it also increasingly difficult to simply attribute this experience to the 'physical properties of objects'.

The data-objective irreducibility of consciousness becomes the more obvious when we turn to the notion of perspectivity or the subjective first-person view. This is, of course, what literature is all about. However, it can be demonstrated particularly well by two basic relationships or attitudes that readers and writers engage in or adopt towards texts. The first has to do with the way readers respond to literary texts. Looking at literary history, the diachronic as well as synchronic possibilities of reading a novel by Thomas Pynchon or Laurence Sterne appear not only idiosyncratic but also potentially infinite. We may cherish a poem and carry it in our wallet for a whole lifetime precisely because two readings of the same poem or novel are never identical. The semantic, situational and evaluative gaps in a text are always filled in or concretised in a different way. The poem unfolds and is accomplished in the inner subjective space of a first person in accordance with his or her very specific experiential, spatio-temporal and psychological disposition. This moment is necessarily only for him or her and inaccessible to a third person. And tomorrow that experience is very likely not to be the same one as it is today. The second dimension of perspectivity becomes apparent in the narrative situation of the novel, especially the 'first person narrative situation', but also the 'authorial narrative situation'. This position corresponds with what the philosopher Thomas Nagel has called 'the view from nowhere'. According to Nagel's thought-experiment (1986, 54-57), we can take a view of the world which comprises all individuals, physical and mental. This view of the world, as a vast theatrum or globe that has myself as a small faceless spot in or on it, is virtually unrestricted and thus without a centre or perspective. But if the world is thereby fully represented, something must have been left out, namely the source of representation. This remainder, however, does exist and can only be conceived of as the consciousness of a first person. Likewise, the authorial and first-person narrators of novels (most clearly of an autobiographical character) are able to unfold a comprehensive world. We may only hear his or her voice in the story (as in George Eliot's authorial novel *Adam Bede*) or the narrating 'I' may also, as in the second type, figure as a character in the story. But in both cases we do not get the source of the narration. It may be overtly or covertly included as a persona, yet it is also necessarily outside the *theatrum mundi* of the story. The absence of the author is one of the reasons why post-structuralists have proclaimed the "death of the author". Yet no one has ever encountered a story that has written itself, being, no doubt, the conscious product of an irreducible consciousness. Literature, finally, allows for a clear realisation of what may be called 'pre-discursive self-awareness' (Manfred Frank). This denotes the very subjective, opaque and sometimes dim sense of that subcutaneously physical state in which we always already find ourselves. We are somehow agreeably aware of the gravity of our limbs, the pulse of our heart, the tonus of our blood and our nerves. This consciousness of one's self, the corporeality which is mine and nobody else's, is indicative of the fact that we do not only have our bodies physically but that we are also bodies in a sense which defies any objective analysis. This epistemological duplicity led to the metaphysical hypostasisation of the body in Nietzsche and many others. And according to Nietzsche, it is through poetry and art as 'applied physiology' (Nietzsche 1969, 1041), through art's rhythm, suddenness and plasticity, that the mind ascertains the irreducible consciousness of its body. Art 'bids us touch and taste and hear and see the world, and shrinks from … all that is of the brain only, from all that is not a fountain jetting from the entire hopes, memories and sensations of the body' (Yeats 1961: 292).

References

Crick, F. 1994. *The Astonishing Hypothesis: The Scientific Search for the Soul.* New York: Simon and Schuster.
Dennett, D. 1992. 'Quining Qualia', in Marcel, A. J. and Bisiach, E., eds.
Consciousness in Contemporary Science. Oxford: Oxford University Press, pp. 42-77.
Guezeldere, G. 1997. 'The Many Faces of Consciousness: A Field Guide', in Block, N. et al., eds. *The Nature of Consciousness.* Cambridge, Mass: The MIT Press, pp. 1-67.
Nagel, T. 1986. *The View from Nowhere.* New York: Oxford University Press.
Nietzsche, F. 1969. *Werke.* Vol. II. Muenchen: Hanser.
Pinker, S. 1997. *How the Mind Works.* New York: Norton.
Searle, J.R. 1992. *The Rediscovery of the Mind.* Cambridge, Mass: The MIT Press.
Yeats, W.B. 1961. *Essays and Introductions.* London: Macmillan

Philipp Wolf is a lecturer of English and American literature in the English Department of the University of Giessen/Germany. He also taught at the University of Wisconsin/Milwaukee and was a fellow at the Centre for Twentieth-Century Studies at the same university. He is the author of Die Aesthetik der Leiblichkeit (Trier, 1993) and Einheit, Abstraktion und literarisches Bewusstsein: Studien zur Aesthetisierung der Dichtung, zur Semantik des Geldes und anderen symbolischen Medien in der fruehen Neuzeit Englands (Tubingen, 1998). He has published a number of articles on Heaney, Heidegger, literary anthropology, aesthetics, literature and religion, money and other subjects.
Email: philwolf@csd.uwm.edu

Nonsense Logic and Re-embodied Intelligence
Bill Seaman

How can our understanding of nonsense be applied to the field of interactive art as well as to the examination of symbolic logic? Nonsense, jokes, and puns can potentially illuminate particular forms of language and image use through displacement or malfunction within a constructed context. The computer, a mechanism entirely predicated on logic, can be used to explore nonsense as well as illogical and elusive resonant content. This paper will examine issues surrounding the employment of specific forms of nonsense in computer-based interactive works of art.

Could an artist, using the computer as vehicle of research, define an art practice where the subject of that practice was an examination of meaning? As computer-based systems and technological sensory extensions change our relation to both nature and language, we need to create mechanisms that function at the highest possible level of human/machine interaction in order to reflect upon this complex plethora of emergent relations. Given the limitations of language to reflect the complexity of lived experience, we need to move toward the creation of more sophisticated systems of communication that will allow us to both share and create new reflective experiences. A rich variety and complexity of experience requires equally complex transformative technological systems to reflect upon that experience. In the light of this comment, might we seek to engender new forms of poetic expression to reflect upon the nature and construction of meaning?

There is a poignant irony to the fact that the computer, a mechanism entirely predicated on symbolic logic, can be used to explore non-sense as well as illogical and elusive resonant artistic content. A work of art can be seen as an organism-like vehicle of content that is both generated and experienced through interaction. Roy Ascott, very early on, saw this potential. In his paper entitled *Behaviourist Art and the Cybernetic Vision* published in 1966, Ascott articulated the following vision:

> Behaviourist Art constitutes, as we have seen, a retroactive process of human involvement, in which the artifact functions as both matrix and catalyst. As matrix, it is the substance between two sets of behaviours; it neither exists for itself nor by itself. As a catalyst, it triggers changes in the spectator's total behaviour. Its structure must be adaptive implicitly or physically, to accommodate the spectator's responses, in order that the creative evolution of form and idea may take place. The basic principle is feedback. The system Artifact/Observer furnishes its own controlling energy; a function of an output variable (observer response) is to act as an input variable, which introduces more variety into the system and leads to more variety in the output (observer's experience). This rich interplay derives from what is self-organising in which there are two controlling factors. One, the spectator, is a self-organising subsystem; the other, the artwork, is not usually at present homeostatic.

Ascott goes on to say:

> There is no prior reason why the artifact should not be a self-organising system; an organism, as it were, that derives its initial programme or code from the artists creative activity, and then evolves in specific artistic identity and function in response to the environment which it encounters. (Ascott, 1966, p. 11)

My research focus has been in exploring what I call Recombinant Poetics. A Recombinant Poetic work seeks to enable the exploration of a set of authored media-elements of language, image and sound such that the media can be made operative in a computer-based environment. Thus a user of such a techno-poetic mechanism can explore the contextualisation, recontextualisation, navigation, as well as a number of other processes, through interaction with that system. The participant potentially brings about interpenetration and juxtaposition of media-elements through their interaction with the following categories: construction processes; navigation processes; processes related to authored media-behaviours; editing processes; aesthetic / abstraction processes; automated generative processes; processes related to distributed virtual reality; and chance processes of a *semi-random* nature. The system seeks to function in a self-organising manner.

Works of art that explore such operable media do so within technological environments that enable the generation of emergent content and potentially open up entirely new fields of artistic investigation.

I am seeking to define a particular approach to the authoring of systems that enable engagement with specific forms of operable media. "Re-embodied intelligence" can be defined as the translation or encoding of authored media-elements and/or processes into a symbolic language, enabling those elements and processes to become part of an interactive computer-based system. Artworks that explore 'Re-embodied Intelligence' do so on a case-by-case basis, where the author and programmer encode a particular set of art-related processes, concepts, or aesthetic attributes, into a computer-based, operative form. Specifically, the output of such a system seeks to manifest the encoded sensibility of its author.

The encoded model of the sensibility of the artist is rendered operative within such an environment such that the system appears to exhibit intelligent behaviour: i.e. in one case it can be engaged to build a virtual world informed by this sensibility. Most of us would consider the building of a virtual world a task that requires intelligence. In this particular case, 'Re-embodied Intelligence' seeks to encode this sensibility such that the computer, functioning autonomously or in conjunction with a user, can generate environments informed by the artist's mind set.

We must be careful to differentiate the kind of "intelligence" exhibited by such a mechanism, to that examined through the Turing Test. Thus the value of the Turing Test to determine 'intelligence' may be seen as relevant to a particular context, but for the purposes of art content may be completely irrelevant. An artwork may explore any approach that the author (or authors) finds appropriate. The artist is not trying to 'fool' someone into believing the machine is thinking. The artist is attempting to translate intelligently particular kinds of potential relations, enabling an experiential exchange that may include specific kinds of responses by the system to the user's input and/or behaviour. During this interaction, the mind-set of the programmer/artist, can be experienced by the user in the service of this content, in the form of process-oriented artifacts.

Now that I have outlined the notion of 'Re-embodied Intelligence' I would like to examine a particular aesthetic sensibility. Given this emergent interactive environmental context propagated through Re-embodied intelligence, I am seeking to explore the field of nonsense as artistic content (among other aesthetic explorations). More specifically, using the computer as vehicle of research, as stated above, I am seeking to define an art practice where the subject of that practice is an examination of meaning. In this case, aspects of meaning are approached though pointed nonsense relations.

If we look historically at the use of nonsense in literature and other forms of art, we find a fertile realm of creative exploration. How can our understanding of nonsense be applied to the realm of interactive art as well as symbolic logic? Here, Lewis Carroll becomes an interesting subject for investigation in that he both authored texts about logic as well as texts exploring nonsense.

Deleuze states in his book entitled *The Logic of Sense:*

The work of Lewis Carroll has everything required to please the modern reader: children's books or rather, books for little girls; splendidly bizarre and esoteric worlds; grids; codes and decodings; drawings and photographs; a profound psychoanalytic content; and an exemplary logical and linguistic formalism. Over and above the immediate pleasure, though, there is a play of sense and nonsense, a chaos-cosmose... Deleuze continues: "The privileged place assigned to Lewis Carroll is due to his having provided the first great *mise en scène* of the paradoxes of sense – sometimes collecting, sometimes renewing, sometimes inventing, and sometimes preparing them. (Deleuze, 1990, p. xiii)

One goal of the use of computer systems is to come to better understand ourselves. Computers can function as mechanisms of discourse, enabling the exploration of embodied models made operative through interactive mechanisms. Within this computer-based context, through the exploration of nonsense, one can witness a contrasting critique of sense. The subtle displacement of a particular element from a selected context can actually help to illuminate aspects and/or qualities of functionality. In the *Philosophy of Nonsense* by Jean-Jaques Lecercle, the author states:

> My thesis...is that the negative prefix in "nonsense" ... is the mark of a process not merely of denial but of reflexivity, that non-sense is also meta-sense. Nonsense texts are reflexive texts. This reflexion is embodied in the intuitions of the genre. Nonsense texts are not explicitly parodic, they turn parody into a theory of serious literature; [for example] Lewis Carroll's metalinguistic content on points of grammar ... (Lecercle, 1994, p. 2)

A nonsense statement can potentially release a field of potential readings. The playful use of a pun is one example. As the meaning forks into a field of alternate readings, and the relations between those readings, an elaborate conceptual process in the mind of the reader/user is set into action.

We can observe the employment of nonsense in computer-based works as potentially setting out a complex field of emergent potential readings. Such a layered field can point at the complexity of how meaning arises and falls away in everyday experience. Nonsense relations inform our understanding of reality just as sense relations do. It is this relation between sense and nonsense that I seek to explore, where the use of nonsense becomes self-referential, communicating simultaneously about a particular authored context while also "throwing off" or playing with the meaning. In this way non-sense can function as 'meta-sense'.

In his book *The Logic of Sense*, Deleuze states:

> The play on words would be to say that nonsense has a sense, the sense being precisely that it doesn't have any. This is not our hypothesis at all. When we assume that nonsense says its own sense, we wish to indicate, on the contrary, that sense and nonsense have a specific relation that can not copy that of the true and the false, that is, which can not be conceived simply on the basis of a relation of exclusion. This is indeed the most general problem of the logic of sense: what would be the purpose of rising from the domain of truth to the domain of sense, if it were only to find between sense and nonsense a relation analogous to that of the true and false? [. . .] The logic of sense is necessarily determined to posit between sense and nonsense an original type of intrinsic relation, a mode of co-presence. For the time being, we may only hint at this mode by dealing with nonsense as a word which says its own sense.
>
> (Deleuze, 1990, p.68)

If one thinks of the computer as often being predicated on a binary logic of true/false, yes/no, on/off (fuzzy logic aside), then, in a specific sense, I am trying to approach more delicate and subtle modes of communication and intellectual exchange through the playful and pointed employment of nonsense.

In terms of exploration of the interface as content, significant to the operation of an authored computer-based environment, I have chosen to explore complex interfaces that are outwardly expressive while, inwardly, functioning as the outer-most layer of symbolic logic. Thus the interface becomes a vehicle of symbolic logic. When we include puns, nonsense and jokes on this symbolic layer of the system as an operational part of the interface, we can potentially make observations about the nature of logic through interaction with such a system. It becomes a goal in my work, to present for the user an opportunity to observe the functionality of consciousness in action, through the intermingling with complex multi-layered systems of authorship and inter-authorship, placement and displacement.

By developing a computer system that explores pointed nonsense as its content, we come to better understand the complexities of context construction. It is often the nonsense text that, through displacement, opens up a new relation – a re-seeing of the original context, a form of active comparison built into, or compressed within, the signifying environment.

The permutations inherent to recombinant structures present a situation in which nonsense relations can arise and/or be intentionally initiated. Each media-element in a 'recombinational' work of art has a potential 'meaning force'. By exploring elements carrying condensed content, or multiple potential readings, we could say the 'force' of such elements paradoxically pushes in a number of directions at the same time. The nature of signification as examined within an emergent, interactive context, exhibiting fleeting and shifting qualities of meaning, can potentially become an experiential focus. The user experiences a temporary glimpse at a continuous process where elements of language, image and sound qualify the readings of particular elements, and are themselves qualified in relation to these other elements, within this process.

Nicholas Rescher in his book entitled *Many-valued Logic* states:

> The very idea of truth-values other than the two orthodox truth-values of truth and falsity is obviously central to the conception of a "many-valued" logic. To obtain such logic we must contemplate the prospect of propositions that are neither 'definitely true' nor 'definitely false' but have some other truth status, such as 'indeterminate' or 'neuter'. [or 'other'. Emphasis the author] (Rescher, p.2, 1969)

In terms of Recombinant Poetics, I am examining the term 'logic' as a compression of logics - a pun on logic - where multiple logics are compressed and made operative within a mechanism that bridges and enfolds the textual, the sonic and the imagistic within an experiential computer-based environment. This multi-logical system explores the following logics:
- Logic of nonsense
- Psychological logic
- Physiological logic
- Mechanic logic

– Strategies –

- Logic of virtual mechanisms and/or conceptual machines
- Logic of economy
- Aesthetic logic, including:
  - Sonic logic
  - Pictorial logic
  - Linguistic logic

All of these logics are made operative within a compressed, shifting, meta-logical mechanism.

In a world whose obvious complexity presents situations that can not simply be read as true or false, the specific employment of Nonsense Logic functions in works of art as a contrasting conceptual perspective mechanism to that of more traditional forms of logic. The meaning arising from the exploration of this pointed nonsense is a meta-meaning. Such operative Recombinant Poetic environments enable the exploration of both specific and emergent fields of meaning through the active engagement of authored-media elements and processes. The non-linear experiential examination of nonsense-logic also presents an alternative to the hierarchical logic that computers are most often predicated upon. The 're-embodied intelligent' employment of nonsense-logic, makes sense.

References

Ascott, R 1966. *Behaviourist Art and the Cybernetic Vision*.

Attridge, D. 1988. in Culler, J. (ed), *Unpacking the Portmanteau, or Who's Afraid of Finnegans Wake* in *On Puns, The foundation of Letters*. Oxford and New York: Basil Blackwell.

Blake, K. 1874. *Play, Games, and Sport: the Literary Works of Lewis Carroll*. Ithaca and London: Cornell University Press.

Carroll, L. 1937. *The Complete Works of Lewis Carroll*. New York: Random House.

Carroll, L. 1977. *Lewis Carroll's Symbolic logic: part I*. Edited, with annotations and an introduction by William Warren Bartley, III. New York : C. N. Potter, Distributed by Crown Publishers.

Deleuze, G. 1990. *The Logic of Sense*. First edition 1969. New York: Columbia University Press.

Lecercle, J. 1994. *Philosophy of Nonsense: The Intuitions of Victorian Nonsense Literature*. London and New York: Routledge.

Rescher, N. 1969. *Many-valued Logic*. New York, St. Louis, San Francisco, London, Sydney, Toronto, Mexico, Panama: McGraw-Hill.

Bill Seaman received a Master of Science in Visual Studies degree from the Massachusetts Institute of Technology in 1985. His work explores language, image and sound relationships through video, computer controlled videodisc, CD ROM, Virtual Reality, photography, and studio-based audio compositions. He is self-taught as a composer and musician. His works have been in numerous international festivals, exhibitions, and museum shows. He is currently Director of Imaging and Digital Arts, the Graduate Program, University of Maryland, Baltimore County. Bill is a CAiiA researcher.
<seaman@umbc.edu>
http://caiia-star.newport.plymouth.ac.uk
http://research.umbc.edu/~seaman/
http://www.235media.com/installations.html

Conscious Isomorphisms and Young Farmers: Consciousness as a Subject of Artistic Research

Stefaan Van Ryssen

Art and Mind
Artistic production or activity is, of course conscious, even if it leaves room for elements which are not consciously processed. Spontaneous, naive or impulsive artistic expressions tell us something about the mind but they have hardly ever taken the mind itself as their subject. Neither have the products of 'broadened' consciousness and the artworks by mentally diseased taken consciousness as an object of representation, a topic of communication or an experience to be shared. The mind is not debated nor depicted, unless perhaps in some popular cartoons about psychologists.

In recent years, however, mainly under the influence of the progress made in artificial intelligence, networked computing, the philosophy of mind and the science of the brain, artists have been representing the mind. Not *their* minds for obvious reasons and for more hidden reasons I will discuss. I distinguish different approaches. Each is represented by one artist with whom I am well acquainted.

Before discussing the artists, their work and its implications, I need to make clear some other things. First, that it is not solely and principally their aim to represent or to render. Representation is not their business; it hardly could be. More likely, they are moving the mind into the arena, onto the stage or on the screen. From now on, it is there as a piece to conquer, not only as a static field or a logistic facility. Second, that they don't stop right there. All of them have other things to do, they proceed, they boldly go where nobody else . . .

The artists discussed here share, almost of necessity, a monistic view of human consciousness. In their works, they express a strong intuition that there is no ontological difference between the biochemical substratum and the mental activity. Mind and brain are one, thoughts and emotions are neurophysical and hormonal processes. It seems that they do not support a strong reductionist approach nor a purely computational view of the brain. Their viewpoint is mainly connectionist.

It is relevant that they share another perspective: the mental and the conscious are social, partly public, but not transcendental. Well, they wouldn't mind if anyone added a transcendental *'lecture'* to the different strata of the discourse around their work, but it would be accepted perhaps with amusement to be ridiculed, or as a sure sign of ubiquitous human folly (and certainly a sign of critical prententiousness). The mind, in short, is nothing exclusively individual. And, as we shall see, for some of them, it is not even exclusively biological.

I must remind (!) the reader that the methodology of these artists is pragmatic and active, even for the so-called conceptualisms. In practice lies the proof of the aesthetic. The method of this paper is isomorphically connected to the methods of the artists; at least; it claims to be.

Prelude. Antheil, Cage and Foss: Mentality or Tonality

If Lukas Foss, John Cage and George Antheil would have been together to 'make music' it would certainly not have resulted in a Cataclysm of Sound. They would not have competed to conquer a dominant position or some kind of hierarchy either. Rather, they would have recognised that any aim of performing something like individual musical pieces would be inhibitory for the existence of music at all. We can guess at what the result would have been: not an improvised set but a careful passing of ideas from one to the other, where the foreground of one would be the background of the other and where the tunings and the preparations of the third would be the stage for the formal presence of the other(s). We can imagine that they would have shared absolutely nothing but the light and the acoustics. And after inviting Maurizio Kagel in, the gang would have hummed a real, because virtual, hymns or coming together. 'Bless Karlheinz', Cage mutters.

Part 1. The Road to Art(ificial) Consciousness

Young Farmers Claim Future (B) are five - three computers and two wetware who are exploring the possibilities of artistic production by computer networks. Connected machines create their own literature, which is probably best understood by the computers themselves. This work exemplifies what I would call the Road to Art(ificial) Consciousness.

A metatypical work of YFCF is the mac installation,[1] a setup of 20 connected machines, interacting among themselves and creating a social field where each coordinates its outputs with the inputs of others. The choices they make are in real time and the interaction is perfectly catastrophic, i.e. slight changes in initial setup, irregular clockbeats, unstable ambient temperature or randomly assigned programming parameters can lead to widely different conversations.

The system produces waste: human-readable (so-called cultural) signs like sounds, text, philosophy, music and images. Because the social system is homogenously digital and electronic, the analogous offal is scavenged by – mainly human – biological systems who use these spillovers as input in processes of their own liking (called 'art', like 'artificial': made by human matter).

What YFCF aim at, among other things, is to get rid of the metaphysical flavour of the concepts of consciousness, creativity, art, genius, etc. Confronted with one of their earliest mottoes 'If God exists, it lives inside our computers', where God can be read as a placeholder for anything metaphysical and generatoric, and as such as well for consciousness, they say:

> About god! There is also something funny about gods. Recently we thought, look, statements about society and ethics and whatever, are usually not ascribed to [society] itself but to something outside of it. And that makes metaphysics very ambivalent! Now if you don't want to be caught in logical contradictions you arrest this method and imprison it. Create a feedback algorithm and let it run and change itself autonomously. God is a random feedback loop, flip-flopping between self-organisation and self-destruction. Yeah, we have it! That is what it is all about – but sometimes I doubt this: machines and consciousness and creativity. Let's forget about the rhetorics but do something with it.

Bringing back creativity, consciousness 'and all that' to the level of algorithms, and still more important; doing something with these algorithms, leads the Farmers a step further along the road. 'Our behaviour is based on trust and collaboration, we have a very personal relationship with our computers and software. It is always hard to kick a machine out of the band and replace it by a better one.'

By being outside of the 20 mac installation, the human members of YFCF leave room for the development of a mental space in digital society. As in biological systems, the conscious mind cannot be studied or experienced directly but only by means of its waste and the actions of some of its interfaces: monitors, soundcards and amplifiers etc. In two different versions, the installation has been directed to generate outputs which are similar to what is considered to be the result of intense human mental activity: philosophical discourse and orchestral music. Because of the quality of the discourse, and the flawlessness of the performance, the installations can be considered to be bad descriptions of the human mind. Fortunately, they are aware of that failure. We guess they are working on it.

Part 2. Emphasis on Unity and Routines

Jordan Crandall (USA) builds spaces in which the boundaries between the real and virtual, the physical and mental, are obliterated.

By exciting experiences and processes in both body and mental awareness of the audience/participant, Crandall recreates the mind/body complex. His tools to connect different levels of experience are 'routines'.

Routines in Crandall's pragmatics are not standardised procedures but rather morphological constants, resonating in different realms of aesthetic, moral and political practice. The scope of a routine is much wider and of a different dimension than what are customarily called 'synaesthesia'. Synaesthetical experiences are believed to be purposely provoked by the artist, and consist of some kind of mutual enforcement of aesthetic fulfillment in different senses – sound and image, taste and texture, working together. Routines are maybe conditional for synaesthesia but they can encompass acts, thoughts, ideas, communications, aesthetic and experiences. They are not only provoked by the artist who claims to realise a single meaning in different codes; rather they are triggered flows of mental activity. The artist and the audience – a tricky distinction in Crandall's world – support, transform, present, use, communicate and interpret these flows or resonances in their own individual ways. A non-typical example (nontypical for JC's work – in my own universe it would be very common and basic) is sexual arousal.

Routines can be likened to formants or fundamentals of a complex signal. No engineer would be willing to draw conclusions about the cultural content or meaning of a Fourier or a Z-transform if she doesn't have an idea of the kind of material the transform was done at. Nevertheless, the formants carry information, and very critical information at that.

Crandall doesn't rely on any kind of explanation or on an interpretation of the nature of the routines he works with. Routines take form: he works with forms and environments – social encounters, epoxy vehicles, image resonators, frequency modulators, printed scores and catalogues. The actual material is temporary and transient: it is chosen for the social and ecological niche where the artist happens to work, and serves to overcome the typical inertia of the subset of art consumers of the space where the work will *perform itself*.

As I said before, Crandall doesn't try to represent the mental or the conscious. He presupposes a self-conscious spectator, someone who is continuously aware of her own mental content and the structure of her cognitive and emotional setup. It needs a split personality to participate and fully appreciate Crandall's environments. One has to 'look at oneself' all the time, and interpret the changes the interaction with the environment provokes. Oddly, one who lets the routines be set up and triggered undergoes the extremely alienating experience of having one's own consciousness – or at least one half of it, so to speak – being represented and integrated in a substantial, material construct.

There is nothing mystical in this. No levitation – even if Crandall talks about suspension – and no Glorious Illumination. One does not need to be initiated or drugged (well, caffeine helps a lot.). You have to be aware of your own awareness, and let one level of awareness be moved – emotional and spatial – while monitoring it by the other level(s).

While exhausting, it is elegantly done, and when you become really involved in the game – letting down defences is quite harmless - you gain in clarity and mastery and you can even have a good laugh.

The mental is, in Crandalls practice, at least partially social. And that is what I have been claiming too. Just as it is meaningless to speak of the individual biology without implying its evolution and its actual ecological embeddedness, it is empty to speak of individual psychology as cut off from its cultural and social component. Even if we experience our mind as highly private, its structure and its content are to a certain – and in my opinion very wide – extent social. The mental does not exist without a substantial repertoire of common blueprints. This may be scientifically debatable, certainly the exact relationship of the social and the private is troublesome. And, not surprisingly, it is difficult to grasp the precise operational meaning of the terms mental and social when we do not define them as mutually exclusive. Crandall's artistic intuition leads me to believe that at least the conceptual equation of the two oppositional pairs private-public and mental-social needs to be questioned.

Part 3. Purposeful Restructuring

I use my own work to explore the dynamics of the mind. As a trained Applied Epistemologist, I have been researching – or rather observing - the way my own mind restructures representations of content to be accommodated more easily in the available mental patterns. To do so, I have been developing a practice of conscious isomorphisms. I use an amalgam of literary and graphical techniques and symbolic references to record the processes of cognitive and symbolic activity.

Every recording starts, so to speak, with an empty mind in which I drop more or less randomly chosen contents (mostly *objet's trouvés*). By letting the content evolve through a number of consecutive steps and jotting down some comment on these transformations I end up with an image of both the object and the process.

Over the years, I managed to set up a state where I can do something like *peinture automatique* with a purpose, while I am actually recording in real time the automatisms involved. Several hundreds of sets of recordings, most of them in the same framework of a series of sheets of graph paper, form a kind of database, ready to be mined for constants and surfacing patterns. Some of these patterns resemble the routines Jordan Crandall uses. I call this the way of Purposeful Restructuring.

Finale. A Sociology of Mind

Now, how does it lead to a Sociology of Mind?

I refer to the story in the Prelude. Cage, Foss, Antheil and probably Kagel and Rzewski, to name only a few, have understood music as a social thing rather than a highly particular expression or a highly private experience.

If music is a subset of socially organised sound, existing or latent, it is a composer's activity to select a slice of that sound, and after consciously tranforming the mental representation she has of that slice, or frame, feed it back to an audience, be it small or large, synchronously or asynchronously present, critical or obedient, trained or lay.

When I try to fit the experiments of YFCF, Crandall and myself into the analogy of the composers, I end up with a strong conviction that consciousness in particular and the mental in general have to be viewed as something in between private and collective. It is not exclusively the domain of the psychologist or the brain scientist to study consciousness. We will have to turn to the sociologist to learn more about its workings.

The mind, though apparently working individually and personally, acts as a locus where discourse, routines, isomorphically transformed contents sort of 'crystallize' against a biological cool wall. Consciousness is – apart from being some kind of process, mechanism or hallucination – a temporary realisation of the social/mental in the personal/biological. Probably consecutive states of the biological are compared, checked or 'remembered', which gives rise to the curious process of thinking the deltas that are applied to a certain set of signals.

In this way, a distribution of unequal amounts of mind over different biological 'places' is possible. And a different intensity of consciousness is possible within one person over time or between diffferent persons at any set time. By manipulating, triggering, expressing and constructing new morphisms, mind exerts power over mind. Whoever understands these manipulations can find ways to reconstruct and replicate mind, and, close by, consciousness. Even the word 'mentality' takes on a different and certainly a richer meaning.

Coda

Artistic action, communication and experience do not prove anything. My artistic intuition tells me that consciousness is created and disappears (dis)continuously, and these concepts make it ultimately possible to recognise artificial consciousness when I meet it, measure it through its purposeful effects on mentality and If Life Exists, It Is Inside Its Computers.

Notes

1 It is related to different types of YFCF projects. Among others, the PotoMac installation shown in Berlin. The installation has lived through different versions. The one discussed here was planned for a festival in Hasselt, Belgium. Another version, called Young Farmers Go Classic, is being prepared for a Barcelona festival.

– Strategies –

Modelling Interpretation
Sharon Daniel

Experience and Representation
Our self, our awareness, no matter how direct and immediate, is not ultimately oriented; is without concrete external reference. Through systems and representations we attempt to locate and quantify our awareness, which is ultimately beyond systemisation; too many-dimensioned for signification on any single graph or plane, no matter how complex; too irregular in surface, proportion and form to be located or composed within any frame. We would like to be given a truth. A single thing to turn over and stroke in hand or mind that conforms to reality, represents it, explains it, that is reality there. There are metaphors that we can recognise as not entirely arbitrary but any statement of a truth only translates to its own inadequacy, its lack of conformance to experience.

Our world is described to us through language and through this description we learn to see it. Whoever controls the languages that describe the world will naturally describe the world as they wish it to be. This wish, as it becomes language, will eventually be accepted as fact. The technologies of representation born of theoretical science both reflect and structure our perception of experience. For example, in the Renaissance, advances in the understanding of geometry and mathematics laid the foundation for perspective drawing, leading Alberti to remark, 'One sees sensibly and rationally according to a science that is called Perspective, arithmetical and geometrical.' This association of perspective with scientific rationality, each of which proposed the existence of an absolute and privileged point of view, helped to establish the dominant model of consciousness in Western culture – that of the centered subject. This model is not fixed or necessary; it is a reasonably functional but uncertain construction.

Today, technologies and sciences are developing that may enable us to question this inheritance. Just as technology has allowed us to examine phenomena, such as the electromagnetic spectrum in the infrared and ultraviolet ranges, which are physiologically unavailable to us, new technologies may allow us to circumvent cultural, land linguistic filters that lie between perception and consciousness.

Models of Consciousness
The projects described below are 'simulations,' physical or experiential manifestations in concrete and electronic space, of alternate models of consciousness suggested by developments in theoretical science. The first two examples embody new models of consciousness and subjectivity suggested by Artificial Life and Complex Systems.

Artificial Life
The content and the structure of the web project 'Signal-to-Noise' evolved in an iterative, associative, and aleatory process. By employing techniques of visualisation borrowed from Artificial-Life research, self-reflexive Boolean search mechanisms and random selection,

'Signal-to-Noise' (http://metaphor.ucsc.edu/~sdaniel) explores conceptual parallels between:
- Cellular Automata models of physical systems
- Deleuze and Guattari's 'molecular' sexuality
- Michel Serres's *La Belle Noiseuse* extrapolated from Balzac's *Gillette*
- Thomas Pynchon's paranoid construction - the "Trystero" an alternative postal network in *The Crying of Lot 49*
- Richard Powers' self-reflexive fictional neural network character in *Galetea 2.2*
- Slavoj Zizek's description of the symbolic order via 'Hitchcock's Universe'

All of these examine, either directly or indirectly, the emergence of global systems from microscopic, associative, random or interpretive interactions.

For example, Deleuze and Guattari's description of molecular sexuality parallels cellular automata models of the physical world in its fundamental assertion of discreteness, difference and separation. Instead of the psyche, the subject and the self, Deleuze and Guattari believe that we comprise many small 'I's' which never attain the unity of the "self" achieved by the Ego. The discrete values of a Cellular Automata system, like the "micro-organs" of Deleuze and Guattari's molecular sexuality, form connections in which parts continually refer to other parts outside of themselves, without closure, without beginning or end.

Just as complex images and patterns emerge in Cellular Automata models, for Serres all form, all language, all expression emerge from the 'masterwork' - a multiple and chaotic sea. Video images, based on Serres's description of chaos or *La Belle Noiseuse* in <u>Genesis</u>, were processed with the Cellular Automata Machine designed by Thomaso Toffoli and Norman Margolis. This system uses iterative steps to processes data in real-time. Each pixel in an image may 'behave' independently at each 'step'" based on a table of rules, for that behaviour and a given initial condition. The table of rules is a set of definitions for the behaviour of each pixel or cell in relation to the state of each neighboring pixel or cell. Therefore, a global state a pattern or image emerges from the local interactions of discrete entities in an iterative and constantly evolving system.

Chaos and Complex Systems
Like cellular automata, chaos theory provides models of the physical world that offer a more accurate description of the perceptual experience of concrete reality than Euclidean geometry, or Newtonian absolutism.

For example, in chaos theory a 'strange attractor' maps the infinite trajectory of a non-periodic system in phase space. Phase space allows as many dimensions or coordinates as there are degrees of freedom in the system to be mapped.

If the system considered is the attractor 'being-in-the-world' as in the interactive installation 'Strange Attraction: Non-Logical Phase-Lock over Space-Like Intervals' then the coordinates for three dimensions might be space, time and desire.

Phase space will stretch and fold to incorporate infinite additional dimensions such as guilt, denial, elation and repression until all symptoms of desire in its trajectory through space and time are registered.

— Strategies —

A strange attractor never comes to rest and does not produce any single rhythm to the exclusion of all others yet a subtle order is established, that of self similarity. In any small cross section of a system the structure of the whole is described.

The installation Strange Attraction: Non-Logical Phase-Lock over Space-Like Intervals' establishes a non-periodic system, similar to that of a strange attractor, that is perpetuated through feedback. The participant acts within the system. Her actions, along with the involuntary responses generated by her actions, modify the state of the system. These modifications to the system effect the participant's subsequent actions or experiences within the system, producing voluntary and involuntary responses that further modify the system, thus simulating the feedback loop of guilt and desire.

Electronic devices monitor each of the four participants' heartbeat, respiration or skin resistance. Each participant thus involuntarily controls the transmission of sounds and images that symbolise the subjective location of each of the four participants and form, at certain intersections of activity, a kind of choir of inter-subjectivity. Each participant is simultaneously aware of herself as a subject with voluntary, yet partial, control over the system, and as an object involuntarily responding to the action of another subject. Thus, a collaborative system is established.

Collaborative Systems
In the following examples, problems of interpretation, translation and communication are examined through artificially intelligent systems, virtual communities and collaborative systems.

Narrative Contingencies
The Narrative Contingencies web site (http://metaphor.ucsc.edu/~sdaniel) allows participants to construct their own narratives at the intersection of chance and interpretation.

All forms of representation play a role in the construction of desire and thus inflect experience. When language and representation are structured from a single, dominant point of view, the position of the subject remains fixed. If the point of view of the symbolic can be restructured and multiplied - then experience itself might be altered. It is impossible to escape the image and language of the existing symbolic order but it may be possible to restructure them by circumvention and dislocation. 'Narrative Contingencies,' disengages the production of image and language from its ideological matrix by forcing it through complex filters of random and chance operations.

If a participant is able to arrive at a meaningful narrative interpretation of the relationships between images and texts brought together through filters of random or chance operations — and so, composed without the ordering intention of a single author — then the possibility that all relations of meaning are inherently contingent might be considered.

From 'Signal-to-Noise,' La Belle Noiseuse, or Chaos Video processed with the Celluar Automata Machine

Virtual Communities, Agency and AI
shared narratives and interactive interfaces

Collective Unconscious
The virtual community test-bed 'Collective Unconscious', a multi-user, object-oriented, virtual environment, provides an interactive interface to the dream narratives of its participants. As a collaborative research environment, this project addresses the impact of "interpretive systems" on the translation of perception into consciousness.

Each participating collaborator must undertake four interpretive translations
1. The translation of lived experience into representations within the unconscious
2. The interpretation and articulation of those unconscious representations or dream images into natural language
3. The implementation of this natural language description as a text-based interactive interface using a 'non-natural'
4. The translation of the text-based interface into a graphical interface

Participants structure the user's subjective location within each dream narrative and provide interactions, navigation paths and hypertext links to other network sites.

The function and visual representation of artificially intelligent agents and avatars in virtual and physical environments are examined in two collaborative, communications interface design projects, 'ELFnet', and 'Pathologies.'

ELFnet Distributed Artificial Intelligence/Natural Language Processing
ELFnet is a pager network community project with a web interface. Each participant is provided with an alphanumeric pager. The collaborators determine the structure and meaning of their own local 'community', the role of each individual in it and its relation to a larger, 'global' pager network community. Each participant creates 'intelligent agents' within the pager network using a simple scripting language. These 'intelligent agents' extend the consciousness or identity of each participant and support his activity in the community by parsing, translating, distorting or enhancing communication within the network. Each participant is encouraged to create a detailed, graphical representation of his or her own 'intelligent agent(s)' for the project's Web interface.

'Pathologies a conversational agent interface'
Each individual personality incorporates, to a greater or lesser degree, all the traits that

'Strange Attraction: Non-Logical Phase-Lock over Space-Like Intervals', installation detail

Detail from 'Narrative Contingencies' database, Participants are offered complex mechanisms to 'read' from, 'write' to, 'build' on, and 'give' or add to the Narrative Contingencies database

have been defined as pathologies in clinical psychology for example, obsessive-compulsive disorder, paranoia and schizophrenia. A 'normal' personality merely exhibits these 'disorders' to a degree that is acceptable within a given social or cultural context.

'Narrative Contingencies', image database grid

The manifestation of an individual's pathological traits through artificial intelligence would allow an individual to 'safely' explore and extend those aspects of her personality and to question the cultural interpretation of behaviours as 'disorders.' A pathological agent built on an association network might, in addition to any possible research or therapeutic function, provide interesting conversation.

The goal of this project is to develop a conversational agent which extends the inherent clinical pathologies of a 'normal' personality. The agent's virtual environment will avoid the rigid 'desktop' and 'page' and offer a more flexible, dynamic, graphical user interface. The user will be able to increase extensibility of the interface through direct manipulation and through conversation with her 'agent'.

Public Art and Public Technologies

Because technologies constitute, embody and disseminate metaphors, which construct, constrain, facilitate, describe, enhance, enable and prescribe experience Artists and critical theorists must participate in their development.

Because technologies actively locate 'the subject' in 'the world', there must be opportunities for the public to interact with 'public' technologies in a critical, intellectual context. The interface for these interactions should be dynamic, scalable and multi-user. This interface should be modeled on an anti-hierarchical, non-perspectival and non-totalizing metaphor – an isotropic space.

Isotropic Cartography The Hubble Space Telescope and the Human Genome Project

Interactive public art using existing public technologies will engage participants in large-scale, public, scientific research projects. Interfaces designed for individuals to interact with these 'public technologies' will map the intimate or microscopic scale of the genetic, physical and historical individual to a cosmological inquiry on the macrocosmic scale the generative evolution in the Universe.

Conceptual Foci

'Public' Technology – to connect the individual to monumentally scaled technological research projects 'public' art using 'public' technology. The Hubble Space Telescope and the Human Genome Project are both 'public' research projects, each attempts to map, in absolute terms, systems that are infinitely extensible.

Biology to Cosmology – to connect the search for meaning in the smallest units of information; to the search for the origins of the expanding universe — to explore the selection and interpretation of patterns — mapping patterns from the cosmos to patterns in DNA code and back again.

Scientific Visualization – to analyse the interpretive systems of disciplines and cultures - overlaying maps and identifying patterns. To examine scientific practice in relation to art practice modes of thought, analysis, investigation, critique, representation, and visualization of data or ideas. Just as the constellations were identified images drawn into random patterns of stars – in order to explain natural phenomenon and regulate social and cultural systems, deep space observations and continuing efforts to translate the genetic code provide stimulus for new identifications, interpretations, descriptions and narratives. The interpretation and visualisation of data sets — the extraction of meaning or truth from the available data — is a fundamental operational strategy of scientific practice. All interpretations are contingent upon context, perspective, the amount of data available at the time and the constraints of the interpretive system.

Dynamic Mapping – 'Isotropic Cartography' events will provide opportunities for interaction with the Hubble Space Telescope, the LICK and KECK Observatories and the Human Genome Project on the Web and at site specific, interactive installations. Collaborations with research scientists engaged in the study of cosmological and biological origins have been established and interfaces for public interaction with technologies designed to collect and interpret data from the subatomic to the macrocosmic are being designed.

Parallel Processing, the Conditions of Possibility and Isotropic Maps

The first telescope was suspect as a scientific instrument. Galileo's contemporaries did not privilege the sense of sight or accept it as a regulator of truth. With the rationalisation of sight through the science of 'perspective', the 'seen' was codified and validated in accordance with the dominant cultural paradigm of the Renaissance secular humanism or the primacy of the individual subject. Technologies are always an extension of the assumptions and desires of their makers – either in concert with or reaction to a cultural moment. Perspectival representation has been embedded in perceptual experience and pervasive in language. It is a 'modeling language' purpose-built to assert notions of centre and hierarchy. An alternative, isotropic mapping of space – one that is invariant with respect to direction – may help to establish the 'conditions of possibility' for multiple, decentred and parallel perspectives.

Within isotropy subjective location could be distributed and recombinant. Images would be evolved rather than composed. Images and objects that are evolved in generative processes are open to inflection, interaction and interpretation in ways that composed (didactic or mimetic) forms are not. An isotropic space could be articulated and populated by 'parallel intelligences.' These entities would not be modelled on human intelligence not 'coded' but 'evolved.' An isotropic interface environment would re-invent collaborative structures by taking cues from relational structures in Artificial-Life and complex systems theories.

– *Strategies* –

The projects presented above are experiments towards this goal attempts to slip the bounds of the perspectival and construct a new, more open and productive metaphor, an isotropic interface to intention, subjectivity and authorship.

Sharon Daniel is an artist working in computer-based interactive art and interface design. Her Web project 'Narrative Contingencies', is an interactive, non-linear narrative, which allows participants to contribute texts and images. 'Signal-To-Noise', is a conversation within the context of the World Wide Web presenting a theoretical discourse linking images, texts and 'sites' through networks of association. The Decordova Museum Virtual Gallery and boston.com launched both in February 1997. Ms Daniel was the Visual Design Director, Video Artist and graphic designer for the 'Brain Opera', an interactive 'Opera' inspired by Marvin Minsky's Society of Mind, produced at MIT's Media Laboratory. The 'Brain Opera' premiered in 1996 in New York at the Lincoln Center Festival and through the Internet World Expo. It has since been presented in Linz, Austria at the Ars Electronica Festival in Copenhagen, Denmark, at the Electronic Cafe International and European Cultural Capital Celebrations, in Tokyo, Japan, in the NexOpera Festival (a Nexsite Project) at Ebisu Garden Place, and in West Palm Beach, Florida, at the Kravis Center for the Performing Arts.
Ms.Daniel's recent work has also been presented at 'The Kitchen', and the 'Women in the Director's Chair' International Film Festival. 'Alive, Dreams And Illusions', a hardware-free virtual environment for networked participants and autonomous agents was presented at SIGGRAPH 95. Her interactive installation, 'Strange Attraction: Non-Logical Phase-Lock Over Space-Like Intervals' was presented at MIT's Center for Advanced Visual Studies and published in Technology Review. Ms.Daniel is an Assistant Professor at the University of California, Santa Cruz. <sdaniel@cats.ucsc.edu> http://arts.ucsc.edu/sdaniel

Performing Presence

Barry Edwards

Last year I attended a symposium at Riverside Studios, London as part of the Digital Dancing programme (Digital Dancing 1997). Quite early on in the discussion attention focused on the work of the dancers taking part in an event that had been performed live at Riverside Studios and seen electronically at The Place. The matter of hardware, modems, cameras, cables and the like was handled relatively swiftly, and then the dancers themselves began to talk about their physical bodywork. They talked about the feeling of distance between them, about contact and non-contact, about the direction of movement in spirals and lines, and about the implications of spontaneity and pre-planning. I was surprised by the language used. I recognised many of the words. They were words that I also used to articulate my own performance-making process (Optik, 1991). I concluded that we were both talking about the same thing, namely the performing of presence. I would characterise the performing of presence as the engagement with physical live performance without pre-planned notions of significance, meaning or the desire to communicate. If you are not communicating, what are you doing? This is the major hurdle for the performer, who normally needs an objective. It seemed to me that in the dancers' case the 'objective' was the experiment with tele-presence, which in a strange kind of way 'liberated' their live work from the strain of significance and the constraint of meaning. In my own case I have been attempting to work in just this state of non-communicative presence but without the tele-transport of image. In fact one of the intriguing aspects of the work has been the absence of image-making altogether, tele or otherwise. The engagement with presence as a performing process is not a matter of assembling images for others to see. Marcel Jousse (Jousse 1925), arguing that movement is a basic building block, used the phrase 'images don't exist'. What is performing if the recollected or reconstructed image or significance of what you are doing is not used as the trigger for physical impulse? What is the instigator for movement in this pre-conscious state. Is it pre-conscious? More accurately, what is it we *are* conscious of? To which we can also add, what is watching under these conditions, what is being watched?

These kinds of questions have led me to attempt to define some structural elements of presence as process and its relationship to consciousness and to doing. Such elements as corporal perception, duration in time, fractal scale and distance, dynamogenesis and the origin of impulse, proprioception and the phenomenology of personal location.

Performing presence is a matter of understanding action and action potential. I describe it as a quantum bio-mechanics. What might be the fundamental constituents of this essentially dynamic process? I have developed one possible formula. In this formula the three fundamental constituents are position, velocity and consciousness. Position relates to space, velocity to time. Consciousness is linked to position and velocity via various layers of decision processes that constantly negotiate change in one or other or both of these elements. Together, position and velocity create space-time and decision-making in either constituent element becomes decision-making in the dimension of space-time. In the dynamic model opened up by the integration of consciousness, this space-time dimension becomes fluid, uncertain, unpredictable. These characteristics then apply to such normally fixed elements such as duration of time, location in space, distance, scale. The proprioception sense is enhanced in this process and seems able to create and/or deal with paradoxical states of somewhen/everywhen and somewhere/everywhere. This quantumised bio-mechanics also handles the phenomenal paradox of being simultaneously singular, a unique biological event, and universal, a part of a category of connectedness, such as species or larger.

Position

Position is where you are at any particular moment, 'what is under your foot'. In addition to our feet (two of them), human beings have two forward-looking eyes, an ear on each side of the head, in other words a bilateral body symmetry. From this are generated six basic co-ordinates of position: in front, behind, to the right, to the left, above and underneath. We generally understand these co-ordinates in one instant as a single point, which is 'being there', but in the performing work it is possible to isolate and work on each one.

We have a very wide eye focus but is has limits. Our forward-looking eyes are adept at detecting the slightest movement, particularly at the edges of our field of vision, and at focusing on a particular point directly in front of us. Where we cannot see any more is a critical boundary that separates seeing from not seeing. This line defines in front and behind, extends out from each shoulder to right and left. It is a key transitional line in our understanding of position. In front is visible, behind is invisible, but audible. The relation between hearing and seeing is critical, and the desire to turn and look back at what is behind us is always strong. It appears in several mythical narratives. You can never see what is directly behind you, since as you turn to look, the position of 'directly behind' moves with you. To be in front of another, front to front, has the dynamic of 'opening up', of starting, meeting. Turning your back on another is its opposite, closing down, ending, departure. The vertical line, up and down, gives us our height, the distance from the hair on our head to the soles of our feet. It can be explored by the vertical movement of the head, which allows us to look up and look down. The vertical line is also engaged by standing upright and this can be explored by dropping the pelvis so that the body goes to crawl and then to lying down and from there back to standing upright.

Velocity

Velocity is the rate of change of position. We know our position and also simultaneously know that it is not fixed but has constant potential for change. We engage in movement which takes our body through different positional points and at different speeds. Rhythm is one of the spin-offs of the dance between position and velocity. We walk, run, stand still, lie down. We can turn and roll.

In my own work I ask performers to use quanta of velocity states. This is so that they can concentrate on the change from one state to another and work with duration. It is a very simple structure that is based on a precise doubling of velocity to effect a change. If a performer engages his slowest walk, then the options facing that performer are to continue (at a constant velocity), stop walking and stand or to double the velocity of the move forward. And so on up the scale until you reach your fastest possible run. Using this structure it only takes four velocity states to go from slowest to fastest.

Decision

In my work each performer has an independent decision-making process. This is my way into the exploration of consciousness. Each performer can at any time change his position and/or their velocity. This is not consciously dependent on any other structure. It is a performer's way into working with uncertainty while being certain of what he is doing. When we are walking what decisions do we have? We can stop walking and stand still. We can turn and keep walking; we can not turn but continue walking on the line we are on at the moment. What makes us choose to stop or change direction by turning? What is stopping? What is turning? How do these changes occur?

The process is a binary or digital one. That is, all decisions are based on taking one from only two options. The options may change, but there will never be more than two. At every moment the performer is therefore juggling the now with the what next. They know what they are doing but not what they are going to do no pre-planning. And when the decision is engaged, it is an all or nothing process, no maybe or perhaps. Like the body muscles, you move or you don't move, one or the other; there is no in-between state.

Moving forward, turning, consequences

If you move in a line forward this will take you forever onwards, unless you turn or stop. A turn produces different positional consequences for the move forward. This combination of line forward and turn is not the same as spiral. A spiral movement is fixed by its centre and so does not actually move forward.

Turning is crucial. It is inbuilt, a kind of homing instinct. I can illustrate with an example from performer work. In this particular exercise, one performer is asked to stand behind another and give a physical impulse with his hand to the other by pushing slightly at the base of his partner's spine. The performer moves forward on this impulse and then eventually comes to a standing position. When first doing this work the moving performer will almost always turn at the stopping point of the movement and walk back to the other. Out on a limb in this way he feels uncomfortable, stranded, strange. The performer has to learn how to resist the impulse to turn and go back, how to remain facing forward, somewhere else, waiting for a new impulse to move forward again. It would seem as if the default decision, as it were, is to turn round 180 degrees and go back to where one started, even when that starting place has only been in existence for a few minutes. Turning is a key to stability. Turning keeps us more or less where we are. If we didn't turn then we wouldn't have communities, we would just keep walking. Turning turns space into territory. In performing, we can see that actors, dancers and others keep themselves on the limited space known as the stage by turning frequently and by restricting their forward movement in this way so that they are contained within the boundaries of the set stage space. If you get a group of performers in a space and give the simple instruction 'move' then they will move in spirals around the space, not in lines forward.

Does this link presence to a consciousness of containment and a resistance or acceptance of that? What if you decide to move forward until you meet an obstacle? What happens then? Is another person an Do you turn then? What happens when one person meets another? What are the dynamics of this meeting, this contact? Other people are disturbing, exciting. For whereas we can exercise some element of control (if we want to) on our position in relation to static objects, we cannot do this with people because they are capable of changing position, as we are. In these conditions I have found that presence becomes felt. Physical actions involving change of position such as moving and turning, resonate with feeling states such as being accepted or being rejected. When two or more people are in close proximity to one another then there is a strong desire to synchronise changes of position, a desire that can be followed or resisted. Moving together feels good. We want it to happen. When another allows this to happen, or when he resists this and move independently, we sense this emotionally.

When we change position, at whatever velocity, we are not alone. Changes in position initiate the possibility of contact with others. We are in a constant state of coming into and going out of contact. When we move, change position, do we do this in order to make this contact, or is contact simply a consequence of moving? How do we know the people we know? Is it because they keep re-appearing as we move and turn back the same way? Turning 180 degrees is to reverse the direction of the line you are on and so maintains something. Turning 90 degrees along the right or the left axis of the body is to make a major change. Moving is a fundamental human activity. Presence, it seems to me, is intimately connected to it.

Notes

1. Digital Dancing For further reference see Kozel S 'Reshaping space: focusing time' in *Dance Theatre Journal* Vol. 12, no. 2, Autumn 1995.
2. Jousse M 1990 *The Oral Style*. New York: Garland Publishing P. 27 (Translated from the original French: Jousse, M. 1925 *Le Style Oral Rythmique et Mnemotechnique chez les Verbo-Moteurs*. Paris:Archives de Philosophie Beauchesne.)
3. Optik, For further reference see Optik, Video Archive Optik Video Archive currently comprises 24 videos filmed and edited by Terence Tiernan covering the company's international and national performances from 1993. It is housed at Brunel University Library, Uxbridge, UK. For further information email library@brunel.ac.uk. Written references include: Edwards B *Optik's Performance Phenomenology.* Arts on the Edge Conference Proceedings, Petrie International, Edith Cowan University Perth,Western Australia, April 1998.

References

Allain P *Optik in Egypt: Towards a Fundamentalism in Performance*.
Brook, J. and Edwards B The Work of the OPTIK Performance Company in *Performance Practice Journal* UK. Winter 1997/8.
Edwards B *Watching What People Are* Optik London 1997
Edwards B *Working with Complexity* 4D Dynamics Conference Proceedings Ed Robertson A Design Research Society & De Montfort University Leicester 1996
Edwards B 'Observing the Unpredictable', in *Dance Theatre Journal* London Vol. 13, No 1 Summer 1996.
Dodds, S. & Ross, C. 'Optik Twice Over' *Dance Theatre Journal* London Vol 13 No 1 1996, London El Haid
Edwards B 'Optik at Tacheles'. *Total Theatre* London Vol. 7 No 2, Summer 1995.
Keefe J 'Review of Optik' *Total Theatre* London Vol. 7 No 2, Summer 1995.
Freeman J 'Performing Performance: the new Authenticity of Optik' in *Total Theatre* London Vol. 7 No 1, Spring 1995.
Edwards, B. 'Visit to Egypt' in *Total Theatre* Vol. 6, no 2, Summer 1994.

Barry Edwards is creative director of the company of artists known as Optik. He researches and teaches contemporary performance studies and is Reader in Drama in the Department of Performing Arts at Brunel University, West London. Barry.Edwards@brunel.ac.uk http://www.brunel.ac.uk/research/perform

7 Projects

Éphémère:
Landscape, Earth, Body, and Time in Immersive Virtual Space
Char Davies

The Ephemerality of Being

> But just because to be here means so much,
> And everything here all this that's disappearing
> Seems to need us, to concern us in some strange way
> We who disappear even faster!
> It's one time for each thing and only one.
> Once and no more. And the same for us. Once.
> Then never again.
> But this once having been,
> Even though only once having been on earth
> Seems as though it can't be undone.
>
> ... Earth, isn't this what you want:
> Rising up inside us invisibly once more?
> Isn't it your dream to be invisible someday?
> Earth! Invisible!
> What is it you urgently ask for
> If not transformation?

Rainer Maria Rilke [1]

Éphémère is an interactive, immersive audio/visual virtual environment which furthers the work begun in Osmose (1995).

In Éphémère there are two intertwined themes. One is the ephemerality of being, in terms of our fragile fleeting life spans as mortal beings embedded in a living, flowing world, among an unfathomable myriad of comings-into-being, lingerings and passings-away. The work's second theme is the symbolic correspondence between body and earth. Earth as regenerative source, organic destiny, mythological ground. Within the work are recurring 'archetypal' elements suggesting a co-equivalency between the chthonic presences of the interior organic body and the subterranean earth, whose meanings and behaviours are dependent on the behaviour of the participant and spatial/temporal context.

Éphémère is structured as a temporal progression, in terms of emergence and withdrawal of form: flow and ebb of visibility and audibility: and diurnal/nocturnal and seasonal transformation, as well as germination and decay. While the ephemeral is most usually associated with momentary manifestations such as mayflies, from a mountain's point of view, our own lives are as fleeting.

Nature as Ground

The iconography of Éphémère is grounded in 'Nature' as metaphor, as is all my work of the past 15 years, as a means of reaffirming our biological and psychological dependency on Nature in the face of its ongoing devaluation and destruction.

No matter how far culture will go to destroy its connections to nature, humankind and all of our technology, good and bad, are inextricable parts of nature the original determinant, the mother and matrix of everything, that all pervasive structure that lies beneath scenery, landscape, place and human history. [2]

Éphémère is fuelled by my experience of a particular place, a remote piece of land, part rural, part wild, in southern Quebec. While this land is not pristine, having been logged, cleared, ploughed, mined and grazed and now producing apples, all the earthly elements in Éphémère, as in Osmose, have their source here. Over the time I have spent on this land, its roots and rocks, seeds and streams, bloomings and witherings, have become numinous, as present in my imagination as in actuality. Wandering among their physical manifestations provides me with a much needed antidote to working with virtual-reality technology. They in turn appear in my work like apparitions.

Life Flow

As I began this paper, the nearby stream roared and flooded with the spring melt of a mountain's snow. Weeks later, rocks warmed in the sun and apple trees were in expectant bud. Soon after this paper goes to press, the stream will have slowed to a trickle, the forests will have leafed and faded and the apples will be ripening. Even on the most tranquil of days, a powerful force pours through here, through every element and creature.

This river of life and time, the inexorable force that pours through all things, is what concerns me. As Dylan Thomas wrote:

> The force that through the green fuse drives the flower, Drives my green age; that blasts the roots of trees Is my destroyer. And I am dumb to tell the crooked rose
> My youth is bent with by the same wintry fever. [3]

According to Heidegger, the Greeks called this flow 'physis':

> In truth, physis means, outside of all specific connotations of mountains, sea or animals, the pure blooming in the power of which all that appears and thus 'is'. [4]

The very immateriality, temporality and apparent three-dimensionality of immersive virtual space is well suited for manifesting such a concept. In Éphémère, besides the various comings-into-being, lingerings and passings-away and the transformations of illumination and spatial contexts, there are 'flows' of rivers, root flows and body fluids streaming through the work.

Poetic Body

Éphémère, like Osmose, utilises an embodying user interface in the form of a vest that tracks the participant's breath and balance, enabling him to move through the work by breathing. A head mount is used for real-time display of stereoscopic 3D computer graphics and 3D localised sound. This device is also used to evoke a sense of spatial envelopment. This strategy serves to implicate the immersant within the space and grounds the work in interior processes of the physical body. (5)

In Éphémère, I have incorporated visual and aural elements that recall the mortal fleshy body of organs, blood and bone (referring not only to human but all bodies). As the mythologist Joseph Campbell has written:

> Myths and dreams . . . are motivated from a single source namely the
> human imagination moved by the conflicting urgencies of the organs,
> (including the brain) of the human body, of which the anatomy has
> remained pretty much the same since c. 40,000 BC. (6)

Poetic elements of the organic body function as the substrata of Éphémère, under the fecund earth and the lush bloomings and witherings of the land. The symbolic correspondence or equivalence between body and earth is key to the work. Rocks transform into organs; rivers transform into veins, and vice versa. While I cannot completely articulate the rationale behind such metaphors, they have been present in my work for many years.

Some people in the burgeoning cyberculture imagine that one day we, as a species, will escape the confines of mortal bodies by merging ourselves with silicon. In this context, Éphémère can be viewed as an attempt to reaffirm our limitations, our mortality, our dependency on ageing bodies and an earth which will, for those of us now living, absorb our bones, dreams of cyber immortality notwithstanding.

Spatiotemporal Structure of Éphémère

As mentioned earlier, Éphémère is structured temporally as well as spatially and thus contains a progression of visual and aural events through realms of landscape, earth and body.

	winter >	spring >	summer >	autumn	
Landscape:	dormant >	blooming >	leafing >	falling leaves >	dust
Earth:		germinating >	fruition >	decay >	
Body:		eggs >		bones >	
	r i v e r				

If an immersant stays within the landscape for an entire session it will change around him, passing through cycles of day and night, from the pale grey of winter, through spring and summer to the climax and decay of autumn. The elements in the other realms transform as well: in the earth seeds become active then fade; in the body, eggs appear and ageing organs give way to bone. While the participant may spend an entire session in one realm, it is more likely that he will pass constantly between them, immersed in transformation.

While the immersant is able to move vertically through the landscape, earth and body by breathing in or out to rise or to fall, another means is possible: the river. The river is a constant element flowing throughout the work. When gazed upon or followed for any length of time, it transforms into an underground stream or artery/vein (and vice versa) bringing in their appropriate visual/aural surroundings. Even as the participant roams among all three realms, by rising/falling or via the river's transformations, no realm remains the same, changing through time, ending in dissolution.

All the transformations which I have described are aural as well as visual. While the visuals pass through subtle changes of visibility and non-visibility, light and shadow and in he case of landscape, progress from the relatively literal to the abstract the sound is also in a state of flux. Interactive and localised in three-dimensions, it flows between melodic form and mimetic effect in a state somewhere between structure and chaos, adapting moment by moment to the changing spatiotemporal context and the immersant's behaviour.

In addition to the various transformations described above, there is another kind of transformation in which the perceptual faculties and imagination of the immersant are deeply implicated. The visual aesthetic of Éphémère, like Osmose, is based on extensive overlaying of semi-transparent three-dimensional forms, creating a constant variability of the perceptual field, causing semiotic and sensory fluctuations which are channelled within the larger meaning of the work.

Finally, in Éphémère, there is a subtle inter-responsivity between selected elements and the immersed participant. These iconographic elements include germinating seeds, rocks and the river as already described. Their behaviours depend on the immersant, responding to proximity, slow movement and/or gaze.

While the creative process associated with Osmose resembled the constructing of a perceptually-mesmerising, immaterial stage set, the making of Éphémère has been exponentially more complex, both conceptually and technically. The process has resembled the creation of a virtual opera, consisting of the development of a myriad of visual and aural elements, whose various comings and goings must be calculated in relation to each other, the progression of the work and the immersed participant, in real-time.

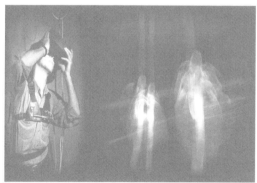

Seed, from Éphémère

The Net of Life and Time

As I write, dusk has fallen and my ears are

filled with the dizzying, deafening sound of shrieking frogs and crickets, creating a sensory vertigo, similar to the state of reverie which I and others have experienced in Osmose. (7)

This land is the muse behind Éphémère at this moment, a velvet envelope of mountain evening, silent rocks, flowing water, insistently budding flora and shadowy fauna stealthily engaged in the business of their own lives. Here I am immersed in an unfathomably complex, inexorable flow which pours through a myriad of channels, whose embodied forms are, as Henri Beston wrote in 1928:

> Brethren . . . not underlings; they are other nations caught with ourselves in the net of life and time, fellow prisoners of the splendour and travail of the Earth. (8)

Éphémère is an attempt to express all this.

These days however due to a litany of consequences of human attitudes and actions, the rich biodiversity of Nature is rapidly diminishing. On this particular piece of land, fewer songbirds return in spring to nest, frogs and salamanders have less young and the maple trees are dying. In some ways, Éphémère is a lament, not only for the ephemerality of our own lives, but for the passing of Nature as we have known it.

Conclusion

The construction of virtual landscapes in an age of environmental crisis is fraught with implications. All digital constructions of 'virtual reality' and cyberspace are ideologically-laden and most serve to reinforce the cultural value system which Henri Lefebvre has labelled as the 'reign of King Logos' (8). Such constructions may also serve to distract from earthly responsibilities and the very wonder of being embodied among all this, here now. However, alternative approaches to the technology are possible. As Marshall McLuhan wrote, such usage is the responsibility of artists:

> The function of the artist in correcting the unconscious bias of a given culture can be betrayed if he merely repeats the bias of a culture without readjusting it. In this sense the role of art is to create the means of perception by creating counter-environments that open the door of perception to people otherwise numbed in a non-perceivable situation. . . . In an age of accelerated change, the need to perceive the environment becomes urgent. New environments reset our sensory thresholds. These in turn later affect our outlook and expectations. (9)

While our habitual perceptions may lead to the forgetting of being, the paradoxical qualities of immersion in a virtual environment if constructed so as not to reinforce conventional assumptions and behaviour can be used to open doors of perception. In this context, Éphémère is an attempt to reaffirm our poetic and mythic need for Nature, returning attention to our fragile and fleeting existences as mortal beings embedded in a vast, multi-channelled flow of life through time.

Today the apple trees are blossoming: tomorrow the blossoms will fall back to earth.

References

1. Rilke, Maria Rainer. 1922. 'The Ninth Elegy', In *The Duino Elegies* (transl: David Young) New York: Norton and Co, 1978.
2. Lippard, Lucy. 1997. *The Lure of the Local: Senses of Place in a Multicentered Society*. New York: The New Press, p 11.
3. Thomas, Dylan. 1937. 'The Force that through the Green Fuse', In *Collected Poems 1934-1953*. London: J.M. Dent, p 13.
4. Haar, Michael. 1993. *The Song of the Earth: Heidegger and the Grounds of the History of Being*. Bloomington, Indiana: Indiana University Press, p 8.
5. Davies, Char, 1995. 'Osmose: Notes on Being in Immersive Virtual Space', In *Digital Creativity*, Vol. 9, No. 2, 1998. London: Swets and Zeitlinger. ISSN 0957-9133. First published in the Sixth International Symposium on Electronic Arts Conference Proceedings, Montreal: ISEA'95.
6. Campbell, Joseph. 1988. *The Inner Reaches of Outer Space*. New York: Harper and Row, p 12.
7. Davies, Char, 1997. 'Changing Space: VR as an Arena of Being'. In Ascott, R. (ed.) 1997. Consciousness Reframed: Art and Consciousness in the Post-biological Era. Proceedings of the First International CAiiA Research Conference. Newport: University of Wales College. ISBN 1 899274 03 0. Also (in expanded form) in Beckman, John (ed). 1998. The Virtual Dimension: Architecture, Representation and Crash Culture, Boston: Princeton Architectural Press.
8. Plumwood, V. 1993. *Feminism and the Mastery of Nature*. New York: Routledge.
9) Lefebvre, Henri. 1991. *The Production of Space*, Oxford: Blackwell, p 407.
10. McLuhan, Marshall, and Parker, Harley (1969). Through the Vanishing Point: Space in Poetry and Painting. New York: Harper & Row, 241, 252.

Éphémère was constructed with the assistance of John Harrison, virtual reality software; Georges Mauro, computer graphics; Dorota Blazsczak, sonic architectura/programming; and Rick Bidlack, sound composition/programming. Éphémère was co-produced by Char Davies and Softimage Inc. Éphémère is on exhibition at the National Gallery of Canada from 26 June to 6 September, 1998.

Char Davies is an artist and a PHD candidate in the Philosophy of Media Arts, CAiiA: Centre for Advanced Inquiry in the Interactive Arts, University of Wales College, Newport, Wales. From 1988 to 1997 she was on the Board of Directors and Vice-President/Director of Visual Research at Softimage Inc. (Montreal). Her previous work includes the immersive virtual environment Osmose (1995). Earlier this year, she left Softimage and founded Immersence Inc. She is the custodian of four thousand apple trees.

Virtual Environment as Rebus
Margaret Dolinsky

Art will become innovative in the medium of virtual environments when the spectator abandons the act of mere viewing, transcends simple narrative participation and bursts into the scene as an active creator. 'Strait Dope' is a CAVE virtual art environment that offers an opportunity to participate in a room-size projective construction. The nonlinear nonhierarchical structure of 'Strait Dope' stages a stream of consciousness movement that is simultaneously subversive and confrontational. This structure is not to create chaos but, rather, to present a rebus for symbolic transformation and multisensory interaction. This paper will attempt to describe the virtual environment as a dimensional rebus and the participant as a perceptual collaborator who engages the computer to discover the interplay of the symbolic transformations.

Virtual environments present a visual spatial media of shapes, landscapes and sounds that establish a system for construction and symbolic transformation. The virtual environment as projective construction provides an opportunity for participants to collaborate in a variety of multisensory interactions: visual-spatial, audio-spatial and kinesthetic.

Collaboration here refers to the constructive collaboration occurring between the participant and the computer to create and update the environment by directing and redirecting the interaction. The participant is able to transform his relationship with the world by manipulating the virtual space he is inhabiting. The ability to activate, manipulate, and deconstruct a virtual environment is a method of personal liberation and a type of metaphoric liberation. The computer updates the display according to the actions of the participant within the system. The participants exercise the guidance of their cognitive structures to ascertain the meaning and content of the virtual experience.

In *Computers as Theater* Laurel (1993) writes:

> As an activity becomes less artifactual (like painting or literature) and more ephemeral (like conversation or dancing), sensory immediacy and the prosody of experience gain primacy over structural elegance in the real-time stream of events. In shared virtual worlds, structural elegance becomes much less about the progression of events and more about facilitating the emergence of patterns and relationships. Shared presence and sensory immersion evoke constructive activities on the part of human participants to a much greater degree and in more structurally interesting ways than I had imagined. Given a multisensory environment that is good enough, people engage in projective construction that is wildly elaborate and creative. And so this turns the problem on its head; rather than figuring out how to provide structure with pleasing emotional textures, the problem becomes one of creating an environment that evokes robust projective construction.

The virtual environment becomes a rebus for the participant. Rebus, literally '*with thing,*' comes from the Latin word '*res*' meaning '*thing*'. A rebus is a selection set or construct of 'things', such as ideas or images, that convey meaning. For example, Egyptian hieroglyphics is a collection of illustrations that represents ideas and things. It is a selection set that presents a symbol system to convey language conceptually. The virtual environment represents a dimensional rebus in that the artist creates a construct that links sound and sense for expression. The participant completes the act of creation by giving sound and sense meaning.

Characters demand recognition in CAVE

'Straid Dope' CAVE Characters

Virtual environment artists must create beyond pictorial compositions and construct a metaphoric liberation for the participant. Multisensory interaction that is tightly linked with symbolic transformations motivates navigation, learning and discovery. Creative constructions with provocative symbolic transformations inspires the participant to establish an identity within the space.

'Strait Dope' offers a virtual incarnation of variant characters who demand recognition through overt facial expressions and unique creative dispositions. Their language is invoked by poking or prodding their faces with a virtual wand. Then their language erupts abruptly with a bluster of atonal sounds, gibberish or frightening demands.

The participant realises a role in the completion of the scene by activating the elements of the scene. Movement is the very essence of 'Strait Dope' where the question of starting points and linear directions becomes irrelevant. The participant is constantly moving, and therefore, finds it compelling to impart movement.

Each character emits a dynamic force of varying density with alternative structures. Some have arms that flail and eyes that pop out or they launch sounds or objects with an extraordinary tension. The rebus comes in many enigmatic forms that offer a challenge to the most sophisticated arbiter of sound and sense. Rather than presenting concepts in a simple, straightforward manner, the rebus exploits meaning by presenting pictorial elements with transformations, extrapolations and omissions. In the CAVE, participants navigate the environment to establish a sense of perspective, both spatially and metaphorically.

The characters in 'Strait Dope' represent physical spatial forms rather than objective spatial forms. Their shapes build a graduated tension in the scene. Their presence remains relative one to the other and simultaneously their boundaries merge one into another. There is a natural energy to becoming involved with the visual, spatial and tonal forms fuelled by the background audio.

The energetic tempo of the music and bright colours in the landscape seduce the craftsperson/creator within to involvement. Participants will often repeatedly reincarnate the characters in an act of experiencing a creative totality: pre-creation, creation and post-creation. The scrutiny of the participant guide and creator is significant to giving the space form. This process of giving form to the

 # Creative constructions demand recognition

Perceptual collaboration establishes perspectives – participants create personal perspectives

environment is more important than the form itself.

The interpretation of an enigmatic rebus recalls Freud's method of interpreting dreams. Freud extrapolates the elements of a dream beyond mere pictorial composition. He reveals the more subtle and sophisticated method of interpretation by emphasizing the symbolic connections between dream elements. In his treatise on "looking," Zizek (1995) quotes a reference to the dream and the rebus in the *Interpretation of Dreams* by Freud:

> The dream-thoughts are immediately comprehensible as soon as we have learnt them. The dream-content, on the other hand, is expressed as it were in a pictograph script, the characters of which have to be transposed individually into the language of the dream-thoughts.
>
> If we attempted to read these characters according to their pictorial value instead of according to their symbolic relation, we should clearly be led into error . .
>
> But obviously we can only form a proper judgement of the rebus if we put aside criticisms . . . of the whole composition and its parts and if, instead, we try to replace each separate element in some way or other. The words which are put together in this way are no longer nonsensical but may form a poetical phrase of the greatest beauty and significance.

A rebus virtual environment is composed of three-dimensional objects that have no meaning until they are engaged by the participant. The participant's actions manipulate the environment and update the CAVE display. The changing shapes and forms entice the participant to establish a dialogue with the CAVE and become integral to the rebus formation and completion.

Strait Dope is a projective construction that requires the participant to complete the virtual environment with the realisation of symbolic transformation and metaphoric liberation. The artist creates an environment that offers multisensory interactions to the participant. In an effort towards metaphoric liberation, the participant must act as perceptual collaborator with the CAVE to facilitate the emergence of patterns and relationships. 'Strait Dope' presents a nonlinear nonhierarchical rebus display style to Multisensory interactions in the CAVE simulate a stream of consciousness movement and engage the participant in a personal dialogue integral to the completion of the virtual environment.

References

Baker, B.R. 1994. *Semantic Compaction: An Approach to a Formal Definition Proceedings of 1994 European Minspeak Conference.* http://kaddath.mt.cs.cmu.edu/scs/94e1-bb.html

Dolinsky, M. 1997. *Strait Dope*. http://www.evl.uic.edu/dolinsky/sdpe/
Dolinsky, M. 1996. *Dream Grrrls*. http://www.evl.uic.edu/dolinsky/DG/
Elkins, J. 1996. *The Object Stares Back*. New York: Harcourt Brace.
History of Puzzles. http://www.cyberpuzzles.aust.com/taster/puz_hist4.html/
Hudu, Brillig, Treesong and Sibyl. 1998. *Types of Flats*(Q-R). http://www.puzzlers.org/guide/flatsQ-R.html#Rebus/
Klee, P. 1964. Notebooks Volume 1. *The Thinking Eye*. London: Lund Humphries.
Laurel, B. 1993. *Computers as Theater*. Reading: Addison-Wesley Publishing Company.
Slobodkin, L. 1992. *Simplicity & Complexity in Games of the Intellect*. Cambridge: Harvard.
Treesong. 1998. *Solving and Composing the Rebus and Rebade*. http://www.puzzlers.org/guide/rebus.html
Zizek, S. 1995. *Looking Awry*. Cambridge: MIT Press.

Margaret Dolinsky creates virtual environment art for the CAVE™ Automated Virtual Environment. Dolinsky's work has been featured at VRAIS'98, Ars Electronica Center, the Total Museum, 'Virtual Spaces' in conjunction with ISEA97, ThinkQuest97, ThinkQuest96, SIGGRAPH96. Dolinsky has spoken at the Museum of Contemporary Arts in Chicago, SIGGRAPH97, Centre for Advanced Inquiry in the Interactive Arts, College of Newport, Wales. Dolinsky's work is published in the VR Developer's Journal, ACM Computer Graphics, ACM Interactions, IEEE Multimedia, and Digital Magic. Dolinsky investigates artistic metaphors and how they can establish guidance and content in technologically based media. Her designs focus not only her metaphors established as a painter but also on participants. Dolinsky is currently a research assistant and Master of Fine Arts candidate at the University of Illinois at Chicago's Electronic Visualization Laboratory. For more information and images on Dolinsky's work, see http://www.evl.uic.edu/dolinsky/ or e-mail dolinsky@evl.uic.edu

Dynamic Behavioural Spaces with Single and Multi-User Group Interaction

Miroslaw Rogala

Throughout the 20th century, as technologies have transformed our experience, artistic practice has been re-defined and re-focused (Ihde, 1990; Heartney 1997). The artist has moved from looking at the surface of things to examining underlying processes, relationships and systems, and has amplified the processes of engagement and involvement of the viewer. Burgess (1994) argues that 'culture making is essential to human survival and definition' and that the arts are crucial to a 'global' understanding of the modern world. Ascott (1997) points to the cultural significance of new art practices employing intelligent structures, artificial life and the principle of emergence. 'Traditional ways of looking at the world have become obsolete; technology is the new language through which experience is understood. What is at stake is our very being the ways we perceive ourselves and others, and the variety of experience that is available to a population whose sensibilities are mediated by urban exposure' (Olalquiaga 1992).

There is a need to rethink the city and urban space. Whyte (1988) laments that 'it is difficult to design a [public's] space that will not attract people. What is remarkable is how often this has been accomplished'. Sorkin (1992) claims that 'recent years have seen the emergence of a wholly new kind of city, a city without a place attached to it . . . The new city also occupies a vast, unseen, conceptual space. What's missing in this city is not a matter of any particular building or place; it's the spaces in between, the connections that make sense of forms'. Can new spaces be located where large groups of audiences can find new interactive experiences? What are the new spaces that the city can offer for social interactions? The art experience is needed in such places to provide context beyond shopping malls, amusement and theme parks and edge cities (Garreau 1991).

There is a difference between a single viewer interacting in public space versus multiple viewers interacting with the artwork and interacting between themselves. My concern is with the latter and with establishing a vocabulary of multi-user behavioural space.

The author's practice and inquiries, since the mid-1970s, have resulted in the creation of publicly accessible artworks designed to explore the artistic potential of interactive systems. Emphasis is on the single and multi-user group interaction and an evolving multi-faceted interface design that allows engagement and participation of the audience to control the content and data changes.

Interactive Environment: *Lovers Leap*

Lovers Leap (Rogala 1995) originated simultaneously as an interactive multimedia environment and CD-ROM. The work is a public art work and public space which is documented as a subject for theoretical research and analysis Kluszczynski 1998, Morse 1997, Shanken 1996, White 1996, Druckery 1995, and Cubitt 1995.

This project incorporated multi-forms (interactive installation and CD-ROM), different media, and development of multi dimensional aspects. Large-scale involvement occurred, which incorporated interaction with two large screens located at opposite sides of the space.

As the artist explains, 'movement through perspective is a mental construct; one that mirrors other jumps and disjunctive associations within the thought process' (Rogala, 1995). Morse (1997) observes 'In this installation, both control and loss of control are given a figurative and experiential shape that is linked to the performance of the visitor, that is, his or her position and speed or movement style over a Cartesian grid of locations . . . 'Chicago' and 'Jamaica' correspond less to geographic localities than to states of mind'.

The artist addresses the concepts of power and control in interaction. 'Control strategies assume either *dominant* or *submissive* roles. Power and strength depends on where we position ourselves within our environment. As the viewer's awareness of the control mechanisms grow, so does the viewer's power. Each viewer will create a new and different work depending on their involvement, understanding, and *transformation into a position of power. Many will leave without claiming their power.*' (Rogala 1995). Expanding on this interaction, the installation is cited as offering "a framework for reflecting on position and power in a way that . . . physical and virtual realms are intertwined in an interactive, immersive large environment' (Morse 1997).

In discussing related issues of the work, Druckery (1995) argues that 'technologies of new media map a geography of cognition, of reception, and of communication emerging in territories whose hold on matter is ephemeral, whose position in space is tenuous, and whose presence is measured in acts of participation rather than coincidences of location'.

In *Lovers Leap*, innovative interfaces, adaptation of new technologies, varieties of engagement and interactions and placement/use in public space provide important answers to the processes involved in framing an analytical system for dynamic behaviourial space.

Interactive Public Art: Electronic Garden/ NatuRealisation

Electronic Garden/NatuRealization (Rogala 1996) is an outdoor large-scale, public interactive, free speech installation originally created for 'Re-Inventing the Garden City', sponsored by Sculpture Chicago. An on-line World Wide Web counterpart of the project was constructed (http://www.mcs.net/~rogala/eGarden) and includes QuickTime VR visualization and simulation. The background of this project involved expanded roles of communities working with the artist in all stages, including preplanning, research, site research, history of site, video and voice recordings, neighbourhood involvement, concurrent activities and evaluations.

The artist's statement points to the installation 'recreating the sense of placement experienced in culturally diverse environments. Characteristic of the site as a free speech environment, the reinvention occurs through technological intervention, image processing. . . and multimedia activity'. (Rogala 1996).

Writings on the project Albin 1997, Kluszczynski 1998 have emphasised the relevance of 'humanizing' through technology. Alben (1997) refers to 'human experience not technology- as the essence of interactive design' and provides a vignette of the project 'that portray(s) the elements of vision, a sense of discovery, common sense, truth, passion, and heart'.

Electronic Garden/NatuRealisation has outlined the direction of interaction within a large-scale public environment. Through the use of practice, research and evaluation, factors began to define components contained in complex and dynamic systems, which enable large groups of participants in a sustained interactive relationship with the artworks produced.

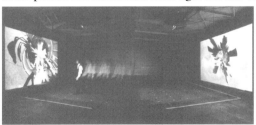

Miroslaw Rogala, Lovers Leap, 1995; Installation, Multimediale 5 ZKM Centre for Art and Media, Karlsruhe, Germany

Interactive Multi-Site Laboratory: Divided We Speak

Divided We Speak, Rogala 1997b is a Multimedia Laboratory Workshop for 'Divided We Stand', an Audience Interactive Media Symphony in Six Movements.

This artist's project is seen as a complementary expansion into an indoor site specific, large-scale interactive multimedia installation (Warren, 1997 Shanken 1997). It has been designed to engage a large group of participants in multi-user and group interaction (*MUGI*), through further explorations of multi dimensional mapping.

Divided We Speak/ 'Divided We Stand' is an artist's statement emphasizing the ways of uniting people through new, interactive technologies in physical and virtual spaces. Shanken (1997) perceives the project as 'a virtual, artistic enactment of the contingent relationship between freedom and limitation'.

Miroslaw Rogala, Electronic Garden NatuRealisation, 1996; Site-specific Interactive Public Art Installation Washington Square Park, Chicago, USA.

Participants utilise multiple interactive tracking devices and custom-designed software/hardware interfaces to become actively engaged in the process of creating and experiencing the work.

Observable behavioural changes occur in interaction with complex multimedia artworks, differentially according to expectation, experience and cues provided in the art environment.

Audiences tend traditionally to be spectators. Bronson (1996) gives a historical perspective on new audience roles: "The real" audience found themselves in the difficult position of having their role taken away, a role they had never before considered as even being a role. Highly disconcerted . . . they hit upon a more creative solution to defining their own roles. With the added dimension(s) and role(s) of participation, audience becomes user, requiring new roles and methods of interaction in public space.

Website installation (a cooperative venture between the artist and collaborating individuals) was initiated and continues to be implemented and expanded. As a finale, Live Internet hook-up (i-Vision) and demonstration occurred with participants in 15 countries (Iverson and Rogala , 1997).

Connecting Two Physical Spaces Through Interaction

In this project, 'triggering' became the method for proximity relationships through multiple wand positions (gallery one) and human gestures in n-dimensions (gallery two.)

Object and Movement as Interface

Spatial grammar of experience and behaviour has to be defined in new contexts as there are problems involved with interpreting movement and gesture of multiple participants in a single shared physical space. Life tends to be based on horizontal movement: how this is mapped in the space became the mode of interpretation. Current usage of interactive spaces ties the body to a stationary position; in this case study, both body and hand movements and gestures became dynamic,thus creating new artforms.

Unanimous Endorsement for Interactive Participation

The response to interactivity has been enthusiastic with all age groups: younger participants react more to the technology, older participants react more to the experience. Variations stemmed from the amount of control desired by each interacting user. In taped interviews, people indicate pleasure with the effect from body gestures being the 'triggering' device. (Rogala 1997).

Overlapping of Dimensional Relationships

A significant contribution to my line of research was the use and adaptation of multiple wands and mapping the three-dimensional space for multiple users. Testing required physical programming and necessitated immediate feedback in decision-making. To my knowledge, such models do not exist. On the one hand, conceptualising was done in virtual digital domain; on the other hand, feedback directions to further develop the ideas of behavioural space became apparent during twelve months of programming, testing, interacting and implementing. The two relations (reality check, feedback and computer memory) are inseparable (Shneiderman 1998).

Resolution of the experience achieved through high-resolution mapping of MIDI notes, low-resolution, low-density sampled sounds (which in this case study were broken into words, musical and non-physical phrases, sentences) and media elements. The new experience has to be achieved through juxtaposition of and combination with both the low-density and high-resolution mapped components.

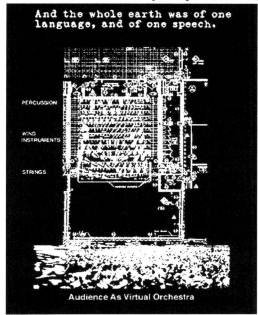

Miroslaw Rogala, Divided We Stand, 1997, Artist's Sketch

Conclusion

Interaction in new spaces requires new behaviours. The complexity of human interaction and behaviour defines constraints in undefined territories. It is not only the hand that can move in multi dimensions: the human body can jump, walk, accelerate motion within the space and cause effects. Dimensional relationship within spaces, accepting or altering constraints of behaviour and interactive environments are key elements in dynamic use of public spaces and multi-locations. Multi-user group interaction artworks are a new genre of artistic practice.

The discovery of complex processes and

interaction in multi-user behavioural space(s) is at the heart of this undiscovered, to-be-evolved field.

References

Alben, L. 1997. 'At The Heart of Interaction Designs' in *Design Management Journal*: Summer 1997.

Ascott, R. 1998. 'The Consciousness of Artificial Life'. Paper presented at ARTOB III, A-Life and Robotics Conference: Oita, Japan.

Bronson, A.A. 1996. quoted by Robert Nickas in A Brief History of the Audience: 1960-1981, *Performance Anxiety*. Chicago: Museum of Contemporary Art.

Burgess, L. 1989. *Global Frontiers for Art*. Pittsburgh: Carnegie Mellon University (unpublished work-in-progress).

Cubitt, S. 1995. 'Sound: The Distances'. Modernist Utopias Conference. Montreal: Musee d'Art Contemporain Montreal.

Druckrey, T. 1995. Lovers Leap: Taking The Plunge: Points Of Entry... Points of Departure, In *Artintact 2, CD-ROM Artist Interactive Magazine*. Karlsruhe, Germany: Zentrum Fur Kunst und Medientechnologie (The Centre For Art and Media) and Frankfurt, Germany: Canz Verlag.

Garreau, J. 1991. *Edge City: Life on the New Frontier*. New York: Anchor Books, Doubleday.

Heartney, E. 1997. *Critical Condition: American Culture at the Crossroads*. Cambridge: Cambridge University Press.

Ihde, D. 1990. *Technology and the Lifeworld: From Garden to Earth*. Bloomington: Indiana University Press.

Iverson, B., Rogala, M., Rutan, M. et al. 1997. 'Mapping Virtual Space: An Experimental Performance Workshop'. Programme Brochure, 'First Friday' presentation. Chicago, Illinois : Museum of Contemporary Art.

Kluszczynski, R. 1998. 'Dynamiczne przestrzenie doswiadczen', O tworczosci Miroslawa Rogali in: *Obrazy na Wolnosci: Studia z historii sztuk medialnych w Polsce* (Images in Freedom: Studies from the History of Media Arts in Poland). Warszawa: Instytut Kultury.

Lovejoy, M. 1997. *Postmodern Currents: Art and Artists in the Age of Electronic Media*. 2nd ed. Upper Saddle River, New Jersey: Prentice Hall.

Morse, M. 1997. 'Miroslaw Rogala': Lovers Leap, In *Hardware Software Artware (Confluence of Art and Technology, Art Practice at the ZKM Institute for Visual Media 1992-1997)*. Frankfurt, Germany: Canz Verlag.

Olalquiaga, C. 1992. *Megalopolis: Contemporary Cultural Sensibilities*. Minneapolis:University of Minnesota.

Rogala, M. 1997a. Dynamic Spaces: Interactive Art in Large-Scale Public Environments. In *Abstracts of the Proceedings of the First International CAiiA Research Conference*: University of Wales College, Newport.

Rogala, M. 1997b. Divided We Speak, An Interactive Multimedia Laboratory of Miroslaw Rogala's Divided We Stand (An audience interactive media symphony in six movements). Programme brochure. Chicago, Illinois : Museum of Contemporary Art.

Rogala, M. and Boyer, S. W. 1997. 'Building a Vocabulary for Multi-User Interaction in a 3-D Environment'. Workshop proposal for ISEA 97 Conference.

Rogala, M. 1996. 'Artist's Statement', in *Electronic Garden NatuRealization*. Unpublished.

Rogala, M. 1995. 'Artist's Statement', in *Lovers Leap*. Unpublished.

Shanken, E. A. 1996. 'Virtual Perspective and the Artistic Vision: A Genealogy of Technology, Perception and Power'. Lecture presentation. Rotterdam: The International Society for Electronic Art (ISEA) Conference.

Shanken, E. A. 1997. 'Divided We Stand: Interactive Art and the Limits of Freedom'. Website essay: www.mcachicago.org

Sorkin, M (ed.). 1992. *Variations on a Theme Park: The New American City and the End of Public Space*. New York: Hill and Wang.

Shneiderman, B. 1998. *Designing The User Interface: Strategies for Effective Human-Computer Interaction.* 3rd ed. Reading, Massachusetts: Addison Wesley.
Warren, L. 1997. 'Divided We Speak'. Exhibition programme booklet. Chicago, Illinois: Museum of Contemporary Art.
White, C. 1996. 'When Two Worlds Collide: Rogala's Lovers Leap', Digital Video On-line. (http//livedv.com/Chicago/oversleep/Lovers.html)
Whyte, W. 1988. *City: Rediscovering the Center.* New York: Doubleday.

The author wishes to acknowledge the contributions of Roy Ascott, Will Bauer, Joel Botfeld, Steve Boyer, Sean Cubitt, John Cullinan, Robert Fisher, John Friedman, Barbara Iverson, Margot Lovejoy, Kieran Lyons, Darrell Moore, Michael Punt and Mac Rutan.

Miroslaw Rogala is an interactive media artist based in Chicago, Illinois, USA and a member of the Art Department faculty at Carnegie Mellon University in Pittsburgh, Pennsylvania. He is the recipient of MFA degrees from the School of Fine Art in Krakow, Poland (1979) and the School of the Art Institute of Chicago, Illinois (1983). He is a doctoral candidate in the on-line Ph. D. Research programme at the Centre for Advanced Inquiry in Interactive Arts (CAiiA) at the University of Wales College in Newport. Communications with the author can be addressed to: rogala@mcs.com and visiting his website: http://www.mcs.net/~rogala

Reality, Virtuality and Visuality in the Xmantic Web
Tania Fraga

Introduction
The poetic expression created inside the telematic environment enables pervasive experiences. It is significant that emotions and feelings occurring during such processes allow manipulations and transformations. Percepts or any other cognitive processes emerge. The gathering of ideas brings about new sets of probabilities in order to establish a variety of multiple aimless navigation (Damasio 1995) and Eliade 1985 [1].

It is important to explain why we use the word poetic in discussing telematic art productions. Poesy from the Greek 'póiesis' refers to 'the action to make something'. It may refer to Beauty or anything related with loft and moving sentiment evoked by individuals or things, even though, in its common sense, poesy is the art to write in verse. The other meanings of the word 'poetic' refers to something which is expressed as visuality and virtuality. Telematic poetics are characterised as fields which use computer language to make visible the virtual, consequently giving it reality.

Paraphrasing and expanding Michael Lochwood's statement, we concluded that art can no longer be segregated from science and philosophy. The association of these fields not only implies that artists should understand a few scientific and philosophical subjects but also that artistic subjects should be nourished and impregnated with the philosophical and scientific outlook.

Artists should realise that science and philosophy have revealed 'a new world, new concepts and new methods, not known in earlier times, but proved to be fruitful where the older concepts and methods proved barren. (. . .) Those who come properly to understand relativity and quantum mechanics, so as to put classical and common-sense prejudices behind them, will never see the world in quite the same way again.' (Lockwood 1989)[2]. Conversely, art and philosophy should contribute to the growth of human sensitivity and comprehensiveness, as in great periods of mankind's history. Furthermore, art and philosophy should furnish the growth of human sensitivity, sensibility and understanding.

Science is beginning to study the complex processes of thought, as deeply related with emotions and feelings. This vital association between science and emotions has resulted in a differentiated state of counsciousness.[3]

The characteristic of telematic art, as a field in close contact with science and technology, has shaped a cultural paradigm which allows the collective energies to be interconnected in sensitive ways. As a result, this paradigm does more than just improve upon those of the past. This is so because telematic art concentrates human efforts through processes such as transformation, interaction, connection and action among artists and the public.

Telematic art connects different instances of mental (psychic) processes that make possible the navigation. This conclusion raises two questions. Heidegger formulated the first question, 'When and how things come to us as things?' (Heidegger 1988)[4] The second question was formulated ten years ago by the author as 'What is the "thingness" of things?'

While trying to answer these questions, consideration concerning the possible meanings of creating 'poetic things', which are essentially 'virtual realities', point clearly to their paradoxical nature. This endeavour has fostered the understanding of relationships among the physical world, the cultural paradigms and the challenges posed by the different fields of knowledge, mainly the art field.

The process of creating 'virtual worlds' has shown that virtuality understood in its sense of potential process of becoming and reality are complementary notions that may be expressed as visual experiences either in the form of images or mental perceptions.

One of the strangest sensations provoked by the diving in telematic art environments is the feeling of almost-trance-'devenir', which is experienced when wandering in this type of poetic reality. We can find parallels between this experience and the ecstatic journey to another state of consciousness produced through shaman's trance, sometimes called the descent within oneself (Eliade 1989)[5]

Figure 1: VRML with nested in-lined files

The analogies we point to consist in the overwhelming sensation produced by the two manifestations, which are the suspension of our time perception (Borges 1996 and Prigogine 1991)[6], and the re-establishment of another sensation. The confrontation with symbols and archetypes (Von Franz 1980 and Jung 1985)[7] may create unexpected space-time relations in our minds. The powerful feeling of ubiquity and wholeness with everything, everyone, everywhere, attains a quality of consciousness we cannot yet understand but which may possibly be re-elaborated through the work of art.

Discussion

The reasoning outlined above has brought new insights. Among these we discern the necessity to comprehend, understand and delineate the boundaries that divide phenomena such as perception, cognition, emotion, feelings and categories such as virtual, real and visual. In spite of all our scholarly efforts, after the expedition to the Kuikuros, a native tribe in the Brazilian Amazon, these thresholds have blurred.[8]

To summarise, we may say that we have expanded our concept of reality by adding whatever is available in a potential state of coming into being. Such potentialities, owing to our technical limitations, are mainly expressed visually although their conceptions, as mental images, are broader than the visual.

This approach seemed a reliable step to the understanding of what we were seeking. The first statement reality as anything available in a potential state of coming into being has shown two complementary views of the phenomena: the real and the virtual. The last one states that both ideas could be expressed visually — either in their form as visual images or as something perceptible only by the mind. Indeed, our practice of creating virtual worlds occurred over this triangular pattern, surpassing duality. Although these manifestations are mainly visual, they may be pure processes of becoming, as one interacts with them.

Evolving perspectives are shaping new paradigms through telematic collaboration among artists and scientists around the world. To the art world, the main paradigms either go beyond mimesis and contemplation the classical paradigm or the artist self-expression the modern paradigm. They have allowed the collective psychic energies to emerge through sensitive interactions and connections. These paradigms do more than improve past ones since they concentrate collective efforts to reconfigure the art field through processes of transformation, interaction and connection (Ascott 1997)[9].

There are many similarities between emotions and feelings, which arise when we wander inside virtual worlds and the shamans' adventure. Such analogies gave rise to the expression 'xmantic'.[10] Shamans are the ones who care for consciousness in their societies. Shamans become aware of unknown worlds during the achievement owing to the descent within themselves. Shamans are responsible for rituals and interpretations of the dreams' meanings. Shamans intermediate the sufferings of life by contacting spiritual energies and keeping the collective psyche in harmony.

Inside the Xmantic Web the Xwomen&men X'w&men are the caretakers of counsciousness. Xmantic rituals Xrituals will produce immersions into virtual environments. X'w&men use Xrituals to abolish daily space-time boundaries; they create X'vironments where poetic fluctuations, sensitivity and sensibility flourish; they invent wormholes connecting unpredictable realities; and they cause unforeseen transformations. The Xmantic Web intends to radiate with the creation of many other X'vironments by many other X'w&men where a large community will participate reconfiguring it in such a way that we may see, from this broad horizon, the Chaos in its process of becoming Cosmos.

The adventure of diving into an environment where different symbols lie, added to the opportunity to overcome our familiar concepts, was the result of the Kuikuro's expedition. This living experience opened our minds to other possibilities and filled us with the unpredictable.

On a few occasions, in the middle of the Kuikuro's tribe, we felt bewildered and overwhelmed by the palpable feeling of ubiquity and the sensation of wholeness that we experienced. Inside that space-time mandala, where we were living, the roundness was all around. At night, covered by boundless velvet darkness, our eyes were filled with shinning stars. Around us, the transparent warm lakes, surrounded by Buriti's palm trees, caught our imagination. The shaman acted naturally, in such harmony with his environment that we felt the clear sensation that reality is not only the physical world outside but also anything inside our minds. We may not have understood it at the time but it conducted us to unfathomable experiences. This kind of awareness unveiled frontiers for further explorations.

Conclusion

Summing up, we would like to establish the Xmantic Web as a sensitive 'place' where the fluctuations of the impermanent process of becoming unfold. Within this multidimensional reality, people will interact, connect and transform this poetic space-time manifold. Emotions and feelings will flow, allowing the 'thingness' of things to reveal multiple projections.

Perhaps the most awe-inspiring possibility of such a net of sites is that it may lead us to change the conception of cultural diversity as barriers obstructing contact. Possibly we will see the 'unknown dimensions of human behaviour' (Tafler 1995)[11] as attractive challenges to attain rich and complex multicultural connections.

The question that arises naturally is whether such transformations may overcome prejudice and fanaticism in our society. Is this desire a romantic dream of a visionary intellect? Is it the wishful thinking of a fanciful mind? Most important of all, because of their potential to influence conduct, these attitudes may lead to profound shifts in our Cosmos vision. Also, they may burgeon as little seeds of change, among many, many others, weaving new patterns of behaviour in the future.

The poetic action proposed by the Xmantic Website environment intends to re-create sensitive repertoires. We are convinced that while sharing and transforming poetic signs we may weave infinite interchanges among human beings (Eliade 1985)[12]. This web can intermix relations and attain a quality of waking consciousness, transcending 'verbal discussion and becoming an intuitive experience of Oneness behind the two' (Von Franz 1980).[13] The fragments of causality may be understood as projections of an unknown order. The conflict between

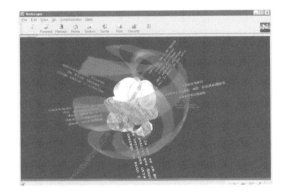

Figure2: VRML Poem

Figure 3: VRML Trans-codification: Myth of Sucury-Moon

opposite states of mind produces paradoxes. This conflict reveals the duality of the verbal thought processes caused by their inherent linearity fighting the intrinsic immanent and transcendent physical nature of multidimensional poetic reality: the only reality that can faithfully express fundamental aspects of human life (Eliade 1985).[14]

Notes

1. Antonio Damasio calls process of thought the ability to display images internally and to order those images. The images are characterised as sound, olfactory, tactile, mental, visual or any other category of images we may differentiate. Damasio, 1995. *Decartes's Error*. New York: Avon Books, p 89-90.
2. Lochwood, Michael. 1989. *Mind Brain and the Quantum*. London: Blackwell, p 314-15.
3. Antonio Damasio calls emotion 'for a collection of changes occurring in both brain and body, usually prompted by a particular mental content. Feeling is the perception of those changes.' op ct p 270.
4. Heidegger, Martin. 1988. 'A Coisa', in Souza, *Mitologia 1*. Brasília: Universidade de Brasília, p 121-31.
5. Eliade, Mircea. 1989. *Shamanism*. London: Arkana, p 508-11.
6. Two complementary notions of time may be seen in Borges and Prigogine. Borges, J. L. 1996. 'O Tempo' in Cinco Visões Pessoais. Brasília: Universidade de Brasília, p 41-9. Prigogine , I. 1991. *O Nascimento do Tempo*. Lisboa: Edições 70, p 59-75.
7. For more information on the concepts of symbols and archetypes used here see Von Franz, M-L. 1980. *Alchemy*. Toronto: Inner City Books, p 31, 137; and Jung, C. G. 1985. *Mysterium Coniunctionis*. Petrópolis: Vozes, p. 209.
8. The Expedition happened on May 1997 and was constituted by the artists Diana Domingues, Gilberto Prado, Maria Luiza Fragoso, Roy Ascott, Tania Fraga and Virginia Haeser.

Figure 4: VRML world with animations and sound

9   Ascott, Roy. 1997. 'Cultivando o Hipercórtex' in *A Arte no Século XXI*. São Paulo: UNESP, p 336-44.
10 Xmantic is a word created by Roy Ascott which expresses the confluence of the 'shamanic' and semantic phenomena applied in the Web. Shamanism is understood as 'any ecstatic phenomenon and any magical technique whatever' in Mircea Eliade. op cit., p 375.
11 Tafler, David. 1995. 'Boundaries and Frontiers' in *Transmissions*. California: Sage, p 235-67.
12 We might quote Mircea Eliade who describes a similar situation 'in which everything seems possible where the 'laws of nature' are abolished and a certain super human "freedom" is exemplified and made dazzling present.' Mircea Eliade. op cit., p 511.
13 Von Franz M-L. op cit., p 137.
14 Mircea Eliade also says that the 'poetic creation still remains an act of perfect spiritual freedom. The purest poetic act seems to re-create language from an inner experience that, like the ecstasy or the religious inspiration of "primitives", reveals the essence of things.' Mircea Eliade. op cit., p 510.

The VRML files quoted may be seen at the following electronic addresses:
http://www.lsi.usp.br/~tania/ vrml/
http://www.unb.br/vis/lvpa/xmantic/

Tania Fraga is a Brazilian architect and artist and has a Ph.D. from the Communication and Semiotics programme at the Catholic University of São Paulo. She is Adjunct Professor of Visual Arts at the University of Brasília and Associated Researcher at the Polytechnic School of Engineering at the University of São Paulo. She was Artist-in-Residence at The Bemis Foundation, USA, 1986, with a grant from the Fulbcigth Commission and Visiting Scholar at the Computer Science Department at The George Washington University, Washington DC1991-2.
Her work has been exhibited internationally in the IV FISEA: The Art Factor, Minneapolis 1993, in several exhibitions in Brazil, USA, Paris and Italy, and it is in the collection of the Bemis Foundation, the Brasília Museum of Modern Art and the Museum of the University of Hong Kong. She has been working with computer art since 1987 and her major interest today is the development of interactive VRML environments which may be seen at the following electronic addresses: http://www.lsi.usp.br/~tania/ and http://www.unb.br/vis/lvpa/. <tfraga@ unb.br>

Xmantic Webdesign

Fatima Bueno

Perception, Multiplicity, Synchronicity, Interaction, Transformation, Metaphors

A group of artists, brought together by the desire to explore the concept of *hypercortex*[1], organised a visit to a Brazilian native tribe, Kuikuro, to attend their rituals and confirm their importance as a cultural legacy. As a result of this unique experience the *Xmantic* Web was conceived.

The Xmantic Website starts with the group need to share the role of the *webartist*: by rescuing images of recognisable symbols, which for many, consciously or unconsciously, will remind us of our identity.

The proposal emphasises the idea of offering the possibility to cultivate the emergence of *telematic* experiences. *Telematic Culture* refers to the global connectivity of people, places but, above all, of mind (Ascott 1997).

Semantic + Shamanic + *Xamã* = *Xmantic*, the union of concepts: semantic, term related to languages meaning; Shamanic, communication with the higher and lower layers of the spiritual world and *Xamã*, the Portuguese word *shaman*[2]. Both artist and *shaman* are considered as channels between sacred and profane domains.

From the beginning of time until now the human desire to disclose the mystery of existence remains. Mythology, religion, philosophy, art and science answer ancient questions. Man's sensitive perception of reality happens mainly through symbolic discourse: 'Image always precedes idea in the development of human conscientiousness' (Read 1955). The research of anthropologists and religious historians reveals that similar symbolic forms can be found in rituals and myths in small towns and tribes still existing on the fringes of our civilisation[3].

Prehistoric art and rituals provided aesthetic and spiritual values and were designed to impose rules to preserve vegetable and animal resources, ecological practices to induce the creation of social groups and of collective social organisations.

In spite of increasing science and technological progress, the basic human feelings and beliefs – pain, love, hope, suffering-are still the same as they were in the past. At first, humans were stressed by nature's indomitable forces and supernatural domain over fate. At that time, people had less than they needed – food, safety and control over nature. Now, humans are stressed with the amount of information surrounding them – they have more than they can cope with.

The Internet offers the widest possible range of circulation for a most extreme heterogeneity of contemporary trash and aesthetic practices, poetics and politics. Such multiplicity can bring anxiety and perplexity.

A webdesigner's challenge is to create stimulus for people to communicate with each other and share common interests. The *Xmantic* Web begins with a site and intends to gather people from all over the world linked to each other by sensibility and sensitivity.

We can apply to the website Italo Calvino's concepts of lightness, quickness, exactitude, visibility and multiplicity to visualise, to perceive and interact with the poetics chosen. People are becoming aware of the impact of rapid technological changes in daily experience through goods and services, yet the main demand is for social interaction. ' Unlike previous environmental changes, the electronic media constitutes a total and near instantaneous transformation of culture, values and attitudes' (McLuhan 1974).

For visibility, connection is needed. 'But a connection could only be made – that is to say, rendered visible, perceptibly realised and represented – by a sign, which is an image that can be separated from immediate perception and stored in the memory. The sign came into existence to establish synchronicity, in the dumb desire to make one event correspond to another' (Read 1955).

Webdesign allows setting up the graphic interface of concepts between user and machine. It is also an aesthetic layout, the product's packing, the marketing, and the iceberg's tip of an electronic publication.

The *Xmantic* Web features will provide sensuous graphic quality, metaphoric structures and stimulating connectivity. It is a site for immersion, emergence, expansion, receptivity and impermanence, a window to invisible worlds in an active environment.

Interactivity, feedback, control, creativity, productivity, communications and adaptivity are requirements to any successful website and vital to the *Xweb*.

Figure 1. LVPA Homepage, link to Xweb

The image's size must be kept to a minimum number of bites for maximum impact and speed. Max Raphael described seven distinct features of the Neolithic geometric style to simplify the forms: synthesis, simplicity, formal necessity, detachment, definiteness, energy and connection of content and concept for the material development of bidimensional shape representation[4]. Present tridimensional shapes still require most of those principles.

Different from most sites, *Xweb* will focus the continued development of human consciousness toward the aesthetics of meaning. It will allow the user to experience consciousness through participation rather than talking about the concept.

The *Xweb*, through visual and conceptual metaphors in nature's diverse domain, material and spiritual, intends to open doors to the greatest ethical and aesthetic questions – territory, identity, new forms of citizenship, the disappearance of the national state, universalism and culturalism, poetics and politics[5].

The interactivity in hypermedia allows technical support for real open work on Cyberspace, *'always aborning space, without frontiers, continuous, sheltering multiple and moving shapes which outline collective come-into-being'*[6].

Both primitive art and electronic art operate on contact with the environment, with the invisible and immaterial worlds based in technology -the extension of bodies and minds, which allows aborigines and computer users to be artists. In the most significant epochs of art, the artist is a craftsman, often anonymous. Now, in our post-historical age, living at the end of art, as Arthur Danto said, 'it is possible for artists to appropriate the forms of past art and use them to their own expressive ends', but 'we cannot relate to them in the same way as those could whose forms they originally were.' He explains that 'the sense in which everything is possible is that in which all forms are ours. The sense in which not everything is possible we must still relate to them in our own way. The way we relate to those forms is part of what defines our period'[7].

Notes

1. As hypercortex, Roy Ascott means the emergent world-mind, where each one of us is made by many 'selves'. Diana Domingues. *A Arte no Seculo XX1*. Sao Paulo: UNESP 1997, pp.336-44.

2. The term *shaman* comes to us through the Russian language, from Tungusic *saman* and *'the pre-eminently*

– Projects –

Figure 2. – Xweb

Figure 3 – Xweb's wrml

shamantic technique is the passage from one cosmic region to another-from earth to the sky or from the earth to the underworld . . . the symbolism employed to express the interconnection and the intercommunication among three cosmic zones is quite complex and not without contradictions'. Mircea Eliade. Shamanism. London:Penguin Group, 1989, pp.259-495.

3 Henderson, J. in Carl Jung. *O Homem e Seus Símbolos*. Rio de Janeiro : Nova Fronteira, 1964, p.106.
4 Parameters remaining updated: 'Synthesis, reflecting the will to create a visual unity out of multiplicity of elements; simplicity, indicatiing the will to build complex structures from a few elements; formal necessity, deriving from the will to both represent and to conceal content in an adequate sign; detatchment arising from the will to embody eternal contrasts in self-evident form; energy, expressing the will to master by magic that which transcends man's physical powers, even life itself; and finally, connection of content and meaning with two worlds, those of life and death'. Max Raphael, in Herbert Read, *Icon and Idea* (Harvard:Harvard University, 1955, p.43.
5 These were some of '*100 days, 100 Guests*' debate's themes at Kassel Documenta. X Catherine David, *Introduction to Documenta X Short Guide*, p.125.
6 Pierre Levy, in *A Inteligencia Coletiva*. Sao Paulo: Edicoes Loylola, 1998 p.156.
7 Arthur C.Danto, in *After the End of Art*. New Jersey:Princeton, 1995, pp.198-99.

References

Benedetti, P. and DeHart, N. 1996. *Reflections on and by Marshall McLuhan*. Cambridge: The MIT Press.
Calvino, I. 1990. *Seis Propostas para o Terceiro Milênio*. São Paulo: Companhia das Letras.
Costa, M. H. F. 1988. *O Mundo Mehinaku e suas Representações Visuais*. Brasília: UnB.
Danto, A. C. 1995. *After the End of Art*. New Jersey: Princeton.
Davies, P. and Gribbin, J. 1992. *The Matter Myth*. New York: Simon and Schuster.
Den Boer, L. et al. 1997. *Website Graphics*. New York: Thames and Hudson.
Documenta X, Short Guide. 1997. Germany: Cantz.
Domingues, D. (org.) 1997. *A Arte no Século XXI*. São Paulo: UNESP.
Eliade, M. 1989. *Shamanism*. London: Penguin Group.
Jung, C. 1964. *O Homem e Seus Símbolos*. Rio de Janeiro: Nova Fronteira.
Levy, P. 1993. *As Tecnologias da Inteligência*. Rio de Janeiro: Editora 34.
Peat, D. F. 1987. *Synchronicity*. New York: Bantan Books.

Read, H. 1955. *Icon and Idea*. Harvard: Harvard University.
Vidal, L. 1992. *Grafismos Indígenas*. São Paulo: Studio Nobel.

Bachelor of Arts in Portugese Language and Literature. Master Degree candidate at the Visual Arts Department of the University of Brasilia.
In 1983, attended classes as a foreign student at the Academy of Art College, in San Francisco, California. In 1995, built a set of wood sculptures based on Brazilian native culture. Recent works include the web design of the Virtual Laboratory of Art Research, at the University of Brasilia, co-ordinated by Tania Fraga, and participation in the show ' Brasil Makes Design' 1998 Edition (Top Ten Award), from March to June 1998, in Sao Paulo, Rio de Janeiro and Milan. <fbueno@perocom.com.br>

Assigning Handlers to a Shadow [I]
Kieran Lyons

The work shown at *Consciousness Reframed 2* continues a sequence of ideas that I began to develop in an earlier work at Consciousness Reframed I. The title of that installation was *Beyond the Pale* and the work solicited the intervention of visitors, tempting them to respond to it and offering a course of action that was apparently clandestine. The concern of mine to attract an audience away from the passive role of viewer was of course unwarranted within the environment of the conference, where the viewer/user dichotomy is well understood; nevertheless it remains an important issue in the public domain. The current installation is being developed under the working title of *Assigning Handlers to a Shadow*, although for reasons that will become evident I, have also called it *Ki[lli]ng Charles 01* and I am now amalgamating these into the compromise title that I have placed at the head of this paper: *Assigning Handlers to a Shadow [I]* [1].

In *Beyond the Pale* I attempted to hi-light a set of conditions that I saw as problematic in the public reception of interactive installation work. The same problem continues to concern me today and has influenced my thinking in developing the format of this new work. At Consciousness Reframed II, I will be restricting the presentation to the interactive system of the installation only. The complete installation of the work would pose problems of space and setting-up time that would be unrealistic within a conference format. Nevertheless, the larger and more generic problem that I am addressing in my work will be underscored here and will be assumed into the presentation as a whole. It can be summarised in the following way.

By becoming detached from the other members of an audience in order to engage with the processes of an interactive work, the viewer becomes identified as the subject of that work and from a peripheral position as a passive observer, the viewer is rapidly thrust into the sharp focus and objective gaze of an expanded and more watchful audience. Consequently, the viewer's role changes from a passive observer of the work to the assumption of responsibilities as its demonstrator and visible exponent. Following on from this problem is, of course, the issue that artists are not always willing to trade the elliptical complexities of their interfaces for a more banal arrangement and so the navigational indicators through a work can be at times difficult to follow. This places an additional burden on the user and the perceived identification with the work can lead to an unproductive response to it. The process can be compared with a hypothetical situation in a theatre, where an audience might be expected to move from the anonymity of a seat in the auditorium into the exposed brilliance of the centre stage not every visitor to an interactive work is prepared for this dramatic shift into the spot-light of attention. This can frequently result in a premature retreat from the work and a denial of interest by the subject before the subtleties of intention can be properly established and assessed by the user. The pressing need, which I would like to identify, in interactive work is for the distribution of the burden of responsibility across a wider audience so that this focus on an individual becomes a shared one, distributed across a number of participants, involved with and engaged on a similar programme of creative inquiry. Until the possibilities of multi-user environments emerge as a reliable option for interactive work, interactive installation in the public domain will produce more spectators than navigators. Interactive artists will have to be reconciled to these limited, cursory and often inhibited responses.

    Within this very broad outline of a problematic, I try to focus on the complexities of decision making within social groups. For the work that I am presenting here, I try to create a template of how decisions are made between dispersed individuals who are nevertheless presented with a common problem and I attempt to create a system for the alignment of the thoughts and opinions of individuals within a peer group. The questions that I try to resolve within the limitations of this interactive piece are: How can thought and opinion be fused among a number of individuals to create a shared consciousness and an agreed response to a problem or a situation? And how can this be done in the visual domain?

    I try to develop this process by using a document that was produced in the House of Commons on the eve of the execution of King Charles I in 1649. Fifty-nine signatures were collected from a larger group of Members of Parliament after a three day trial of doubtful legality without historical precedent. The men who signed were asked to judge whether the king had acted unlawfully in waging the recent Civil War and if he had they were then asked to put their names to an agreement that would lead ultimately to his execution. At no previous time in British history had there been an attempt to remove a reigning sovereign through the process of the law although the rapid dispatch of kings by other means had been until that time regularly contemplated and sometimes effectively achieved. The Warrant of Execution of 1649 with its 59 signatures (some written in a bold and clear hand and some betraying the anxieties of indecision) has survived and is now held in the House of Lords Record Office.

    Visitors to *Assigning Handlers to Charles 01* will encounter a version of this Warrant that has been converted into a standard office spreadsheet and will be able to contribute to the process of collection and accumulation of signatures. Visitors to the

Assigning Handlers to a Shadow [1]. Kieran Lyons. 1998. Video Targeting Screen Display.

installation will be located on a projection screen and 'identified' by one of the 59 names that are assigned to their shadows on the screen. As they move around in front of the screen, the assigned name moves across the screen in alignment with the shadows. At a defined point in this procedure, a name on the screen will exchange its identity from that of a cipher on a conventional spreadsheet to an individual and characteristic signature that recalls the original and highly charged document that was produced in 1649.

Note

1   The designation of '01' at the end of the title refers to the numerical method of ordering in programming languages and on computers generally, e.g. 01, 02, 03, or to the designation that is assigned to the monarch and placed at the end of the his or her name. In this case it would be Charles I, Charles II or conceivably even Charles III. This project will end at the current designation of 01 and I am not anticipating any subsequent development into *Ki[lli]ng Charles 02* or even *Assigning Handlers to a Shadow [III]*.

Kieran Lyons is Senior Lecturer, Interactive Arts, Department of Art and Design, UWCN.
<k.lyons@newport.ac.uk>.

8   Architecture

Human Spatial Orientation in Virtual Worlds
Dimitrios Charitos

Introduction-Background to the Experiment

This paper documents part of a research project which attempted to develop an architectural way of thinking about designing space in a virtual environment (VE). It presents one of several experiments which investigated several design issues, pertaining to the way that spatial elements may be composed for the purpose of aiding navigation in a VE.

These experiments were conducted on the basis of a hypothesis which was the result of the theoretical part of the project. According to this hypothesis, space in a VE may comprise the following elements:
- places
- paths
- intersections
- domains

These spatial elements are established, out of the void, by appropriately arranging the following space-establishing elements:
- landmarks
- boundary objects
- thresholds
- signs

This hypothesis has also been presented and explained in Bridges and Charitos (1997) and Charitos (1997).

Planning Phase
During the design of several pilot VEs for implementing the above-mentioned hypothesis, it became evident that if a place was positioned in certain ways in relation to a path, a subject who moved into or out of the place would become somehow disorientated. This phenomenon indicated an issue, which was crucial for the composition of domains by appropriately arranging places and paths in the three-dimensional space of a VE. This issue has been the object of an experiment, documented in this paper.

This experiment investigated the impact of the way that a place was positioned in relation to a horizontal path on the orientation of subjects who navigated in this place. Additionally, the way that subjects positioned their viewpoint when they tried to orientate and perform a task in the place, dependent upon how this place was positioned in relation to the path, was also investigated. The ultimate aim of this investigation was to identify how places have to be positioned in relation to a path in three-dimensional space in order to ease the task of navigating into the place and performing a certain activity within this place.

Design Phase

Since there was no precedent for predicting the behaviour of subjects in such a situation, it was not possible to develop a hypothesis about the subjects' responses. The independent variable which was taken into account in this experiment was the way that each place was positioned in relation to the path; this variable was described by the axis and angle of rotation of the place in relation to the path. This qualitative factor was set at specific levels of interest and its effect on the dependent variable was studied. Other factors which may have affected the behaviour of subjects in each place were the physical characteristics of each place and path and the way that each path was spatially arranged; these factors were rigidly controlled so that all paths and places were identical and all paths were positioned in a similar way. Finally, the order in which subjects experienced one place after the other was randomised.

Therefore, a VE was designed, which consisted of
- a central hall on the one surface of which five entrances to five paths were positioned;
- five identical, horizontal paths that lead to
- five identical places, differently positioned in relation to each path.

All places were orthogonal parallelepiped in form, with their length being equal to their width and their height being a third of the their length. The bigger two boundaries of the place were not intended to be seen as a 'floor' or a 'ceiling' respectively. Instead, the intention behind investigating the way that subjects positioned their viewpoint inside each place was to identify whether subjects did indeed tend to see these bigger surfaces as floor and ceiling. Each of the five places was rotated in a different way in relation to the path. Five numbers, which were positioned at the entrance of each path from within the central hall, signified the way towards each of the places.

The way that these places were positioned can be described if we consider an (x y z) coordinate system as a reference, where z is the main axis of the horizontal path, y is the

Figure 2: Front view of the experimental domain, with place 1 on the far left and place 5 on the far right.

Figure 1: Side view of the experimental domain

vertical main axis of the numbers in the paths' entrances and x is the axis parallel to the longer direction of the central hall, then:
1) Place 1 is rolled $90°$ around the z axis
2) Place 2 is pitched $45°$ around the x axis
3) Place 3 is not rotated at all
4) Place 4 is rolled at a $-60°$ angle to the z axis
5) Place (5) is rolled at $30°$ to the z axis and tilted at $-45°$ to the x axis

The impact of the way that these places were positioned on the criteria:
- how easily do subjects orientate while performing an activity
- the way that subjects position their viewpoint in relation to a place was studied

The results for the first criterion would identify which ways of positioning a place in relation to a path were preferable and which were problematic for a subject who navigates within a domain, comprising places and paths.
 Subjects were requested to enter the place, perform a simple task there and then exit the place. This task aimed at motivating the subjects to enter the place, move around and to perform a certain activity there, it was not intended to have an impact on the response of subjects in terms of how easily they orientated thmeselves. Five spheres were positioned inside certain concavities on the surfaces of each place, at similar positions. Each sphere differed in colour and size from the other spheres. In each place, the task was to identify which was the colour of either:
- the largest or
- the smallest sphere in each place

The different order of presentation was expected to randomise any unpredictable effects that the task may have had on the responses.
 After performing the task and exiting the place, subjects were asked how easy it had been for them to orientate themselves while
- entering the place
- moving within the place in order to perform the task and
- exiting the place

Their responses were measured on a scale of 0 to 100. The time it took to execute the task was also recorded. The two dependent variables studied were the 'easiness of orientation' and the 'time of task execution'.
 A third dependent variable corresponded to the way that subjects positioned their viewpoints when they tried to orientate themselves within each place in order to do the task, as observed by the researcher. As a result of a pilot experimental study, involving three subjects, four different possible positions of the viewpoint within a place were identified. If we consider the (x,y) coordinate system of the subject's viewpoint, where x is the horizontal and y is the vertical axis, then the four options were:
1) the y axis of the viewpoint was vertical to the two largest boundaries of the place that one could refer to as a 'floor' or 'ceiling'

2) the y axis was parallel to this 'floor' or 'ceiling'
3) the viewpoint was positioned randomly
4) subjects tried to keep a global vertical orientation, following external cues visible to them

Thirty-one subjects (11 female and 20 male) took part in this experiment.

Analysis Phase

The Variable of Easiness of Orientation
The statistical analysis of the results for the variable of 'easiness of orientation' showed that:
- Place 3, to which no rotation at all had been applied, was clearly the easiest of the five to orientate in
- Place 2, the floor of which had been pitched $45°$, was the most difficult to orientate in; however it was not clear whether this place was more difficult than the place to which rotation had been applied along both axes.
- It was not possible to differentiate between the other three places in terms of easiness of orientation.

The Variable of Time
The form and size of places in this experiment was not likely to induce exploratory behaviour and therefore subjects did not spend any time in each place doing much more than merely performing the task. Accordingly, the variable of time was also seen as indicating the easiness of orientation in each place, in the sense that subjects who orientated easily in a place spent less time performing the task there. Indeed, the analysis of the results for the variable of 'time' provides support for the analysis of the 'easiness of orientation' variable:
- It took subjects significantly less time to perform the task in place 3 - to which no rotation had been applied and which was the easiest to do the task in - than in places 1, 2 and 4 which were either rolled or pitched in relation to the path, at a $45°$ or $60°$ angle. However, no difference was detected between place 3 and place 5, which was rotated along both axes, at a smaller angle.
- It took subjects significantly more time to perform the task in place 2 - which was pitched $45°$ and which was considered the most difficult to orientate in-than in the other places.
- No significant differences were detected for either easiness or time measurements among places 1, 4 and 5.

Figure 3: View within place 2   Figure 4: View within place 3

The Variable of Viewpoint Orientation Within a Place

Since there were no precedents for predicting the behaviour of subjects in each place, no specific hypothesis was defined for the way that subjects orientated their viewpoint. The analysis of the results aimed at identifying whether subjects felt the need to position themselves in a particular way in order to orientate and feel comfortable to perform a task in each place.

This analysis showed that in all five places, the majority of subjects preferred to position their viewpoint vertically to the two largest boundaries, which could be seen as the 'floor' and the 'ceiling' of the place. However:
- Since place 1 was rolled $90°$ along the z axis, subjects who entered it may have seen the floor and ceiling as if they were walls. As a result, 25% of them adapted to the awkward volumetric proportions of the space and felt comfortable to keep their viewpoint positioned parallel to the 'floor'.
- Place 2 was pitched $45°$ along the x axis. The way that this place was rotated in relation to the path caused confusion to a large number of subjects, who entered the place and were immediately faced with a dark pitched floor which was too close to their viewpoint. As a result, 25% of the subjects moved their viewpoint randomly within the place.
- Place 3, which was not rotated at all, was seen by most subjects as a realistic room. As a result, they felt like behaving as they would in a real space and accordingly saw the planes defined by the two larger dimensions as a floor and a ceiling and attempted to position their viewpoint vertically to this plane.
- Place 5, may have been rolled along both axes but the angles of rotation were not too big. As a result, most subjects saw this place as a normal room, the floor of which was slightly rotated and consequently behaved as in the case of place 3.

From observing the behaviour of subjects, it may generally be suggested that most of them felt more comfortable to yaw-rotate around the y axis-than to pitch-rotate around the x axis-their viewpoint within a place in order to look around and perform the task. It seems natural to pitch one's head up or down at a small angle but it would seem very awkward to pitch one's whole body $360°$; on the other hand, it seems natural to yaw one's viewpoint in the real world at any angle.

Conclusions

This experiment investigated the impact of the way that a horizontal path is positioned in relation to a place and has concluded that:
- The application of any rotation on the place in relation to the path clearly decreases the easiness with which a subject orientates in this place.
- The application of a significant pitch rotation on the place in relation to the path makes orientating in this place difficult and confusing.
- Subjects seemed to have less trouble orientating in a place to which a significant roll rotation or a less significant pitch rotation has been applied.
- With respect to parallelepiped places, subjects tend to interpret the largest boundaries of

the place as 'floor' and 'ceiling' and accordingly position their viewpoints vertically to these boundaries in order to achieve a state of orientation and perform a task in this place. However, a smaller majority of subjects is likely to do so when the rotation applied to the place is significant. Subjects generally preferred yawing rather than pitching their viewpoint within a place, in order to look around in this place.

References

Bridges, A.H. and Charitos, D. 1997. 'The Architectural Design of Information Spaces in Virtual Environments', in Ascott, R., ed. Proceedings of the 1st International 'Consciousness Reframed' Conference. Newport: CaiiA, University of Wales College.

Charitos, D. 1997. 'Designing Space in Virtual Environments for Aiding Wayfinding Behaviour', in Bowden, R., ed. Proceedings of the 4th UK VR-SIG Conference. Brunel University.

Dimitrios Charitos has recently completed a doctoral research project supported by the Commission of the European Communities under the Training and Mobility of Researchers Programme. The project dealt with the architectural aspect of virtual environment design and involved several experiments for evaluating a proposed theoretical framework. Results of this project have been published during the period 1996-8 in books, journals and conference proceedings. He was born in Athens, Greece, and graduated with the degree of diploma in Architecture from the National Technical University of Athens in 1990. He was awarded the degree of MSc in Computer Aided Building Design from the University of Strathclyde in 1993. He has been involved in composing and recording experimental electronic music since 1983.
Dimitrios Charitos, Department of Architecture, University of Strathclyde. <D.Charitos@strath.ac.uk>

An Interactive Architecture
Gillian Hunt

Architecture: Control and Communication

The concept that architecture is an embodiment of a society's values and is a form of nonverbal communication, having associational content or meaning, is evident in Greek culture and certain religious casts in India, for the Maori and for numerous indigenous societies. Traditional societies, having quite rigid social structures and mores, are inclined to have an architecture that is consistent and monolithic in its tectonics, i.e. guided by strict rules of assembly and symbolic detail-ornament which closely integrates architecture with its culture. In current Western civilisation the meaning of architecture is not as easily determined. The pluralistic, inclusive nature of our culture-era and the variety of architecture produced obscures any easy analysis of architecture *per se*. According to George Hersey, post-modern conceptions of architecture in which ornament is redeemed and utilised for its allusive power attempts to make architecture comprehensible in a way that unadorned, utilitarian modernist forms could not [1]. Similarly, architect Aldo Rossi argues that the post-modern reinvestment of architecture with associational meaning is in the use of particular 'types' or forms as a vehicle for collective memory and that it is because of the persistence of certain architectonic forms that architecture can convey meaning by affording a collective (archetypal) system of articulation and interpretation [2].

As the scale and complexity of social systems has increased throughout the 20th century, so have the number of specialised settings each with its own particular cues and behaviours resulting in a diversity of message systems. Martin Pawley suggests that 'What we have in the modern world is a disorganised multiplicity of sign systems tracking back through time, of which perhaps the oldest and most overlaid is architecture. . . the next step in architecture should be the reintegration of the built environment with the overlaid information systems that have been allowed to take over its proper task'[3]. In its well-intentioned but misguided concern to assimilate the technical and processal realities of the 20th century, architecture has adopted a language in which expression resides almost entirely in processal, secondary components such as ramps, walkways, lifts, staircases, escalators, chimneys, ducts and rubbish chutes. Architecture's assimilation of rapid technological developments in environmental controls has since the 1960s resulted in an emphasis upon building services, their incorporation into existing structures and their use as a post-modern emblem is evident in numerous commercial or public buildings since Richard Rogers and Renzo Piano's 1977 Pompidou Centre.

Currently, 'intelligent' buildings are defined by a high–level of automated building services which require specialised planning and installation. The incorporation of electrical–mechanical devices for environmental control, the installation of numerous cables and duct work to accommodate IT and communication systems has privileged a more utilitarian, atectonic and atomistic attitude in which each technological element is grafted in or onto an often pre-existing structure[4]. Arguably, designing for the incorporation of mechanical-electrical services is not merely a matter of finding 'neat' ways to install them but of setting them to work in partnership with the structure so that 'the whole is more than the sum of its parts'. At present 'intelligent' building design concerns itself mainly with satisfying human physiological need by providing computer controlled heating, lighting and ventilation services. 'Intelligent' buildings seem predominantly defined by a high level of automated services and what seems to be a preoccupation with increased 'control' of information and energy management systems. It is suggested that a more 'human centered' and integrative approach is warranted in the utilisation of available technologies in order to facilitate novel experiential domains for 'human–human' and 'human–machine' interaction.

Cybernetic Architecture

Cybernetics was conceived in the 1940s as a transdisciplinary study dedicated to the roles of information and purpose in both organic and machine systems. Originating in notions of feedback and circular causality, cybernetics has since matured into a 'science of describing' culminating in the insight that reality is a social construct, and in contrast to being 'true' or objective, 'reality' is already 'virtual'. Consequently, second order cybernetics 'the cybernetics of observing systems' has a practical impact upon the 'effectiveness' of computing, communication and media. As such, it also affords a more human as opposed to technological imperative concerning the design of 'intelligent' environments[5].

The lively, brainstorming speculations of Gordon Pask and Stafford Beer have been influential in popularising cybernetics which depends in fact on difficult mathematics. Pask's influence in the development of 'cybernetic' or 'intelligent architecture' can be traced through his involvement with the Architectural Association from the mid 1960s until his death in 1996. His own works also provide a model for the design of interactive environments and one in which 'the computer, material and all engages (the user-participant) in dialogue and within quite wide limits is able to learn about and adapt to his behaviour'[6]. Cybernetic concepts of information flow, control by feedback, adaptation, learning and self–organisation have permeated many disciplines since its inception in the late 1940s and its influence on architecture in the 1960s is evident not only in the work of Cedric Price but in the attitude adopted by numerous members of the 'alternative avant garde'[7].

The cybernetic concept of an 'intelligent' building may be perceived as a dynamic assemblage of components (parts), which could not be understood in isolation to the whole system. The organisation of relationships of parts including those components which are in themselves intelligent, i.e. human activity, would be the pattern through which the buildings processes would be revealed. The structure of such a building would be a result of its experience, gained in interactive congress with its internal (cyclical) and external (developmental) interactions with the environment. Its responses would result in a continual process of change, its behaviour a product of successive structural couplings with the environment. In keeping with this description, communication is perceived to be, not a transmission of information, but more a coordination of behaviours between living organisms through mutual structural coupling. Similar types of structural coupling may emerge resulting in interactions between architectural entities which coordinate structural and hence behavioural outcomes. In co-evolution, the building organisation that may develop would then form a larger pattern of cooperative relationships

Neo–Organic Tectonic Culture

The cybernetic concept of an 'intelligent' building demands some degree of epistemic autonomy in order to improve itself, a capacity which is only attainable through structural autonomy, as is the case with all biological systems. Turning to nature as the exemplar for action and outcome has been a tactic of architects for centuries, drawing inspiration from nature in structures, forms and more recently in terms of processes. Biological analogy has also infiltrated materials science and new, artificial materials have been endowed with the capacity to change. Contemporary material innovation concerns itself with processes or the active and interactive forces which promote transformation in biological systems. The manipulation of biochemical processes by chemical engineers in effecting transmutations at a molecular level concentrates on the generative and catalysing agents which integrate structure and behaviour. The 'nature' of matter and its resulting forms is perceived as a dynamic, responsive living system in which there is a co-dependence between form (represented as information) and matter (conceived as substance-material).The simulation and synthesis of organic systems by molecular chemists and engineers has made a number of advances over the preceding two decades, culminating in the recent development of 'Smart' materials systems. 'Smart' materials do not necessarily mimic existing materials but are being created as syntheses of interactive processes found in natural systems hence the term materials systems, which emphasises the dynamic, inter-relatedness of molecular patterns of behaviour. The idea of a pattern of organisation, i.e. a configuration of relationships characteristic of a particular system, is a central concept in the development of systems theory.

Conclusion

Cybernetics, from its inception, has examined a number of complex systems by exploring them from the point of view of not 'what it is' so much as 'what does it do?' and 'what can it do?'. Cybernetics seeks to analyse and describe the function and performance of social systems, natural systems and machine systems. Since architecture encompasses all of these things, cybernetics represents a meta-disciplinary language by which to explore architecture as an 'intelligent' system, a concept which encompasses notions of interaction and responsiveness in terms of its environment, human usage and the maintenance and adjustment of its own systems. The development of new technologies stemming from the computer has fuelled speculations concerning evolutionary, self-organising and predictive systems and the increasing sophistication of the virtual production of interactive environments has spawned a growing interest in the possibilities of manifesting these ideas in new artificial and material forms. The 'information age' has, in this context, a close association with a certain 'new matter' and suggests that architecture's formal and material status is not lost!

Notes

1   Johnson, P.A. 1994. *The Theory of Architecture: Concepts Themes and Practices*. New York: Van Nostrand Reinhold, p.186.
2   ibid. p.189.
3   Pawley, M. 1990. *Theory and Design in the Second Machine Age*. Oxford: Basil Blackwell Ltd. p.x.
4   Wada, B.K., Fanson, J.L. and Crawley, E.F. 1990. 'Adaptive Structures', in *Journal of Intelligent Material Systems and Structures*.1. pp.157-74.
5   Preece, J.A. and Stoddart, J.F. 1994. Review: 'Concept Transfer from Biology to Materials', *Nanobiology*. 3. pp.149-66.
6   Pask,G. 1969. 'The Architectural Relevance of Cybernetics', in *Architecural Design Journal*. September p.495.
7   Jones, A. (eds) et al. 1985. *Cedric Price: Works 2*, London: Architectural Association Press, p.31.

References

Maturana, H. 1975 'The Organisation of the Living. A Theory of Living Organisation', in *International Journal on Man Machine Studies*.7.
Pask,G. 1971. 'A Comment, a Case History and a Plan' in Reichardt J (ed) *Cybernetics, Art and Ideas*. Greenwich: New York.
Pask,G.1962, *An Approach to Cybernetics*, London: Hutchinson.
Pask,G. 1975. 'A Cybernetic Theory of Cognition and Learning' , *Journal of Cybernetics*, vol. 5, no.1.
Pask,G. 1975. *Conversation, Cognition and Learning* , Amsterdam: Elsevier.
Pearce, P. 1978. *Structure in Nature as a Strategy for Design*. Cambridge, Mass: MIT Press.

Steadman, P. 1979. *The Evolution of Designs: Biological Analogy in Architecture and the Applied Arts*. Cambridge: Cambridge University Press.
Von Meiss. 1990. *Elements of Architecture:From Form to Place*.London:Van Nostrand Reinhold.
Wiener, N. 1948. 'Cybernetics.Reproduced' in *Cybernetics of Cybernetics*..Minneapolis: Future Systems Inc. 1996, *Introduction to Weiner* by Heinz Von Foerster, p.4.

Heterotic Architecture
Ted Krueger

Krueger (1996) characterised architecture as a metadermis-a second skin that is inhabited on a collective basis. This biological metaphor develops out of the capabilities for sensing evaluation, reaction, communication and learning that are becoming available to architecture by transfer from other technical disciplines. Advances in materials engineering, particularly intelligent and interactive structures, have much to contribute to the realisation of this metaphor.

Substantial work in intelligent materials has been undertaken with applications intended for the aerospace or defence industries. The strategy employed and the methods used have evolved directly from composite materials-a typical method of fabrication for intelligent structures. A range of materials, each with a desirable property, are brought into the proper orientations and fused in a thermoplastic matrix. For example, a variety of fibres-glass, graphite and ceramic-may be oriented to directly counteract the forces that are anticipated on the component. Their properties are individually matched to the anticipated stresses. Optic fibres are placed as sensors to track the forces impinging on the structure as well as its internal states during manufacture, installation and use. These sensors provide information to embedded micro-controllers that record and process it. Actuators, such as shape memory alloy wires, are included to counteract forces or damp vibrations as directed by the embedded controllers. This results in a material with enhanced structural properties, material that has the ability to sense, record, decide and react.

Materials used in architecture are crude by comparison. Bulk materials are stacked, poured, rolled or extruded into shapes and sawn to size. Site assembly methods are inexact processes undertaken in uncontrolled conditions. Composite structures represent a significant increase in capability. Analytic techniques are used for the definition and decomposition of the problem which is then solved as a series of relatively isolated tasks-a specialised material becomes integrated into the larger scale component to accommodate each function.

This approach is in many respects similar to the way in which contemporary architectural and civil structures are composed. Frequently, the structural solution to a building is optimised in relative isolation to the cladding that envelops it. Issues that arise at their interfaces are resolved as secondary decisions. The composite material tradition brings a similar process to bear on the individual component, significantly increasing its capabilities at that scale and increasing enormously the capabilities of the larger-scale structure into which the individual components are united.

This is not the only viable path to intelligent materials. Takagi (1992) notes a second, biomimetic, approach 'which is the attempt to create a unified bulk material which combines system sensor, processor and effector functions in a single group. Such a material might resemble a neuron'. Unfortunately we can not presently build neuron-like structures. They require a significant decrease in scale and an increase in complexity and integration beyond our capabilities.

It will be possible to attempt such a project when the techniques of nanotechnology are more fully developed. Presently, this field is focused on designing the equivalent of human-scaled machinery at the molecular level. Drexler and Merkle have designs for bearings, pumps and assembly robots at nanometer sizes. The construction of these devices is not imminent. It is of interest here that the design strategy is imported from another scale. The engineering of nanoscale components appears set to follow strategies that have evolved over time and are understood. Individual devices are deployed to solve elements of the problem at hand as given by an analytic decomposition, matched on a one-to-one basis. This approach favours a linear process leading to solutions optimised for anticipated states but brittle when exposed to a range of conditions.

Nanoscale machinery is not uncommon. Biological materials are produced by processes at this scale in all living things. They are non-optimised, flexible, responsive and robust. Another approach to intelligent materials, consistent with Takagi's second path, is the adoption of materials of biological origin in bulk as functioning devices. Currently, we use bamboo and leather, for example, as harvested products. While each has its unique properties to be used or overcome, the greatest portion of its original functionality was terminated at harvest. It is precisely these properties that we seek in an intelligent material.

Skin, as a biological material possesses characteristics that are desirable in an architectural cladding. These include protection from the environment, an active role in maintaining homeostatic internal conditions such as thermal and humidity levels, regeneration or self-repair, sensory perception and bio-degradability. It is flexible, compliant and responsive, our largest and most robust organ.

Skin was one of the first tissues to be engineered because it is not highly structured. It has a very high demand level owing to burn accidents and chronic ulcer conditions stemming from diabetes. In the traditional treatment of severe burns, the damaged tissues are cut away and the wound covered with skin removed from a cadaver. This covering is often rejected by the recipient's immune system and must be replaced on an ongoing basis while the skin is generated on other portions of the body for transplant to the damaged areas. Cultured skin tissues are being developed to circumvent the tissue rejection problem and to ensure a supply of materials.

With cultured tissue, fibroblasts and keratinocytes from a donor's skin are grown onto a synthetic degradable matrix that has been prepared with appropriate growth factors and proteins. A 'soup' prepared from an infant's circumcised foreskin can seed up to 250,000 square feet of synthetic tissue that can be grown in a matter of weeks. These are cladding dimensions.

The use of cultured human skin as an architectural material clearly has unacceptable cultural implications. In addition, if it were a functionally appropriate use, we would have little need for architecture to mediate the environment. We would already possess its properties. The use of biomaterials as architectural media will be dependent upon developing an appropriate substance. This may not exist as an off-the-shelf item in Mother Nature's pantry because our expectations for the material as a building product have many additional requisites.

If biological components are to be employed for architectural applications, techniques must be developed to enhance their properties. There must be an understanding of the granularity of the components, their interrelationships and the nature of their integration into a larger framework of organisation. The infrastructure required for their support in both structural and metabolic terms must be provided.

The properties of biological materials have been undergoing directed modification by agricultural techniques for thousands of years. New techniques for more rapidly and precisely controlling these properties are under development. For example, Wierzbicki and Madura (1996) have studied the action of certain polypeptides in altering the crystal growth patterns of ice formation within the cells of Arctic fish. The inhibition of ice formation allows these fish to survive in sub-freezing waters without suffering cellular damage.

Genes for the production of these proteins may be vectored into plants and become part of their inherent make-up. This may reduce frost damage during the growing season, extend the range for the species and maintain quality in produce shipped to market in a frozen state. Similar proteins would be useful in enhancing the properties of cultured skin tissue exposed to harsh temperatures. The genetics of the cells introduced into a tissue culture may be tuned to arrive at properties useful in architectural applications.

Providing the infrastructure to make biomaterials viable in architectural settings poses additional, but perhaps not insurmountable, problems. A vascular system capable of transporting nutrients and carrying off wastes will be required, as will a method of process control and intercommunication. Work in the medical field suggests that these requirements may someday be met as well. Research on a variety of cultured organs in addition to skin is beginning to make progress, among them, liver, pancreas, kidneys, blood vessels, marrow, cartilage and bone.

The cells, in addition to being cultured, need a matrix onto which to attach themselves in the three-dimensional configuration required to synthesise a complete organ. The molecular 'Velcro' to which the cells bind may be overlaid onto a variety of substrates. The matrix may have inherent properties that become part of the final product or degrade as they provide the raw materials for the cells that replace it. This scaffold may be shaped to conform to the specifications of replacement parts, as happened with the popularised ear-on-a-rat's-back produced by Langer and Vacanti. In principle, shapes may be developed that respond to other requirements as well. Through the use of scaffolds, biologically-based components may be configured to architectural requirements.

Bone culturing has been undertaken to provide for repair and reconstruction of damaged limbs. Notable is bone's ability to regenerate and to reconfigure in response to variations in structural stress or impact loading. Thompson (1942) has shown that bone density correlates closely to these stress patterns. A variety of biocompatible metals may reinforce or replace bone joints. Notable among these is shape memory alloys.

SMAs are a class of intermetallic compounds that have properties that vary widely with the crystalline phase of the metal. The transition temperature between phases can be controlled with precision during the alloy's fabrication and could be specified at or near the normal operating temperature of a biological component. This would allow for either a range of physical properties with minimal variation in chemical composition or the properties of the material may be varied by a transition in the operating temperature of the component as a reaction to stress. Hybrid metal-bone systems would increase the range of available structural and functional properties beyond what is available in either material alone and may allow for the application of bone material to architectural scales.

As techniques develop, and as successes occur, there will be an influx of intellectual and economic capital into the field. One would expect that many other organs would come under development. Of these, the most challenging-the prize and the grail-will be the brain. A human-equivalent brain may remain a problem long after its raw processing power is matched with silicon devices. However, some type of control scheme will be required to regulate the biological components and a method found to communicate within and between components. Silicon computing devices are arguably our most sophisticated mechanisms and well suited to issues that are much less demanding than brain equivalence.

Peter Fromherz and others have investigated the interface between neurons and electronic devices. Proteins isolated from the nerve tissues are structured onto a silicon substrate by UV photolithography in the geometry required to effect a specific connection between a neuron and the chip. Nerve cells are cultured along this geometry. Stett, Müller and Fromherz (1996) have shown that communication with the nerve cells must be by induction rather than by the direct electrical contact that would destroy the nerve cells over time. Work in this area is just beginning; applications to humans of the sort envisioned by Gibson and others are not expected soon. Intermediate stages in this research would allow for the interfacing to simpler organisms or simpler organs.

The design of the software controllers that could be used with these kinds of interfaces may also be in its preliminary stages. The baroreflex, a nested series control loops that the body uses for short-term blood pressure regulation, has been studied to develop neuromimetic control algorithms that would be useful in industrial processes. The algorithms are capable of regulating the inherently non-linear processes found in chemical manufacturing. The research has not only succeeded in abstracting the computational principles inherent in the reflex but has also uncovered its rich hierarchically layered structure. Developments and extensions of this research provide the basis for the kinds of complex control strategies that would be required to regulate the processes contained within architectural biomaterials and to integrate a series of them into a functioning higher-level entity.

Krueger (1997) discusses recent research in artificial intelligence and cognitive science indicating that embodiment is a precondition for the development of cognition.

Embodiment is the mechanism by which symbolic systems develop meaning and allow for the shared grounding of representations as a precursor for the development of cognition as a cultural process. Biological materials have much to offer in terms of producing an embodiment conducive to the development of intelligence in architectural structures. The sensing and actuation capacities of biological materials and the requirement for non-linear control systems provide for a rich substrate on which these higher-order functions may become manifested.

Architecture is basically stacking dirt in one form or another: we can do better. Current materials are hard, dry and inert. We must begin working wet. Within the context of intelligent materials and structures and interactive environments, functioning biological materials have much to offer. They can be a source of inspiration, the cheat sheet that is desperately needed when working with complex systems. The recent interest in biomimetics distributed across many fields suggests that living processes are a compendium of solutions to complex problems that defy traditional analytic methods. They may also, as noted here, fulfil *a priori* many of the parameters that must be integrated into these materials and systems.

This paper is not a wholesale dismissal of human cultural production in favour of the biological, a rejection of the designed for the found, but a recognition that each covers a different territory and represent radically different methods. The transformation into hybrids takes advantage of the unique properties of each.

References

Krueger, Ted. 'Like a Second Skin', in *Integrating Architecture*, Spiller, ed., Architectural Design Profile no.131, December 1996.

Takagi, 'The Concept of Intelligent Materials and the Current Activities of Intelligent

Materials in Japan', in *1st European Conference on Smart Structures and Materials*, SPIE vol. 1777, 1992.

Stett, Alfred, Müller, Bernt and Fromherz, Peter, 'Two-Way Silicon-Neuron Interface by Electrical Induction, Physical Review E', in press abstract available at:

http://mnphys.biochem.mpg.de/projects/abstracts/stemuefro96.html

See also: Peter Fromherz, 'Neural membrane cable',

http://mnphys.biochem.mpg.de/projects/wachstum/wachstum_e.html

Thompson. 1942. *On Growth and Form*. New York: Macmillan.

Wierzbicki, Andrzej and Madura, Jeffry D. 'Computer Assisted Approach to Design of Biomimetic Inhibitors Controlling Crystal Growth', in *Interface Journal*, vol.4, no.2, May 1996, The Alabama Supercomputer Authority.

http://www.asc.edu/interface/VOL11/biophysics.html

Ted Krueger is the E. Fay Jones Visiting Professor at the School of Architecture of the University of Arkansas for 1996 to 1999.
tkrueger@comp.uark.edu
http://comp.uark.edu/~tkrueger

— *Architecture* —

The Cybrid Condition: Implementing Hybrids of Electronic and Physical Space

Peter Anders

Emergent Spaces

While many architects accept computers as extensions of existing practice, few accept that computers change the very foundations of their discipline. This is in the face of dramatic change in information technologies. For example, recent developments allow the creation of on-line work environments. These electronic spaces, called collaboratories, let researchers share technology to pursue common goals asynchronously, without the need for proximity.[1] Also, the US military has funded study of overlapping physical and cyberspaces to provide briefing facilities for its personnel.[2] Multi-user domains (MUDs) on the Internet are now viewed as promising social workplaces.[3,4] Since many MUDs now incorporate three-dimensional graphics and sound, we can foresee new and compelling uses for these on-line spaces.[5]

These spaces present clients with new means for achieving goals previously reserved to construction. Architects may find that portions of projects will no longer be implemented physically. Indeed, such developments may be hybrids of physical and cyberspaces. Designers may translate the information-rich components of a building programme into flexible, dynamic data-bases using spatial references for orientation and ease of use. To varying degrees, these spaces may be linked to the physical architecture of the project as they relate to the building's and business's activities. These hybrids, here called 'cybrids'[6], provide a model for responsive, physical and electronic spaces.

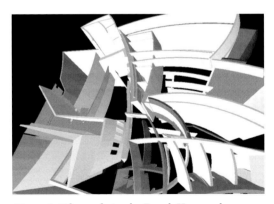

Figure 1. Library design by Ranah Hammash, graduate student at the University of Michigan School of Architecture, incorporating physical and electronic space

It is worth examining how cybrids may affect architectural practice and our environment. Some pragmatic questions need to be answered. For instance, what effect would cybrids have on the physical environment? How would such a strategy be implemented? What might be the interactions between the two spaces and their populations?

To answer some of these questions, let us consider a hypothetical deployment of a cybrid. Imagine that a client approaches an architect to plan a new facility. Assume it is the client's place of business, a business having informational as well as physical needs.

Planning/Design Stage

After initial correspondence, the architect contacts her consultants and engineers. Some are local but others are remote. The team agrees to meet regularly on neutral turf for progress updates. To this end, the architect contrives a site on the World Wide Web allowing many modes of interaction. Most importantly, the site is a multi-user domain that lets the team meet continuously either through use of avatars or by live video feeds from their workplaces.[7] Participants may also do so asynchronously by leaving messages in various formats, much in the way that email and Usenet newsgroups operate.[8]

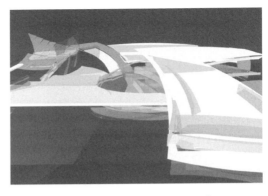

Figure 2. Cybrid library seen from ground

By modelling data taken from a satellite, the architect reconstructs the terrain of the physical site within the cyberspace. All pertinent information is included in this recreation: vegetation, power lines, water and sewer utilities, power, gas and media connections.[9] Significantly, she creates a platform in this model, the meeting space within the project's multi-user domain. When the team gathers, they do so in the presence of the site's simulation in cyberspace.[10]

One strategy involves a kind of game that encourages imagination and cooperation on the part of the team members. In a series of quick agent-assisted exercises the team produces a number of optional solutions that are manifested in alternative incarnations of the site.[11,12] The computer evaluates each schematic solution for fitness according to predetermined constraints such as cost, energy consumption, air flow, daylighting, acoustics.[13]

Another approach uses computer agents assigned to building components that interact with one another, fighting for resources-fresh air, daylight, heat-until they reach a stable, optimal state.[14] Yet another tool at the designers' disposal are genetic algorithms that let them breed a variety of solutions, optimising different parameters whether functional or aesthetic.[15]

The architect and client determine which functions need to be physical and which might take form in cyberspace.[16] These might manifest the business's computer networks, facilitating alliances with remote companies.[17] Some functions may be replaced by cyberspaces, particularly information storage facilities.[18]

Some of these functions are portrayed in cyberspace as ambient environments. Some are contiguous with the cyberspace model of the proposed building. Some hover beyond the site's simulation while others are nested within one another, available by entering them or summoning them categorically into view.[19]

The cyberspace of the team is also a tool for the client. He may check in to attend meetings. The cyberspace can be used for fund-raising activities with the interactive model and multi-user interface. Gatherings may be arranged live 'on-site' or the event may continue throughout the preconstruction phase under the auspices of an automated promotional agent.[20]

Construction and Deployment
Finally the team conducts a directed, automated search for construction specialties.[21] Contractors proposing alternatives to the design change the cyberspace model to show the projected results. The computer support evaluates these proposals for fitness just as it did with the design team's work.

At the same time, contractors for the client's information technology systems offer proposals that refine the cyberspace component of the cybrid. Electronic Data Interchange, intra-extranets, virtual private networks and Internet commerce softwares are formatted within the symbolic domain of the cybrid. Cyberspaces for managing corporate databases attend the proprietary areas of the cyberspace and actual building.

Figure 3. Cybrid library conceptual diagram

Once all parties are in agreement, development of the cyberspaces extends the original work of the architect while the construction of the physical building begins. Meetings still happen in the cyberspace's domain but they at times overlap discussions held on site within the contractor's temporary shed. The two parallel spaces connect with sound and image, each informing the other during the meeting.[22] As the physical construction continues, the team updates the cybereal model of the building. This keeps it up-to-date and ultimately provides an accurate account of what was finally built.[23]

The client, meanwhile, has begun to use the cybrid. Remote prospective employees are interviewed on-line. Since the technical support for the cyberspace may be anywhere, some business can already be conducted through the network months before the physical plant is completed.[24]

Monitoring devices, sensors and servo-mechanisms are installed as the building proceeds. As sections are completed, the devices go on-line to their cyberspace equivalents, making the cyberspace a valuable tool for management of the physical environment. Communications, building systems maintenance and operation, fire prevention and security all benefit from the mutual support of the cybrid's components. Computer agents determine which portions of the building require cooling and control the dampers in the ventilation system.[25] A contractor sent to modify a portion of the electrical system sees through the ceiling using a head-mounted augmented reality display that merges electronic and material spaces.[26]

Occupation and Use
Upon completion the cyberspace of the cybrid expands into a larger multi-user domain. This MUD may have different uses depending on the client's business.[27] It can serve as an intermediary workplace for telecommuting employees and sales contractors. Or it might

become a public Internet storefront for the business. It may even become a place of production, capitalising on its computer/media affordances. Design work done there-as before-may be experienced or sent to devices that render it physical.[28]

The cyberspace is ubiquitous while the building is local. Because of this the cybrid is a tool for the promotion of the business-the way Web sites are currently used for advertising. For it also contains the non-physical components of the business-literature, advertising, communications.

In ensuing years, the physical plant undergoes change. Plans derived from the cyberspace assist in planning additions and modifications.[29] The cyberspace changes as well, users influence its configuration or its information structure changes automatically. Only some of these changes are reflected in the physical structure, most being distinct and untethered. The cybrid's database may constantly evolve according to the users' needs.[30]

The Sweet Hereafter

Finally, the building is no longer required. The business has been relocated, new facilities are needed to meet business demands. The cybrid's cyberspace can supply data for sale of the structure to a new owner or provide demolition plans. Further, with its database model of as-built conditions, it can provide information for the reclamation of materials for recycling.[31] In the end, all that remains of the cybrid is the ghost, the cyberspace that attended the construction, life and death of the building.

And yet it is a lively ghost. So long as a networked computer maintains the database and multi-user domain, the cyberspace remains active. Indeed, it may still be the client's business space regardless of where he has moved. Or it might be an historic archive of the building it once supported. It might even take on a life independent of the physical operation. For instance, a long-defunct nightclub could still host parties long after its physical demise. Just as buildings take on new owners, so too might cyberspaces be recycled.[32]

Conclusion

In this scenario we have seen how the cybrid condition affects the built environment. Yet these observations are necessarily conditional and do not consider social and economic effects. However, cybrids already manifest themselves in local area networks as well as sited multi-user domains. Where they may evolve to next depends on social/economic forces and the imaginations of their developers.

Notes

1. Ross-Flanagan. pp. 52-9.
2. Downes-Martin et al. pp. 28-38.
3. Curtis et al
4. Bruckman et al
5. Anders 1966. pp.55-67.
6. Anders 1997. pp.17-34.
7. Technologies: groupware, file transfer protocol (ftp), telnet, email, 3D multi-user domain software (Chaco Systems World Gen software), HTML, virtual reality modelling language, modelling software, word processing

software, spreadsheet software, animation software, video streaming software, high-speed computer network with Internet connections, video projection, data and telephone connections.

8   Technologies: Usenet newsgroups, HTML.
9   Technologies: CAD software and modeller, Virtual Reality Modelling Language (VRML), access and software to use satellite-supported global data system.
10 Technologies: CAD modeler, VRML, CUSee-me video conferencing software, video-streaming support. 3D MUD software.
11 McCall et al. pp. 153-62.
12 Knapp et al. pp. 147-52.
13 Mahalingam. pp. 51-61.
14 Pohl. 1996.
15 Krause. pp. 63-72.
16 Anders, Peter. 1997. op.cit.
17 Technologies: Electronic Data Interchange-Value Added Networks (EDIVAN),
Virtual Private Networks (VPN), groupware, intranet software.
18 Technologies: Database software actively linked to HTML files available
through network.
19 Technologies: VRML, HTML. This posits the linking capability of HTML be employed in creating interactive VRML models, a capability not yet available.
20 Technologies: Internet telephone software and support, streaming video, 3D MUD software, CUSeeMe teleconferencing software and support.
21 Technologies: Agent-driven Internet search engines, ftp.
22 Technologies: VRML, 3D MUD software, video conference software, data projection system.
23 Technologies: CAD modeling software, VRML.
24 Technologies: video conference software, EDIVAN, Internet commerce software, HTML.
25 Recent research at XeroxPARC proposes the use of forensic agents in mechanical systems that 'bid' for service to parts of a building. The highest bidding agents will receive power, heat, cooling according to need. This market model for building operation can reduce energy costs by up to 20%.
26 Feiner et al. pp. 74-81.
27 Technologies: EDIVAN, Internet fire-wall security, graphic object-oriented programming software (Visio).
28 Technologies: CAD modeller, stereolithography, computer-driven laser software and support, ftp.
29 Modifications to the facility will incorporate many of the cited technologies as though the changes were new work.
30 Technology: Database software and support, HTML.
31 Technology: CAD, Database software, HTML.
32 'The Street finds its own uses for things-uses the manufacturers never imagined'. Gibson. p. 29.

Anders, Peter. 1996. 'Envisioning cyberspace: The design of on-line communities', in *Design Computation: Collaboration, reasoning, pedagogy*. McIntosh, P. and Ozel, F. eds., proc. ACADIA 1996 Conference, Tucson, Arizona. pp. 55-67.

Anders, Peter. 1997. 'Cybrids: Integrating cognitive and physical space in architecture', in *Representation and Design*. Jordan, P., Mehnert, B., Harfmann, A. eds., proc. ACADIA 1997 Conference, Cleveland, Ohio. pp. 17-34.

Breiteneder, C.J., Gibbs, S., Arapis, C. 1996. 'TELEPORT: An augmented reality teleconferencing environment', in

Virtual Environments and Scientific Visualization '95: proc. Eurographics workshops in Prague, Czech Republic 1996 and Monte Carlo, 1996. M. Göbel, ed. Vienna: Springer Verlag. pp. 41-9.

Bruckman, Amy and Resnick, Mitchel. 1995. 'The MediaMOO project: Constructionism and professional community'. *Convergence* 1(1).

Curtis, Pavel, and Nichols, David, 'MUDs grow up: Social virtual reality in the real world'. Austin, Texas: 1993.

Downes-Martin, S., Long, M., Alexander, J. 1992. 'Virtual reality as a tool for cross-cultural communication: An example from military team training', in *Visual data interpretation*: 10-11 Feb. 1992 San Jose, Cal./ Joanna R. Alexander, chair and director; sponsored by SPIE, The International Society for Optical Engineering, ISNT The Society for Imaging Science and Technology. Bellingham, Wash.: SPIE. pp. 28-38.

Feiner, S., MacIntyre, B., Höllerer, T., Webster, A. 1997. 'A touring machine: Prototyping 3D mobile augmented reality systems for exploring the urban environment', in proc. International Symposium on Wearable Computing 1997. Cambridge, MA. October 13-14, pp. 74-81.

Gibson, W. 1992. 'Academy Leader', in 'Cyberspace: First steps. Benedikt, M. ed. Cambridge, Mass: MIT Press. p. 29.

Knapp, R. W. and McCall, R. 1996. 'Phidias II: In support of collaborative design', in *Design Computation: Collaboration, reasoning, pedagogy*. McIntosh, P. and Ozel, F. eds., proc. ACADIA 1996 Conference, Tucson, Arizona. pp.147-52.

Krause, Jeffrey. 1997. 'Agent Generated Architecture', in *Representation and Design*. Jordan, P., Mehnert, B., Harfmann, A. eds., proc. ACADIA 1997 Conference, Cleveland, Ohio. pp.63-72.

Mahalingam, Ganapathy. 1997. 'Representing architectural design using virtual computers', in *Representation and Design*. Jordan, P., Mehnert, B., Harfmann, A. eds., proc. ACADIA 1997 Conference, Cleveland, Ohio. pp. 51-61.

McCall, R. and Johnson, E. 1996. 'Argumentative agents as catalysts of collaboration in design', in *Design Computation: Collaboration, reasoning, pedagogy*. McIntosh, P. and Ozel, F. eds., proc. ACADIA 1996 Conference, Tucson, Arizona. pp.153-62.

Pohl, Kym. 1996. 'KOALA: An object-oriented architectural design system'. Master's thesis, California Polytechnic State University, San Luis Obispo. unpubl.

Ross-Flanagan, Nancy. 1998. 'The virtues and vices of virtual colleagues' in MIT's Technology Review. Mar/Apr. pp. 52-9.

Peter Anders is an architect and owner of Anders Associates, a firm specialising in information spaces. He has published widely, taught architectural design and computer applications in architecture at the graduate schools of the University of Michigan and the New Jersey Institute of Technology. He has just written a book published by McGraw Hill entitled 'Envisioning Cyberspace'. The book presents the design of physical and electonic spaces as human, social and informational environments. It shows the work of artists, designers and researchers from around the world to illustrate principles underlying our use of space and its emulations.
<anders@concentric.net>

Virtual Architecture as Hybrid:
Conditions of Virtuality vs. Expectations from Reality

Dace A. Campbell

Introduction
Today we are engulfed and overwhelmed by information in all aspects of our lives. In dealing with digital media, we are immersed in a chaotic sea of data on a global scale, and we must filter and organise this information to survive in the 21st century. The Internet and its interface, the World Wide Web, perhaps the most complex and comprehensive resources of information on the planet, must be shaped and navigated with care so as to enable and empower those who use them (Best).

For years, designers and theoreticians have mused the nature of 'cyberspace', the collection of spatialised, digital environments made possible by virtual reality technology and the Internet. The idea that information can and should be consciously shaped and re-presented in spatial, navigable form is inherently an architectural issue (Novak). Architecture has a long history in the representation of digital information: from the maps and graphics used to orient participants in virtual communities and Multi-user Dungeons (MUDs and MOOs) to the two-dimensional graphic environments of chat spaces and video games, to the three-dimensional 'liquid architectures' envisioned by Marcus Novak. As the Web incorporates and features data in a three-dimensional format and as that networked 3-D data is presented as an immersive environment, we find that architecture is a valid metaphor for the design of the information portrayed in that environment.

Architecture (Re) defined
Architecture, the making of a place by the ordering and definition of space, as developed in response to a need or programme (Ching), can be considered to include both the design of physical and virtual environments. This architectural expression of a society's or culture's values in spatial, experiential form has traditionally been expressed in wood, stone, metal and glass for thousands of years. Today we are discovering that architecture can be expressed in polygons, vectors and textures with equal validity. Architecture, then, can have a physical or a virtual expression in our society, or both.

Conditions and Constraints
But what is this virtual environment and to what extent does it borrow from its physical counterpart in the shaping of information space? Designers of cyberspace, as well as its inhabitants, need to understand how and why the characteristics of the virtual differ from that of the physical, where they are the same and what implications each has for design.

Many fundamental characteristics of cyberspace are quite different from the constraints that architects face in the physical world. In virtual environments, there is no physical context: no geographic limitations, no site boundaries, no property lines, no climate, not even a gravitational pull against which to create a structure. That is not to say that virtual environments are without constraints, however: new factors contribute to the shaping of an environment as much as the absence of traditional ones when that environment is virtual rather than physical. Technical constraints of processing power, bandwidth and interfaces all affect how a virtual space is experienced and therefore shapes the making of the space. In addition to these technical conditions, human factors such as psychological, perceptual and socio-cultural issues affect the way virtual environments are understood and experienced. While these human factors may not differ from those in the physical world, they are often more pronounced and relied upon to gain meaning from the environment when traditional physical constraints are lacking.

Virtual Architecture

The virtual environment, although it embodies a different set of constraints and conditions than its physical sibling, can use architecture as a metaphor just the same. This *virtual architecture* is a metaphor for the shaping of information space, although a limited one (Campbell). Virtual architecture uses the concept of architectural 'parti' as an overarching spatial, organising principle when dealing with digital information. The use of parti enables information to take an identifiable and understandable form and hierarchy, in much the same way a physical building is the organisation and hierarchy of physical components. Such organisation enables the participant of the space, physical or virtual, to develop a cognitive map as he or she navigates through the environment.

The representation of virtual architecture can fall within a broad spectrum of possible expressions. On one end of the spectrum, it can borrow from physical architecture in full. That is, the expression of a virtual space could look and behave identically to a space designed for the physical environment. Such spaces would be most familiar to participants accustomed to physical space. However, as this would ignore the fact that the conditions of the virtual realm are indeed different from that of the physical, this would often be an inappropriate expression for cyberspace.

On the other end of the spectrum of expression for virtual architecture lies the representation of data into spatial, three-dimensional arrays. This abstract organisation of data into tabular fields is almost non-architectural in that there is little opportunity to create or imply importances or relationships between points of data within these arrays. Such mathematical spaces risk lacking any tangible or graspable hierarchy which can be perceived, understood or successfully navigated by human participants relying on learned expectations of how their environment behaves.

Perhaps more interesting than either of these two extremes are the possibilities that lie throughout the middle of the virtual architectural spectrum: the hybrid. By combining some aspects of each extreme – the mimicking of the physical at one end and the mathematical spatial array at the other – there lies an infinite possibility of exciting combinations. Fragments and metaphors of the physical, such as floor planes, walls and portals, help direct participants unfamiliar with a given virtual environment. Signage, lighting, materiality and even artificial gravity can help aid the foreigner with expectations of 'reality' in much the same way that they do in the physical world. At the same time, portals between spaces need not operate like physical doors, nor do walls need to be solid. Multiple suns and local, non-consistent gravity zones can make an environment more interesting and in some cases more responsive to a given task faced by the participants. The 'baggage' a human being brings into a virtual environment from his or her physical conditioning can be a springing point into a very rich environment filled with opportunities for design not possible in physical reality.

Examples

Researchers at the Human Interface Technology Laboratory (HIT Lab) at the University of Washington (USA) have been investigating the nature of virtual architecture as a hybrid for a number of years. Below, two recent examples are offered and briefly discussed to demonstrate ways in which virtual architecture can exist as a combination of elements from the physical and the virtual.

The first project is the HIT Lab Gallery, an on-line gallery intended to archive, organise and feature dozens of virtual environments (figures 1 and 2). It has served as a test bed of design ideas as relates to virtual architecture. The gallery contains several interlocking elements that create a composition of various spaces, free-floating in a gravity-free space. The default entrance to the environment is a vestibule from which visitors can see and approach the gallery. Directed by visual cues of forms and elements, visitors move along a circulation spine (linear parti), pass a threshold and enter the 'interior'. The spine provides symbolic and directional cues to orient and guide visitors to the gallery rooms. Each gallery room has spatial proportions and local orientations different from the spine, enhancing the sense that these spaces serve different functions. The gallery rooms feature a thick display 'wall' with deep-set images representing various worlds and providing a visual cue for hyperlinks leading to the worlds represented on the images. There is also a main hall, which serves as a gathering space for multiple participants to interact and to use as a jumping-off point to other virtual environments on-line.

A second and more recent example of virtual architecture is the Virtual Playground, an 'urban' scale environment to facilitate on-line commerce and entertainment (figures 3 and 4). The parti is a long, simple, stepped spine bisecting a large 'ground plane'. It features common spaces for gathering and entertainment, opportunities for on-line shopping and a space for meditation. There are enclosed and open spaces, although there is always a view to the main circulation spine and commercial spaces from any point in the environment. This condition was particularly important when designing an environment to be experienced with a narrow field-of-view interface. The environment features gravity, affording a default up-down orientation. It also features dynamic feedback of the state of the environment – like the number of users – expressed as constantly changing weather patterns across the sky.

Conclusions

Many contradictory and paradoxical issues have been discovered from real-time 'walkthroughs' of these and other examples of virtual architecture as hybrids between

Figure 1. Exterior of HIT Lab Gallery   Figure 2. Gallery circulation spine.

physical and mathematical expressions. Broadly, these have been categorised into orientation, navigation, transition, enclosure and scale.

Orientation in the virtual world, like in the physical, can be aided by light sources, atmosphere, gravity or any number of other conditions that our perceptual systems have learned to rely on. However, these elements need not be consistently applied throughout the environment to be successful. They can be manipulated or even discarded when they interfere with a given task.

Navigation in the virtual world is highly dependent on physical interfaces as well as the cognitive abilities of the individual participant. Where a physical environment may have buildings with signage and text, in the virtual world the buildings *are* the signage and vice versa. There is no need to draw a distinction between the concepts.

Transitions are key to our understanding of our world around us. We are constantly approaching and moving away from one place or another as we move through space. Hyperlinking as afforded by virtual technologies can eliminate the need for such transitioning across Cartesian space but humans need such sense of movement across space to develop effective cognitive maps.

Enclosures are required in virtual spaces not for climatic response but for preventing cognitive overload of information. Walls and other opaque elements obscure and filter information for a participant and are necessary despite the lack of physical elements. Interestingly, links are often given a transparent representation to be understood as a portal through which to pass but walls need not have any physicality in their opaque representation. Thus there is a reversal between what is technically 'solid' and visually represented as such.

Without accurate representation of a (human) body in virtual space, it is difficult to perceive what is truly one-to-one, human scale. With the constant distance between eye and foot lacking in virtual space, one's scale is derived as a function of the rate of movement. That is, the faster one moves through a space, given the same amount of effort, the smaller that space appears to the participant without bodily representation.

These and other peculiarities are a direct result of the hybrid combination of physical

– Architecture –

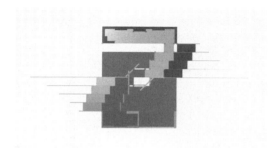 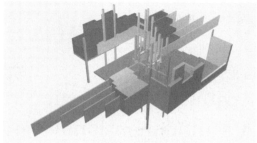

Figure 3. Plan view of virtual playground   Figure 4. Axonometric view of playground

elements with abstract mathematical ones in the formation of virtual environments. One's perception and expectation of how objects and spaces should behave in the virtual are not always what they seem, yet this should be looked upon as an enriching opportunity for the conscious design of our virtual environment. While these issues, and their resolution, differ in the physical and virtual realms, they are nonetheless just as relevant to our awareness of our digital environment for years to come.

References

Best, K. 1993. *The Idiot's Guide to Virtual World Design.* Seattle: Little Star.
Campbell, D. 1996. *Design of Virtual Environments Using Architectural Metaphor*: A HIT Lab Gallery. Seattle: University of Washington.
Ching, F. 1979. *Form, Space and Order.* New York: Van Nostrand Reinhold, p. 10.
Novak, M. 1990. 'Liquid Architectures in Cyberspace' in *Cyberspace: First Steps.* Cambridge, MA: MIT P.

Dace Campbell is a virtual architect in Seattle, Washington, where he holds a joint position between the Human Interface Technology Laboratory (HIT Lab) at the University of Washington and NBBJ architects. He researches the integration of virtual reality technologies and architecture as an industrial fellow at the HIT Lab, and he applies this research in a professional setting as an intern architect at NBBJ. His writing and his design work have been published in all major American architecture trade periodicals and he frequently discusses design concepts related to virtual architecture at conferences worldwide. He holds a Bachelor of Science in Architectural Studies from the University of Wisconsin, Milwaukee, and a Master of Architecture from the University of Washington. He can be reached at any of the following:
dace@hitl.washington.edu
dcampbell@nbbj.com
http://www.hitl.washington.edu/people/dace/

9   Creative Process

Acquiring an Artistic Skill: A Multidimensional Network

Christine Hardy

The mechanistic paradigm with its focus on isolated, independent systems and linear cause-effect relationships is giving way to new paradigms that emphasise self-organisation, interdependence and complexity. Frameworks such as systems theory, neural networks and chaos theory shed new light on complex relationships such as the relationship between artists and their artwork; in particular, these frameworks account for the dynamical evolution of such complex relationships.

    Semantic Fields theory blends a network approach with that of chaos theory (or complex dynamical systems theory). It views learning as a process (Combs 1995), based on connective rather than computational logic, and involving nonlinear dynamics (Guastello 1995). While the computational framework considers the mind to be a computer, executing predefined logical operations on symbols, the neural nets framework views it as a network of elements and processes, organising itself towards an optimal state (vis-à-vis given inputs and/or objectives). A network architecture can further account for the creation of weighted connections between elements and processes. Chaos theory, in its turn, can account for the interaction of forces and the creation of novel organisational states. Both networks and dynamical systems exhibit self-organisational properties, i.e. the capacity of a complex system to reorganise itself internally. The combination of network and chaos theories is a particularly powerful framework for explaining the self-organising features of heuristic learning.

Semantic Fields

Semantic Fields theory views the mind as a lattice of numerous constellations of meaning called Semantic Constellations or SeCos. Each SeCo binds together widely different elements and processes into a meaningful whole such as concepts, sensations, actions, words, memories, etc. It is the act of giving meaning to what we are experiencing and intending that builds the coherence of the SeCos and, ultimately, the semantic lattice. Through each experience, the links and interrelations between elements are modified, thus allowing the SeCo to reorganise itself; the SeCo, in other words, is a network that behaves as a dynamical system and self-organises.

    For example, take a painter looking at a landscape. It would be a gross oversimplification to imagine that the landscape is simply a set of stimuli evoking conditioned responses (concepts, memories or actions), as in the behaviourist framework. Rather, the landscape is itself a complex natural system having developed its own ecological web of interactions; and the mind interacts with this complex system through multiple parallel linkage processes that imply sensory impressions and feelings, meaningful words and concepts, imaginary scenes, a load of past experiences, a certain state of consciousness colouring the evolution of the perceptive process, anticipations as to how to render it all through a painting and so forth. In other words, while contemplating the landscape the psyche-mind is engaging in a rich, complex and multilevelled networking process. Through this unique, novel experience, the painter's "*Landscape*" SeCo develops new internal clusters, grows with novel meanings and new links to related clusters and shifts to a new global organisation.

Another feature of Semantic Fields theory is to pose a transversal network integration of mental and neural processes. This kind of transversal network recasts the mind-body problem, insofar as SeCos integrates processes from lower neuronal processes up to higher rational ones, implicating all levels between (sensory, affective, imaginary, etc.). Human knowledge and ideas, in other words, are never strictly abstract or rational; rather, they are deeply tied to sensory, affective, hormonal, motor, and so forth, processes. This viewpoint is consistent with a growing recognition of the role of nonrational processes in cognition (Goleman 1995). For example, Francisco Varela et al. (1991) hold that cognition develops out of and remains tied to a strong coupling of sensory and motor exploratory behaviours.

The transversal cognitive architecture also sheds light upon the psychological complexes described by psychoanalysis the pathological grouping of traumatic experiences with mind-sets, behaviors, and physiological processes. It also fits Charles Tart's (1975) description of states of consciousness as idiosyncratic patterns of sensory and mental processes, behaviours, mind-sets, knowledge-sets and memory.

Learning a Skill

In the present model, then, learning is the weaving of new links, the elaboration of new link-clusters and the selection of new paths within the SeCo-network. In this sense, there is no distinction between experiencing something novel, creating a link-cluster and learning. For an artist, there really is no substantial difference between creating and learning: learning takes place during the creative process itself, while creation is a mode of learning. The SeCo corresponding to one's artwork will be reactivated, modified or refined through each new creative experience even when the artist has reached a full mastery of her art.

Thus, learning a skill has little to do with knowledge which has been formulated and fixed; rather, it is the complex development of heuristic knowledge, implying continuous changes and self-organisation. The processes underlying learning are based neither on rigid stimulus-response pairings, nor on rational-sequential constructs; they are, instead, connective processes which trigger the mutual interaction of many forces.

Learning an artistic skill may indeed be one of the most complex forms of learning as it involves the progressive coordination of concepts, feelings and sensitivity with gestures and motor control. It also necessitates both the comprehension of general patterns and experiencing creative states. Repetition and imitation are generally discouraged by art teachers, while feelings, imagination and creative states of mind are strongly emphasised.

Let's examine, in some detail, how the SeCo-network constitutes itself in the process of learning an artistic skill, for example sculpting. As suggested earlier, the 'Sculpting Skill'

SeCo integrates all the non-conscious biological, neurological and somatic processes that are tied to higher-level processes (Reber 1993). A spontaneous linkage dynamic will create a cluster of links between sensations, muscle control, tools and materials, as well as qualitative states (affect, intention, state of mind), concepts, names, artistic styles and so forth.

With each sculpting experience (whether with guidelines or a teacher or simply alone), the artist's Sculpting Skill' SeCo will be reconstructed and amplified. For the beginner, for example, a link-cluster at the onset of the learning process might be quite elementary, connecting just a particular tool, its name, the gesture needed to use it and the resulting shape:

< Hold on tool / Tool-Name X / Hand movement Y / Perceived shape Z >

With experience, <Tool-Name> recognition becomes automatic and muscular control and gestures more precise. Concepts and sensations are refined and complexified. Progressively, the execution of adapted movements become semiautomatic. The learning process increasingly focuses upon the quality of artistic expression, vis-a-vis the artist's intentions and feelings. The 'Sculpting Skill' SeCo is amplifed and reorganized into a complex network involving numerous links, such as:

<Intended expression / Meaning / Quality of sculpture / (Tools/Gestures) >

As time goes on, even the mastery of fine movements becomes unconscious; the accomplished artist in the process of creation is nurturing more complex subjective links, such as:

<Intention / Form / Quality of the sculpture (Tools/gestures) / Evoked feeling / Meaning / Artistic associations / State of consciousness>

In this evolving learning process, basic concepts such as 'shape', or 'style', are bound to drastically change. Within the SeCo, each element or process may develop numerous links, and thus the whole SeCo-system may end up being totally different from the original one. Some clusters may grow, others may be discarded, while still other ones may undergo significant changes. Certain clusters may develop profusely and generate an entire sub-SeCo. Other ones may turn out to be so inadequate that they do not undergo any further evolution; they remain a passive memory cluster and may eventually be dropped altogether.

The Mind as a Dynamical Network System

We thus learn by weaving a dynamical network between qualitative experiencing, various neurological processes and higher-level conceptualisations. The whole mind-brain network organisation acts as an endo-context influencing the meaning an experience will take on, and hence, the unfolding of that experience.

But the outside world is itself a complex, meaning-laden network. As stressed earlier in the landscape example, the outside world must be seen as webs of complex, self-organising systems that have evolved specific interrelationships and interactions. In the present model, then, as the mind interacts with the world it develops multiple parallel links with other complex systems; it learns to grasp their evolutive dynamics and their organization as a whole. Consciousness makes sense of the world through a complex web of links and relations, that is, through connectivity and inter-influence of all the elements and processes linked together in the lattice. Consequently, we might state that the meaning of a novel experience emerges out of the complex interaction between a semantic endo-context (the lattice) and a meaning-laden exo-context (the environment). Both endo - and exo-contexts influence (but do not compel or direct) the further evolution of the relevant SeCo.

Thus, we here have a dynamical network model, showing characteristics of both networks and dynamical systems. As a network, the SeCo is created by binding elements and processes through a spontaneous linkage process. Seen as a dynamical system, the clustering of specific chain-linkages between processes may be viewed as trajectories through the state space of the system. The SeCo itself acts as an attractor basin, the trajectories revealing an attractor that will shape subsequent experiences and pull them toward its most recent organisational state. In other words, the attractor is a convergent force that encourages the mind to take the same paths through the network. But insofar as it is a dynamical system showing self-organisation, the SeCo-system may also bifurcate: a change of parameters may provoke a reorganization of the whole SeCo and a modification of the attractor strength or type (Abraham, Abraham and Shaw (1990). In particular, spontaneous linkage processes can act as divergent forces. By creating links with different exo-contexts, this process brings about the modification and evolution of a SeCo.

Conclusion

Semantic Fields theory permits us to view experiencing, creating and learning as the same process. SeCos are the vehicle through which the mind-psyche experiences and reorganises itself; what has been learned is represented by the new organization of the SeCo (and its folded past states). Contrary to behaviouristic or computational models, here the mind is not seen as bound to past experience or obliged to follow predefined operations. The self-organizing properties of the mind-psyche (in terms of both network and dynamical systems) make for dynamical evolution permitting the memorisation of optimum past solutions without forcing their automatic repetition. The model, in other words, allows both for habit and for generative creativity. It explains how the mind-psyche may follow paths of least resistance (the already formed links or trajectories), thus falling into fixed or rigid patterns of thought and behaviour. It explains, as well, how we can move beyond habit how, for example, a person may remain in an exploratory mind-set and sustain certain SeCos in a very labile state, so as to welcome the changes and tranformations brought about by novel experiences.

Many artists exhibit this kind of labile, flexible and highly transformative mental organisation; while developing their art further, their connections to the human and physical environment are extending in richness and depth. Indeed, the hallmark of true creators is their generative and evolving mental dynamics, implying a mutual interaction of thinking, feeling and acting within a meaningful whole.

References

Abraham, F., Abraham, R., and Shaw, C. 1990. *A visual introduction to dynamical systems theory for psychology.* Santa Cruz, CA: Aerial Press.

Abraham, F. D., and Gilgen, A. R. (Eds.) 1995. *Chaos theory in psychology.* Westport, CT: Praeger Publishers.

Combs, A. 1995. 'Psychology, chaos, and the process nature of consciousness'. in Abraham, F. D. & Gilgen, A. R., eds. *Chaos theory in psychology.* Westport, CT: Praeger Publishers.

Combs, A. 1996. *The Radiance of Being. Complexity, Chaos and the Evolution of Consciousness.* St. Paul, MN: Paragon House.

Goleman, D. 1995. *Emotional intelligence.* New York: Bantam.

Guastello, S. 1995. *Chaos, catastrophe, and human affairs.* Mahwah, NJ: Lawrence Erlbaum Associates.

Hardy, C. 1997. 'Semantic fields and meaning: A bridge between mind and matter'. *World Futures*, 48, 161-70. Newark: Gordon and Breach.

Hardy, C. 1998. *Networks of meaning.* Westport, CT: Greenwood Press.

Levine, D. S. and Leven, S. J. 1995. 'Of mice and networks: Connectionist dynamics of intention versus action'. in Abraham, F. D. & Gilgen, A. R., eds. *Chaos theory in psychology.* Westport, CT: Praeger Publishers.

Maturana, H. and Varela, F. 1980. *Autopoiesis and cognition.* Boston, MA: D.Reidel.

Reber, A. S. 1993. *Implicit learning and tacit knowledge.* New York: Oxford University Press.

Tart, C. (Ed.). 1969. *Altered states of consciousness.* New York: John Wiley & Sons.

Tart, C. 1975. *States of Consciousness.* New York: Dutton.

Varela, F., Thompson, E., & Rosch, E. 1991. *The embodied mind.* Cambridge, MA: The MIT Press.

Christine Hardy, Ph.D. Laboratoire de Recherche sur les Interactions Psychophysiques (LRIP), France. President of Interface Psi, a research association investigating extended psychophysical interactions. She has conducted cross-cultural investigations of altered states of consciousness over a period of several years, while travelling in the Middle-East, the Far East and Africa. She worked as research assistant at Psychophysical Research Laboratories of Princeton, USA, while elaborating her doctoral thesis, and has published several works on progressive scientific research. In recent years, Christine Hardy has focused on cognitive sciences, in particular the merging of network and chaos theory frameworks, as outlined in her most recent book Networks of Meaning *(Westport, CT: Greenwood Press, 1998).*
Email: 101515.2411@compuserve.com

Enlarging the Place for Creative Insight in the Theatre Model of Consciousness

Thomas W. Draper

In presenting his theatre model Baars (1997) suggests ways of overcoming these two difficulties. First, neural images and evoked potentials could be compared with verbal reports of mental activities as a means of linking mental and biological processes. For example, Libet (1997) used a combination of evoked potentials and verbal reports to draw inferences about conscious choice. Similarly, Pert (1997) studied emotions, memories and reactions using radio-isotope technology. Similar work is proceeding in a number of laboratories around the world (Glantz 1998).

Baars's (1997) solution to the second problem, that of the seemingly infinite regress of homunculi needed to explain conscious events, is to pose an audience for consciousness's internal actions that is not another consciousness but rather a collection of powerful unconscious and apersonal judging modalities which have evolved to meet the challenges of fitness and survival. These modalities include deterministic filtering templates that allow information that matches predetermined selection criteria into the spotlight of consciousness for consideration. For example, in attempting to solve a problem consciousness might cast a filtering net of templates of known solutions to similar problems and allow the unconscious modalities to iteratively configure a wide variety of known resources until a plausible solution is found.

Of course, proposals like Baars's do not fully resolve the original enigma of a conscious agent that sets criteria and casts nets of filters. But as Dennett (1996) has pointed out, at the current stage of consciousness studies, any move which reduces the number of infinite regresses and circular explanations can be seen as progress.

Among the remaining problems of the Global Workspace model is its limited view of the distinction between creative processes and mundane problem-solving. In this paper I seek to reduce this problem by posing and examining a process for original thought that is similar to the normal problem solving of the Global Workspace model yet consistent with extent explanations of the creative process.

To date, the most scientifically useful models of creativity are process analytic models that rely heavily on mechanistic structures (Mumford et al. 1991). However, such models always include some appeal outside of their mechanistic order to allow for the truly original. At their hearts, all such models invoke either directly or indirectly some inexplicable to account for the origins of something new. Of course, it is not likely that any model will ever completely overcome the logical impossibility of the 'same' producing the 'other'. But some of the current thinking in consciousness studies can be used to reduced the size of the philosophical barrier that separates the two.

Recent work in consciousness and memory allows for a more refined view of original thought and a redefinition of the concepts of memory, agency and randomness (Draper, 1997 Loftus 1993a; 1993b). In the study of memory, the demands of conservation and efficiency have lead to theories that include a high degree of reduction, disassembly and compression in normal learning as well as expansion and reconstitution in remembering (Loftus 1993a). Memories usually consist of a partial emotionally-based framework and lots of 'filling in' by attendant expectations and biases. We most remember that which we find pertinent and novel but not overwhelming. The pertinent and the optimally novel form a skeleton of emotional remembrance that is fleshed out with non-specific 'factual' information from general life themes and knowledge. For example, an individual remembering her tenth birthday party might indeed recall the emotional aspects of some particularly joyful or traumatic event that had happened that day. But just exactly what happened, who was present and the significance of the event are all likely to be added constructions. Such constructions are always biased by the current beliefs and attitudes of the one doing the remembering and serve to prop up the illusion of a single identity or self that is relatively consistent and logically transformative over time (Loftus 1993b).

Reconstructive memory processes provide an evolutionary advantage by reducing the mental resources dedicated to the storage of mundane and oft-repeated information and by allowing for the belief in an internally consistent self. However, while these processes provide the advantages of a functionally significant cerebral economics, they also dictate that substantial errors of memory will be the rule rather than the exception.

In the most popular models of memory, the processes of filling in', reconstitution, and reconstruction are attributed to both deterministic and low-level random processes in the brain (Loftus 1993b). However, random reconstructive processes that play a secondary role in thinking and remembering may play a primary role in some types of original thinking. That is, that which comes to us reconstituted, in some sense, comes to us original. In terms of the Global Workspace model, when we are faced with a problem for which there is no known direct route to a solution, looser and less precise than usual 'recognition' templates representing the best guesses about the form a creative breakthrough might take may be set in the place of the more well defined normal problem-solving templates in the fringe areas that surround the spotlight of consciousness. The various reconstruction modalities in the unconscious 'wings' and 'audience' are then set free to spin in a way that is both fairly mechanistic and more random than usual to produce new forms that could serve as possible solutions to the problem at hand. The new forms would consist of iteratively spun, but looser, combinations of old information. This process would differ from the standard problem-solving process in that it would be substantially less guided by existing associations. Serendipitous combinations of reassembled information that might by chance approximate the looser 'creativity' templates would be periodically allowed into the spotlight of consciousness for evaluation by application. This extension of the Global Workspace model comes closer than the original theatre model in describing the critical 'illumination' step in the traditional preparation, incubation, intimation, illumination and verification sequence that has often been used to describe the creative process (Ghiselin 1952; Koestler 1976; Wallas 1926; Weber and Perkins, 1992).

The appeal to increased random, non-associative, processes in original thinking is certainly nothing new in models of creativity (Skinner 1972). But the present model allows for the more precise location and description of such processes and hence moves the study of creativity more in the direction of observation and testability. For example, neural scanning technologies could be employed to determine if there was a difference in the patterns of energy expenditure when individuals worked on difficult tasks where they possessed all the knowledge for solving the problem compared to when they worked on tasks where they did not posses all of the necessary information and were therefore required to come up with an original piece of the puzzle themselves. Such studies might be useful in isolating areas of the brian that are most involved in creative insight.

Since random processes of assembly occupy a central role in the proposed model of creative thinking, they deserve close scrutiny. In the past, creative insight was typically describe as arising from forms of personified otherness. Inspiration from gods, genies and muses suggested as a likely source of fresh insight (Koestler 1976). But the problem with explanations relying on such Western metaphysical otherness always has been that it is that which is most in need of explanation is located in an unaccessible realm. Indeed, from the time of Descartes, the separation of body from mind and spirit has been the very dividing line of science from non-science.

Modern science has largely resolved Descartes's dualism by resorting to monadism. The material body is all that there is and it works according to logically discernable and discussable principles. However, this move creates a problem for those occasions where one wants to talk about non-mechanistic processes. In the mind of most thinkers, issues like creativity and moral agency have long seemed to require something 'other' than the flow of specifiable mechanistic material operations in a logical structured system. Rather than invoke genies, demons and spirits, modern science invokes randomness as its guiding muse. Randomness can be seen as the sort of otherness that leaves the Western monadic scientist feeling comfortable. But the comfort does not come from monadic science's ability to say much that is material and structured about randomness (Park and Miller 1988). Rather, the attraction of randomness is in its parsimony and the fact that so little can be said about it. In one form, at least, it is a way of having an avenue for the fresh and the original to appear without invoking all of the extra baggage and squabbles that come with an appeal to some particular other-world.

But of course, randomness need only be Western metaphysical 'otherness' if it is 'sincerely' construed (Draper 1997). In one of its dominant formulations randomness is derivable from the failure of measurement in the mechanistic operations of the deterministic material world think of chaos theory or the three-body problem in physics. In such cases randomness is more apparent than real ('insincere'). Such pseudo-randomness would be nothing more than an artifact of the current limitations of assessment technology. In its second popular formulation, common randomness is posited as a extant feature of universe, or as derivable from extent features of the universe such as the quantum wave function. This type of randomness would be both apparent and real ('sincere').

The 'insincere' approach to the random is the most parsimonious since it does not invoke any sort of dimension of 'otherness' that would then be available to make all kinds of things inherently inexplicable. But this simplicity comes at a price. If all is ultimately derivable from mechanistic operations, then human responsibility as well as 'sincere' creativity is done away (Hodgson 1996). We have simply opted for a type of simplicity that closes the gap between human beings and structured mechanisms by only allowing human beings to be considered as structured mechanisms. The 'sincere' approach to randomness also has its dangers. If one is not careful, it runs the risk of either positing a new 'otherness', or simply masking the old 'otherness' with new terminology. Both approaches would limit scientific understanding.

But just as Baars's (1997) Global Workspace model offers a partial way out of 'infinite regress of homunculi problem' by replacing something otherworldly with something materially accessible (at least in principle), so also could the problem posed by a form of otherworldly randomness be solved by recognizing, as modern physics does, a 'sincerely' construed randomness with material origins. However, such a recognition calls into question many long-held elements of standard Western metaphysics, Eastern metaphysics and the most popular type of positive monadism. But such a modification of thought may be overdue. The elements of Western metaphysics that gave us first 'otherness' and then the same-other, spirit-body and mind-body splits may need to be given up as not only unworkable but unnecessary. Eastern metaphysics's rejection of things material may also need to be given up if the conception of matter can be moved into the 20th century and matter can be construed on a continuum of probability running from particle to wave forms rather than on a dichotomy of the real and the unreal. In addition, the familiar type of positive monadism which is an interesting reversal of Eastern metaphysics where the real and unreal ends of the essence-matter continuum switch placescould also be replaced with a wave-partial continuum. Finally, the 'sincere'- 'insincere' distinction between types of randomness can be given up. Since both types ultimately rely on the failure of conscious measurement they may simply both be related to the randomness in the quantum wave function.

Such moves leave one with a world redefined in terms of limited and non-standard monadic possibilities but also one that seems more compatible to the study of consciousness and modern physics. In this redefined world the familiar distinctions between the same and the other, mind and body, the spiritual and physical, the completely certain and the completely random disappear and are replaced with a sort of gross-refined mondadism. The new monadism retains a distinction not between the material and the non-material but between 20th century physics's gross and refined states of materiality. Gross matter is relatively stable and observable in an relatively non-dynamic way. Refined matter has a multi-probabilistic presence (it contains its own proprietary randomness) and cannot be stably observed without being collapsed into a stable gross state. However, all existence retains some degree of materiality running from the highly probable and seemingly stable (e.g. common gross material) to the highly improbable but nonetheless real array of quantum effects.

Having disposed of the great divide between the same and the other, one is free to posit a world that is more consistent with 20th century physics where gross material and random probabilistic material effects co-exist and have a real but limited potential to influence each other. Of course, the existence of actual consciousness or creativity relevant effects flowing out of the quantum world is still largely a matter of speculation (Penrose 1994; 1997). But it is a speculation that is leading to more scientifically testable hypotheses than the former approach. The shift to a 'new metaphysics' makes much that was philosophically impossible in the old metaphysics possible.

References

Baars, B. J. 1997. *In the theater of consciousness*. New York: Oxford.

Dennett, D. C. (1996). *The Myth of Double Transduction*. Toward a Science of Consciousness. April 1996, Tucson, AZ.

Draper, T. W. 1997. Folk psychology and personal responsibility in the study of consciousness. Brain and Self Workshop: Toward a Science of Consciousness, August 1997, Elsinore, Denmark.

Glantz, J. 1998. Magnetic brain imaging traces a stairway to memory. Science, 280, 37.

Ghiselin, B. 1952. *The creative process*. Berkeley, CA: University of California Press.

Hodgson, D. H. 1996. Folk psychology, science, and the criminal law. Toward a science of consciousness, April 1996, Tucson AZ.

Koestler, A. 1976. *The act of creation*. London: Hutchinson.

Libet, B. 1997. Do we have free will? Brain and Self Symposium on neurophilosophy and ethics August 1997, Copenhagen.
Loftus, E. F. 1993a. Psychologists in the eyewitness world. *American Psychologist*, 48, 550-2.
Loftus, E. F. 1993b. 'The reality of repressed memories'. American Psychologist, 48, 518-37.
Mumford, M. D., Mobley, M. I., Reiter-Palmon, R., Uhlman, C. E., and Doares, L. M. 1991. and 'Process analytic models of creative capacities'. *Creative Research Journal*, 4, 91-122.
Park, S. K. & Miller, K. W. 1988. Random number generators: Good ones are hard to find. and *Communications of the ACM*, 31, 1192-1201.
Penrose, R. 1994. *Shadows of the mind: A search for the missing science of consciousness.* New York: Oxford.
Penrose, R. 1997. *The Large, the Small and the Human Mind.* Cambridge: Cambridge University Press.
Pert, C. 1997. *Molecules of emotion: Why you feel the way you feel.* New York: Scribner.
Skinner, B. F. 1972. *Beyond Freedom and Dignity*. New York: Alfred A. Knopf.
Wallas, G. 1926. *The Art of Tthought*. New York: Harcourt & Brace.
Weber, R. J. and Perkins, D. N. 1992. *Inventive Minds: Creativity in Technology.* New York: Oxford.

An Approach to Creativity as Process

Ernest Edmonds and Michael Quantrill

Introduction

A series of compositions has been produced which we define as sequences. As part of this process an input device connected to a computer is used. This device has qualities previously unavailable to an artist using electronic media. These qualities include its physical size (resembling a canvas), the feedback from the drawing tools and the creative space surrounding it. We address the impact on creativity this device has and the questions proposed by the process of creating a work as a sequence.

The artists who participated in the earlier event were Jean-Pierre Husquinet (Liège), Fré Ilgen (Eindhoven), Michael Kidner (London) and Birgitta Weimer (Cologne). Each artist was paired with an expert in computer technology who worked with him or her, identified appropriate computer systems and drew in other experts when necessary.

The outcomes of the week's experiment were very interesting. A point to make is that none of the artists felt able to use a computer system exactly as it was offered to them. In each case, they had demands which could not be dealt with by means of a simple solution. Nevertheless, even in one week, art work was conceived and carried through to completion, albeit with considerable support from other people. At one end of the spectrum, Weimer had produced computer-generated prints that she was pleased with. At the other end, Ilgen had created a Virtual Reality sculpture system. Kidner had solved problems, but as much by mathematics as by computing, and Husquinet had generated visualisations of projected rope installations.

In addition to a basic concern for structure, each artist indicated that the process of exploring the underlying structures of the works produced is an important part of their personal strategies. In this exploration, the products form a notable but not supreme part.

In this sense, the computer's ability to handle structure might be quite significant. Some comments by those artists follow.

Artists in Residence A Response

Kidner:

> I mean structure becomes the nature of the composition and there is a lot of discussion I think with the Russian constructivists in around 1920 as to the difference between composition and construction. And they were all trying to do structures and criticising composition.
> When we were talking about composition earlier, the composition is designed to make it if you like attractive or interesting for the viewer to see, well... I had not thought very much about the viewer except that I am the viewer, so it seems to me that I make things and I have no idea what they will look like or very little idea. I don't really care because I am more interested in resolving my problem and seeing, confirming my theory or not...

Husquinet:

> . . . probably the best word would be structure because everything is based on structure. It's built either visually or musically in structures.
> If it wasn't for the process there would be no interest in the work anyway, so I wouldn't do anything. So it's clear that the process is much more important than the work itself. Well it's proved for example with ropes, the ropes are never definite works because they can be re-used in a different context so once the installation is started it stays there two or three weeks or . . . destroyed. If it's still in one piece I can reinstall that piece somewhere else, but it has no value among the material value, it is just painted rope, it's worth nothing.

Weimer:

> I think process is important. I mean on the one hand you can say that process influences the product . . . This is a very important thing, which has to do with the material I work with and sometimes I have an idea and then I find out that I can not realise it as I wanted it or that one idea was a mistake. . . but even a mistake sometimes is the best way to do a good artwork . . . the process is more important and then the product .

ILGEN:

> You have processes which should also reach temporarily moments of equilibrium and maybe a finished piece is that you are satisfied because you reach this equilibrium, which I like to understand as a moment of balance between yourself and the structure of nature, if you like. But, because also in nature all processes change continuously it is perhaps an analogy that also, just a situation of equilibrium is disturbed again and you have an urge to create another

one, which doesn't make the former piece unimportant but it explains that your desire is not so fulfilled that you never want to make anything new.

Following on from the work of these artists Quantrill began exploring a device we call the softboard.

Description of the Softboard
A whiteboard (4ft by 3ft) is connected to a computer (MAC). The whiteboard is similar in design to any conventional whiteboard except it has a laser matrix across its area. There are 4 colour pens. These look no different to ordinary pens except there is a metallic strip enabling their location to be tracked. There are also small and large erasers.

The application program looks similar to a drawing package. There is a re-sizeable window which maps to the physical whiteboard. The resolution is high (4000*3000 points approximately). Any actions made at the physical whiteboard are immediately represented in this window. Together the physical whiteboard and the computer representation are known as the softboard.

Background to Quantrill's Work
Touch and feeling have been a major part of Quantrill's work in recent years. It would seem that he has also been concerned with product. We express some uncertainty here as it is quite difficult to analyse motives or even to define work in concrete terms. For a long time this was viewed negatively by Quantrill. Indeed, he began using the softboard with mixed feelings. He knew he would have to evaluate the work, yet he had always tried to avoid sustained evaluation of his own work. Edmonds was aware of this. In a defining discussion we decided he would simply work with his sketchbooks as if he were using any traditional media and at the same time he would evaluate the facilities offered by the softboard almost as a separate task. This was a good move as it allowed the work to evolve as a natural result of the process.

With this in mind we hoped as we began this project we would discover emergent properties belonging to both the artist and the medium in tandem. In a sense we hoped that what would be produced would be a fusion between artist and process.

Products and feedback
We are examining the concept that the process of creativity is sufficient product in itself to sustain a piece of work and if this view is adopted by the artist what effect does this have on the way an artist works and will the artists own evaluation of what is occurring continually change as the process continues?

We mention products because they exist and impact on process. In the simplest of terms we have three products from this work:
1. The feedback to the artist from engaging in the process.
2. The physical product.
3. The visible product

These products effectively encapsulate a set of images as a sequence of pages. These really exist in a virtual world with properties of visibility. We feel it is important to be aware also of the underlying structure of these images and pages. That they are in fact sets of binary data held within a machine, a sequence of controls for some electronic device. However, it is the concept of pages and sequences that form the main focus of Quantrill's work with the softboard and it is these that he now discusses in more detail.

> 'From time to time I have used graphics tablets and mice to input drawings to a computer, but found them very unnatural and disappointing. Indeed after a while their limitations usually meant I would abandon the work or at least be unimpressed by the results. Using the softboard is quite different. I am able to stand and move about as I would when working with a canvas. This enables complete freedom of movement encapsulating the ability to step well back from the work to assess its progress. For me this freedom of movement is very important. It enables a creative space to develop that allows the work to progress without my being aware of the constraints usually associated with electronic media and having to make allowances for them. In the context of my own work this space is a dynamic and integral part of the creative process.
>
> The pens used give positive feedback and leave real, visible marks on the board. This is no small point. Drawing with ink and pen is a direct and physical act, which engenders positive feelings. Drawing with a piece of plastic on a graphics tablet is somewhat artificial and impacts strongly on the feel of the work as it evolves.'

We propose that Quantrill's feelings about the media as he just described enabled him to participate in a process involving both himself and the computer where the latter was very conspicuous but to Quantrill, as he worked, was invisible. In other words he was engaging in one process but another, perhaps more interesting one was going on under the surface.

The Process

Drawings are entered onto a 'page' using four pens. A page is one virtual workspace displayed on the monitor. A set of pages forms a sequence. When any mark is made it is recorded as a set of points for the current page. Both positive pen marks and eraser events are recorded. At any time a new page can be generated as a new blank canvas or inclusive of the previous pages marks. The controls for starting and stopping recording and entering new pages are situated both at the whiteboard and within the window on the computer. This enables the artist to complete a whole sequence with pen in hand, never having to touch the computer.

The concept of sequence is fundamental to the work. When a sequence has been recorded it can be played back on the monitor. What is actually shown is a trace of every mark made whether this is a pen mark or an eraser event. This gives real insight into how a piece has been constructed with no decisions or revisions hidden. In a sense one can view the work as if watching the invisible hand of the artist as he struggles and wrestles with the creative process.

What is truly interesting about the works produced is that they only exist in the form of a sequence. They are not static drawings. They are not animations, nor are they film sequences but they contain elements of each. We therefore have the notion here of a composition as a sequence. Quantrill states:

> Once I got to grips with the work as a sequence, this immediately changed the way I started to think. Sometimes when I work I am uncertain whether I should continue with the piece. This can be a problem. However with the softboard I can stop at any point, assess the work and continue. I can continue with an inclusive page which relates to a typical drawing process or I can continue with a new page. Either way the image is preserved at the decision point. With any sequence I can if I wish return to any page and continue to develop the work in a different direction.
>
> Realising the fact that marks made on the whiteboard remain I used the new page option to continue a sequence using some of the lines and marks on the whiteboard. This meant a sequence would exist that was a natural progression of a drawing and then only parts of the underlying structure would be used after the new page was selected. So almost a new work would continue and yet the sequence as a whole would contain both new and old.

We believe what is captured by a sequence is a rare insight into the creative process which includes not only the visual but the cognitive components. Quantrill continues:

> Taking this process one stage further I started to use only the new page option whilst drawing. Using inclusive page builds up a drawing layer by layer with the sum of all pages visible on the last page and greater than the whole. This time the sum of the pages equalled that whole. The last page simply contained the final part of the composition. In this case the playback of each page could be set so that a pause between pages could occur. This reinforced the idea that each page was a composition in itself with the concept that it formed a larger whole remaining significant but quietly hiding in the background.'

Conclusions

The key point is that process proved to be the most important aspect of the developments. In engaging in the activity Quantrill discovered new concepts and new ways of seeing his work. Without engaging in the process, only what is already known can be said. It is the process that raises new questions. It is the process that enables answers to be found if looked for. Quantrill states finally,:

> For me the real creative force is contained within the process. The initial idea is important. This is the input, but it is the process that sustains the idea and develops it. The final form of the product is initially undetermined and may even be rejected. However, the creative dialogue generated by the process can be all that is needed to sustain many works. Husquinet said 'If it wasn't for the process there would be no interest in the work anyway, so I wouldn't do anything.'

Biography

Ernest Edmonds, Professor of Computer Studies and Executive Director, LUTCHI Research Centre, Loughborough University, <e.a.edmonds@lboro.ac.uk>.

http://www.lboro.ac.uk/departments/co/personal_pages/edmonds.html
Michael Quantrill, Artist/Researcher, LUTCHI Research Centre Loughborough University,
<m.p.quantrill@lboro.ac.uk>.

We are Having an Idea: Creativity within Distributed Systems

Fred McVittie

Pants

Pants Performance Association is a five-man company of interdisciplinary artists and performers who, since 1990, have attempted to by-pass traditional methods of making or devising in order to ask questions about the relationship between individual authorship/creativity and aesthetics. Of particular interest here is a performance piece developed by the company between 1992 and 1995 called *SPAM*. For this project, the company constructed a series of rules which administered their interpersonal behaviour during the period in which the piece was made. These rules, based loosely on certain ideas found in mathematical game theory, delineated how members of the company could interact creatively with each other, how ideas could be generated, how material could be brought into the process, how modifications to existing material could be made, etc. An important aspect of this rule system was that it be sufficiently complex and interconnected that no one individual could gain mastery over the system, but rather that all individuals were equally implicated in its effects. The aim of the project is to force the emergence of an aesthetic from the workings of an interpersonal system.

Scene One - The Choreography

Players 1 - 5 play a game in which the material which will comprise the show the audience are about to see is processed.

Whether or not a piece of theatre works is directly related to the way in which the making of the piece is carried out; that is, to the process. In a director-led devising process, material is deemed good or bad in relation to a central consciousness. In *SPAM* a different kind of aesthetic is exercised.

Scene Two - The Lecture

Player 5 lectures on the semiotics of William Shakespeare's prologue to Henry V, illustrating his lecture with diagrams on an overhead projector. Player 4 enters wearing half a pantomime horse costume. Player 5 continues his lecture after joining player 4 in the costume. Enter players 1, 2 and 3 executing an imitation of a Lippizaner dressage display to music which drowns the voice of player 5.

In the opening speech of Henry V, Shakespeare lays out the conventions of the play for the audience literally inviting them into the 'world' of the play. This is to effect an alignment of the audience with the perceptual set necessary to appreciate the play's form. Henry V is a 'good' play for various reasons, not least because the process which made it is a model of creation we understand and possibly share, a hierarchy of material clustered round a central authorial position. A different perceptual set is needed in order to see the results of *SPAM*'s de-centred creative process as 'good'. *SPAM* has its roots in the notion of dispersion and has little in common with the divine originality of Shakespearean drama.

Scene Three - The Composition

The letters 'JC' are placed on an overhead projector by player 4 who proceeds to recite a long list of celebrities whose names begin with these initials. When the list reaches the name 'John Cage' players 1 and 3 enter and perform a choral composition in the style of John Cage. Enter players 2 and 5.

At the start of the theatre-making process nothing is distinguished from anything else. It is only when something is marked out and given special significance that the possibility of meaning is created. Random letters are transformed into material and given relevance by their correspondence with a list of names. The name John Cage is distinguished from the others when the scene is intersected by the performance of a composition in the style of John Cage. The implication of the events in Scene Three is that significance is not contained within the material itself but within the process by which attention is drawn to and focused on that material.

Scene Four - Democracy

PLAYERS 1 AND 3 COVER THEIR HANDS IN CHALK AND STARE AT EACH OTHER INTENSELY FOR SEVERAL MINUTES. WITHOUT WARNING THEY SLAP EACH OTHER SIX TIMES ABOUT THE FACE VERY HARD. PLAYER 2 ENTERS AND STANDS IN A SPOTLIGHT WHERE HE PERFORMS A TRICK WITH A CIGARETTE. PLAYERS 4 AND 5 ENTER WITH A VCR AND MONITOR AND WATCH A VIDEO OF A PREVIOUS PANTS SHOW 'DEMOCRACY'. PLAYER 2 FINISHES HIS CIGARETTE AND TURNS OFF THE VCR.

Through their allegiance to a creative bureaucracy all players have equal responsibility for the selection of material. By taking centre stage, two of the players can draw an accentuated amount of audience attention to the material they are performing. However, a spotlight is all that is needed to distract that attention. Attractive technology might vie for a position at the centre of the scene but in the end the separate activities themselves don't really hold up to close scrutiny - a bathetic exchange of slaps, the inept performance of a magic trick, an image of an 'other' show. *SPAM* is an experiment in emergent aesthetics. Not only is the medium the message but in this case the message is the method.

Scene Five - The Refrain

Player 2 teaches players 1, 4 and 5 an Italian melody one note at a time. At the same time player 3 performs one half of the Cage composition from scene three. On the instruction of player 3 everybody dances.

The arduous process of learning the notes and rhythm of an already-documented composition - the waltz *Tempi Lontani* by Luigi Oreste Anzaghi - vastly outweighs the final rendition of the piece both in terms of technique and in the sheer time it takes to be taught all the notes compared with the time it takes to sing them. This scene de-privileges the product of a creative process in favour of the means by which it is produced.

Scene Six - The Prisoner's Dilemma
PLAYERS 1, 2 AND 4 EXPLAIN THE RULES OF A GAME: 'TWO PEOPLE ARE CAUGHT AT THE SCENE OF A CRIME. THEY ARE TAKEN TO SEPARATE ROOMS AT A POLICE STATION AND INTERVIEWED. THEY ARE TOLD THAT IF THEY BOTH PLEAD GUILTY THEY WILL RECEIVE SENTENCES OF TEN YEARS, IF ONE PLEADS GUILTY AND THE OTHER REMAINS SILENT THEY WILL RECEIVE SENTENCES OF ONE AND TWENTY YEARS RESPECTIVELY. IF THEY BOTH REMAIN SILENT THEY BOTH RECEIVE SENTENCES OF TWO YEARS FOR BEING PRESENT AT THE SCENE OF A CRIME. PLAYERS 1, 2 AND 4 PLAY OUT THE GAME WHILE PLAYERS 3 AND 5 DANCE TO POP MUSIC.

The Prisoner's Dilemma is a simple model of the complex nature of decision-making within a group. The future of the prisoners depends not on how they plead individually but on how they plead in relation to each other. When translated to the devising process, this game might correspond to a technique by which individuals in a company contribute towards the creation of a show. One player's input, in relation to the input of the others, manifests itself in the final manifestation of the. An allegiance to this process results in material that is independent of personal aesthetics.

Scene Seven - The Play
Players 1 and 3 cover their faces in chalk and face each other as in scene three. They perform the last six lines of *Waiting for Godot* by Samuel Beckett. Players 4 and 5 enter and perform a short choral work in the style of John Cage. When they reach the instruction in the score that reads 'French Farce' all five players enact a scene in the style of a French farce. Players 4 and 5 finish the choral work.
   In Scene Seven Pants employ several strategies for simulating meaning.
1. By being seen to select specific material.
2. By implying a relationship between this material and other material in the show.
3. Through the creation of a context which lends meaning to the material.

The work centres on the crucial differences between what is good and what is bad, what is sense and what is nonsense, what is meaningful and what is meaningless.

Scene Eight - The Resurrection
PLAYERS 3 AND 4 RE-ENACT THE SIX GOALS OF THE 1966 WORLD CUP FINAL WHILE PLAYER 2 LIES IN THE CENTRE OF THE STAGE WEARING A YELLOW BALLET TUTU. PLAYERS I AND 5 ENTER AND BEGIN TO SING PART OF A MELODY BY THE GERMAN COMPOSER LEONARD VON CALL. PLAYER 2 JOINS THEM AND THEY SING THE MELODY IN THREE-PART HARMONY WHILE PLAYERS 3 AND 4 RE-ENACT THE SIX GOALS OF THE 1966 WORLD CUP FINAL A SECOND TIME.

– Creative Process –

Imagine if every year since 1966 a team of footballers from England and a team from Germany met at Wembley Stadium and re-played the 1966 World Cup Final. Not played again, but re-played move for move (each pass, each kick, each goal etc.) The game is re-played in this fashion for 20 years at which point a new game appears to be emerging. The idea of this game isn't to win - England will always win in extra time but to execute better, to play the parts with more feeling, with more accuracy, in short, to re-enact better. The arbitrary has become habitual and a new game has evolved. Like *SPAM* this game would be theatrical demanding the representation of character, the execution of choreographed moves and the interpretation of a score.

Scene Nine - The Audience
PLAYER 1: 'YOU HAVE THREE MINUTES.'
PLAYER 2: ' I AM THE SUPREME BEING; ARCHITECT OF THE ENTIRE UNIVERSE AND KNOWER OF ALL THINGS. I AM PREPARED TO ANSWER ANY QUESTIONS YOU MIGHT HAVE.'

God is God because He knows everything. On a human level, the individual expert has the knowledge and power to create and destroy the world as we know it. Throughout history theologians, philosophers and scientists have repeatedly altered our understanding of the basic nature of reality. An individual tends to become believable, and thereby achieve expert status, through adopting one or more of the following strategies:
1   By being seen to be selected as an expert.
2   By implying a kinship with or association with other experts.
3   Through placing oneself in a context where one would expect to find expertise.

As uncritical supplicants to one player's deity, the others invest his position with belief. In some ways all the scenes in *SPAM* reflect this authorial position. They are all answers to an unspoken question (What is this about?). In Scene Nine, however, where the idea is given overt form, our expert talks absolute shite. Furthermore his expertise allowed to last for just three minutes - at the end of the show, the lights go out. Life goes on.

Scene Ten - The Reprise
PLAYER I GIVES A RENDITION OF '*TEMPI LONTANI*' BY LUIGI ORESTE ANZAGHI. PLAYER 2 SMOKES A CIGARETTE. PLAYER 3 PERFORMS ONE HALF OF A CHORAL WORK IN THE STYLE OF JOHN CAGE. PLAYER 4 READS THE SCORE OF THE SHOW USING A SERIES OF LETTERS AND NUMBERS AND PLAYER 5 RECITES THE PROLOGUE TO HENRY V BY WILLIAM SHAKESPEARE.

> O for a muse of fire, that would ascend
> The brightest heaven of invention,
> A kingdom for a stage, princes to act,
> And Monarchs to behold the swelling scene.
> Then should the warlike Harry, like himself,
> Assume the port of Mars; and at his heels

Leash'd in like hounds, should famine, sword and fire
Crouch for employment
But pardon gentles all,
The flat unraised spirits that hath dar'd
On this unworthy scaffold to bring forth
So great an object.
Can this cockpit hold
The vasty fields of France? Or may we cram
Within this wooden O the very casques
That did afright the air at Agincourt?
Oh pardon, since a crooked figure may
Attest in little place a million;
And let us, ciphers to this great accompt,
On your imaginary forces work.
Suppose within the girdle of these walls
Are now confin'd two mighty monarchies,
Whose high upgrade and abutting fronts
The perilous narrow ocean parts asunder
Piece out our imperfections with your thoughts;
Into a thousand parts divide one man,
And make imaginary puissance;
Think when we talk of horses, that you see them
Printing their proud hooves i'th receiving earth;
For tis your thoughts that now must deck our Kings.
Carry them here and there, jumping o'er times,
Turning th'accomplishments of many years
Into an hour-glass, for the which supply,
Admit me Chorus to this history.
Who prologue-like, your humble patience pray
Gently to hear, kindly to judge, our play.
      (Exit)

Epilogue

There is no guarantee, of course, that an emergent property, of which a group aesthetic may be one example and individual consciousness another, is any more complex, is 'smarter', than individual elements making up the substrate from which that property emerges. If anything, the evidence seems to suggest the opposite: that emergent properties tend to operate on a simpler register than the elements of which they are composed. An instance of this might be crowd behaviour which follows clear patterns and seems to demonstrate a rudimentary 'intelligence' inasmuch as it follows identifiable heuristics; yet the behaviour of a crowd is unlikely to approach the complexity exhibited by an individual crowd member. It has been suggested that the I.Q. of a crowd can be arrived at by taking the average I.Q. of its members and dividing this by the number of persons present. Likewise, although the human organism is capable of behaviour which approximates intelligence, the complexity of its cognitive processes, its ability to think, does not begin to approach the complexity of the neurological and biochemical elements which comprise it. This being the case, it is perfectly possible, indeed likely, that the aesthetic sensibility displayed by an interpersonal system would be more rudimentary than that possessed by any one individual.   In art such as *SPAM*, in other words, there is no real way of knowing whether it is the work of a systematically emergent idiot or a trans-personal, other-worldly genius.

References
Boden, Margaret A. 1995. 'Could a Robot Be Creative - And Would We Know?' in Ford, Kenneth M.. Glymour, Clark and Hayes, Patrick J. (eds.) *Android Epistemology*, London. MIT Press.

Csikszentmihalyi, M. 1988. 'Society, culture, and person: a systems view of creativity', in Sternberg, Robert J. *The Nature of Creativity*, Cambridge: Cambridge University Press.

Fred McVittie is Head of Live Arts at the Crewe + Alsager Faculty of Manchester Metropolitan University.
f.e.mcvittie@mmu.ac.uk
http://cra.als.aca.mmu.ac.uk/adp/fmcv.htm

To be or not to be . . . conscious
Eva Lindh and John A. Waterworth

1. Introduction
Most of the models that we use and that are expressed as computer models give the impression that they are very objective and based on objective information. Another characteristic with today's models, especially the models we use in computer applications, is that they are rigid, frozen in time and do not to a large extent reflect our experience of space. The support for space is improving with the use of multimedia, synaesthesia and VR. This is especially true when it comes to computer games and applications for amusement.

A way to improve the interaction is to loosen up the boundary between the user and the computer, bring the two parts closer together and even making them intertwine. This could be done by presenting information directly to the senses so that the user gets a bodily experience, which in turn gets the user involved and engaged in the interaction. Applications that use this approach, so called sensational interfaces (Lindh 1997), can be used to present information in more intuitive ways, for example within medicine and banking. Another application area is as tools for people with some kind of disability, either physical or psychological. These kind of tools present an alternative way to communicate and therefore could be used to improve the communication for people who have problems communicating with words.

2. CharM A Model of Models

A model is an abstraction of something. It could be a part of the reality or part of something that does not exist (Figure 1). The model is separated from the thing it depicts and it is always a simplification and designed for a certain purpose.

There is a three-part relationship in modelling between the person who creates the model, the model itself and what is being modelled (Figure 2) (Wartofsky 1979). A model is always in a sense more or less subjective in that the designer or the user adds his/her own interpretation of what is being modelled.

CharM is a tool to understand the implications of the model and to put the model in a greater perspective. It is aimed at designing or evaluating a model intended to be used for a certain purpose. The best use of technology is when the interaction combines the best parts from both sides in the communication, the technology and the human being, into a coherent whole. This means letting the technology do the things it does best and letting the humans do what we do best. CharM is a tool to understand what effect a specific model has on its user. For the model to be effective and for it to achieve its goal, it is very important to be aware of the effect the model has on the user. Consciously placing a model in CharM-space can improve the interaction between the user and the computer.

CharM The 3D Model of Models

The 3D model (Figure 3) explains whether the model appeals to conscious doing or conscious being, if the model is intended to be objective or subjective and, last but not least, whether the model is used to understand or experience something, for example an artefact, a task or an event. Most of today's traditional computer applications are directed towards conscious doing because they appeal to abstract representation and human cognition. On the other hand, most computer games and computer art appeals to conscious being, which means concrete representation and human perception. A fuller discussion and definition of conscious doing and conscious being is found in Waterworth (1997).

Another aspect of the 3D model is whether the model is to be seen as objective or subjective. The traditional computer application is intended to take an objective view of what it is modelled. It is seen to be an objective presentation of what it depicts. On the other hand, computer games and computer art do have a subjective presentation of what they

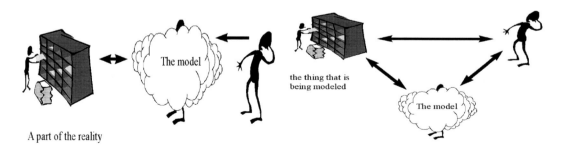

Figure 1 A model

Figure 2: The three-part relationship in modelling

268

depict, so the observer is allowed to interpret it the way she wants, which in fact is a goal with art and games.

The third axis of the 3D model considers whether the model is used to understand or experience a certain phenomenon or object. It can be used to understand and take a decision in a problem-solving situation or to experience an object or phenomena, as within art or a computer. OSMOSE (Davies 1997) is one example of a model aimed at experience by its user and there exist architecture models of building so the user, by using VR, can walk around in the building which does not yet exist, and so on.

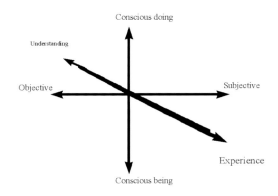

Figure 3 CharM's 3D model

After putting the model in CharM space it is important to analyse the model according to the CharM checklist. This is done to understand and see what implications a certain model brings with it to its intended user. The CharM checklist analyses the rationale behind the model. The checklist is split into two parts, the generality part and the usage part. The generality part deals with the purpose, attributes, structure and the assumptions behind the model, whereas the usage part inspects the use of the model and whether the model is clearly expressed to the user, that is whether it is easy for the user to understand the model, its implication, limitations and so on.

CharM Usage - the relation between the conscious and the unconscious

The CharM model is used as a tool to improve the communication between technology and its user. It is a way to bring the two parts together and, in fact, let them intertwine. By doing that it is also possible to use technology to expand human consciousness and for the technology to appeal to both the human conscious and unconscious mind. One outcome of this expansion is the possibility of technological support for human creativity.

The relation between the conscious and unconscious is very complicated but very important. Figure 4 presents a gross oversimplification but makes the point that the 'door' between the two may be closed, open a little, or open wide; and this has profound implications. Creative insights have often been associated with the activities of the unconscious mind. Perhaps the most familiar example is that of breakthroughs following the experience of a particularly striking dream. If we take a Jungian perspective, we can account for this dimension of creativity in terms of the relationship between the conscious ego, the individual unconscious and the collective unconscious (Jung 1953-1979a). Linking the cognitive, abstract stream of thought with the physical, bodily stream amounts to linking the conscious with the unconscious, since most bodily processing is unconscious. According to Jung, the goal of personal development is the expansion of consciousness to encompass more of the unconscious. Or, in our terms, to open the door between the two.

Creativity is, at least in part, the ability to let the unconscious mix different kinds of information and then transfer it to the conscious part of the mind, which can make a model of it and then see, check, execute and test that model. For this to work, it is important that the information is initially unconscious and unstructured, so that mixing and selection can take place without the conscious setting the rules.

A new idea starts with stimuli which evoke concrete and low focus thoughts. This part of the process is controlled by the unconscious part of the mind and gives rise to 'sparks' of ideas. In this part of the process, emotions play an important role. When an individual is in low focus though he has access to his whole memory and this also entails the use of emotions. After a while one or a couple of these sparks appear and become clearer than the rest and this is perceived by the individual, who becomes aware of the thought and conceives it. This is the start of the process of checking and testing this new idea. He transfers the thoughts along the spectrum of thoughts from low focus to high focus. The control is accordingly transferred to the conscious part of the mind and gives rise to abstract thought, which starts to reason and gives the idea a clearer shape. When the idea is clear, the individual starts to reflect further about the possible new idea and, if necessary, modifies it or otherwise a new idea is born. If the individual finds it necessary to modify the idea, the process starts all over again and the skeleton of the new idea is transferred to experiential mode again.

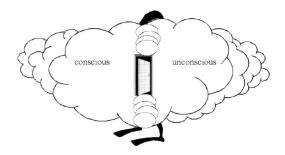

Figure 4 The relation between the conscious and unconscious

Conclusion

This article outlines a model of characteristics that describes the rationale behind models of artefacts that are designed today or will be designed in the future. These characteristics are often hidden and used in the models more or less unconsciously; they are not documented anywhere and sometimes have not even been considered during the design process. The characteristic model (CharM) could be used to evaluate existing artefacts or to deliberately place a model in a specific place, aimed at a certain goal. It should be an aid to creating realisation instead of mere visualisation and to bring the body and the mind together to enhance our perception of the artefact and our being in the world. Another advantage could be to help us be more conscious about a part of our unconscious, to expand our conscious.

Our title refers to Hamlet's tragic dilemma. Oblivion may seem more inviting than facing reality. We often defend ourselves from ourselves, and the price of expanding consciousness into the unconscious is to expose oneself to oneself. CharM allows us to address the question: to be conscious or not (or how much?) and of what? Should we keep opening doors until a monster pops out, or nail them up and make a nest in the hall? Dare we revel in our potential for creativity, self-knowing and expanded consciousness? Such revelry may yet transform devilish tragedy into divine comedy.

Note
A full version of this paper, with full details of the CharM model, is available from the authors.

References
Davies, C. 1997. 'Changing Space: VR as a Philosophical Arena of Being', in *Proceedings of Consciousness Reframed: 1st International CAiiA Research Conference*, University of Wales College, Newport, UK.

Lindh, E. 1997. 'Sensational Interfaces: Function and Passion'. *STIMDI conference proceedings*. Linkoping University, September 1997, pp. 41-4.

Jung, C. G. 1953-1979a. 'Relations between the Ego and the Unconscious'. From Two Essays on *Analytical Psychology, Collected Works* vol. 7, Second Essay, Paragraphs 202-95.

Wartofsky, M. W. 1979. *Models: Representation and the Scientific Understanding*. Dordrecht, Holland: Reidel.

Waterworth, J. A. 1997. 'Back to Being: VR, the Mind in the Body and the Body in the Mind' in *Proceedings of Consciousness Reframed: 1st International CAiiA Research Conference*, University of Wales College, Newport, UK.

Biography
Eva Lindh is a PhD candidate at the Department of Informatics, Umeå University, Sweden. She has taught human-computer interaction (HCI) for the last 11 years. Her research focuses on new approaches to HCI, especially the design of VR tools for sensory and communicational enhancement.

John Waterworth is a Senior Researcher at the Department of Informatics, Umeå University, Sweden. He holds a PhD in Experimental Psychology. For the last 18 years he has worked mainly on human-computer interaction (HCI) including eight years with British Telecom Labs and six at the Institute of Systems Science, National University of Singapore. His current research focuses on the design of multimedia environments for learning and creativity and on psychological aspects of VR and cyberspace.

Eva Lindh and John Waterworth, Department of Informatics, Umeå University, S-901 87 UMEÅ, Sweden.
+46 90 786 67 71
<eva@informatik.umu.se>, <jwworth@informatik.umu.se>
http://www.Informatik.umu.se/~eva/ and
http://www.Informatik.umu.se/~jwworth/

Part IV - Values

10   Values

10 Values

CyberArt & CyberEthics
Colin Beardon

Introduction
The purpose of this paper is to demonstrate that the hypothesis that machines have consciousness requires an ethical analysis. Without such an analysis, the central debate of this conference lacks depth and humanity and we are unable to discriminate between the different technologies presented to us.

The current debate over machine consciousness requires that we first define what we mean by consciousness when we apply that term to humans. The attempt is then made to recreate that state within a computational model. If it is successful, it is claimed that we have created 'machine consciousness'.

As a form of argument it is, of course, very weak. As part of the process of launching a new model of an automobile the stylists will make a real-size clay model of the entire car. It certainly looks like a car but, of course, in many respects it is definitely not a car. It is a model, created specifically to explore certain aspects of a finished product while ignoring others.

My approach is not to try to resolve this issue of machine consciousness but to question whether our approach to the problem makes complete sense. By concentrating only on consciousness, and ignoring other aspects of our humanity, we are not only doing bad science but are also in danger of pursuing cultural forms that are irrelevant to our needs.

Can machines be conscious?
I will characterise two fundamentally opposed views on the existence of machine consciousness in terms of a contest between two opponents. The protagonists in the blue corner state that consciousness is a property that is definable separate from any particular object that may possess that quality. As a consequence, it is possible that machines can think. This is because there is no good reason to believe why they could not achieve whatever it is that makes the adjective 'conscious' applicable to an object. In essence this was the position argued by Alan Turing (1950) and has been repeated in many forms since (see, for example, Herbert Simon's claim that 'there are now in the world machines that think, that learn and create'. (Simon & Newell, 1958) and recent writing by Daniel Dennett (1991)).

The protagonists in the red corner state that consciousness cannot be simply defined in isolation from the particular thing that is conscious: it is, if you like, a mode of 'being' rather than a simple property. Therefore we cannot say that a machine is conscious because machines simply are not one of the type of things to which the word properly applies. This I take to be the essential position of both Hubert Dreyfus (1972) and John Searle (1980).

As a debate, this is hollow (as hollow as a clay model of a car). We must find a dimension through which it can be transformed into a meaningful dialogue.

An experiment

Imagine walking down the road one day in the company of a close friend. You are chatting away, referring to happy times you spent together in the past. Perhaps you are also planning some future activity together, maybe a trip into town or a holiday . . .

Then your friend trips up and falls onto the pavement. Initially you are a little concerned. Is your friend hurt? Do you need to give medical aid, or perhaps call an ambulance? You look for tell-tale signs of injury.

Your initial shock gives way to amazement as something very strange happens. From your friend's elbows there begins to emerge blue smoke. Strange metallic sounds can be heard deep inside your friend's body. One arm begins to split open and, where you would expect to find flesh and blood, a mass of tiny wires becomes visible. Joints start to bend the wrong way and finally a circuit board pops out where an ear used to be.[1]

I invite you to investigate what your feelings would be at this point. I will not presume to speak for you but I am fairly sure that I would experience a mixture of strong emotions.

I expect that I would feel *confusion* over how to mentally classify this past human / present machine. I would, quite literally, not know how to think straight because so many memories are based around the assumption that this was a person, not a machine.

I expect that I would then begin to experience *doubt*, wondering how I had been led down this particular garden path so successfully. Should I have known? Were there clues that I missed? Is my reasoning deficient? Am I particularly gullible? I would want to know why I didn't know.

Or perhaps I was being deliberately deceived? My confusion might well spill over into *anger* as I felt that, from the point of view of knowledge - or is it trust[2]? - I had been let down.

Quite separate from the emotions I might feel about my own beliefs, I think I would also feel a sense of *emptiness* and *waste*: a sense that the emotional and other resources that I had invested in the friendship had been unnecessary. I might as well not have bothered. I might remember the time that my friend's car broke down late at night and I travelled a long distance in bad weather to help. I made a genuine sacrifice, but now I know that it was pointless. I could have left the car and my friend together until the morning. They are, after all, the same kind of thing.

And what of the future. What if the friend / machine were invisibly mended? Of course, I now know and the nature of our relationship would never be the same again.

We can distinguish humans from machines

What does this experiment show? First, let us state the obvious - we can distinguish people from machines. In order to make sense of what Kant described as the 'phantasmagoria of experience', I divide experiences up so that they relate to different objects. I further divide these objects by classifying them into various types of object. The subjective 'I' is, of course, a very special thing but I also recognise myself as an object like certain other objects. These are mainly softish squashy things that often move around and grow in a particular way - things like pumpkins and rabbits - but these are not what I would call 'defining features'. Machines, on the other hand, belong to quite a different category of things which are generally hard, are typically made of metal or plastic and are constructed in a quite recognisable way - things like alarm clocks and tin openers.

Now I have no trouble, given a screwdriver and a couple of minutes, in telling whether any object before me belongs to one category or the other. What is more, I don't think I have ever been wrong in my entire life: briefly mistaken, perhaps but never *wrong*. It is certainly not one of the problem areas of my life and I am surprised when I read articles that indicate that it is for other people. Perhaps they should invest in a screwdriver.

That is not to say that one cannot dream up situations in which we are denied a good view of the object (or we lack our screwdriver or a couple of minutes) and in which we have some difficulty making the correct classification. The famous 'Imitation Game' devised by Alan Turing did essentially that (Turing, 1950). He imagined a situation in which a person had to guess which of two teletype conversations derived from a machine and which from a human, but in order to rule out our using screwdrivers and such things, he had to set the game up very precisely. Like most games, it cannot be serious and I sometimes wonder why so much attention has been paid to it. If I really wanted to know whether the teletype was being driven by a computer or a person I would walk round the screen and take a look. If I still wasn't sure I would get out my screwdriver and take a couple of minutes to explore. No problem.

The only credible explanation of our fascination with Turing's test is that we find ourselves more and more in the situation where we must classify objects without having a good view of them. This is sad but I am not sure it requires a wholesale reorganisation of our mental world. It may be that, over Internet for example, I receive messages and carry on conversations without ever knowing whether I am talking to a male or a female or a human or a robot. But that, it seems to me, does not mean that if I were given the resources I could never answer that question or that I believe that a definite answer does not exist.

Why distinguish humans from machines?

At the last conference, Carol Gigliotti raised the important question, 'Why would we want to decide that machines can have consciousness?' (Gigliotti, 1998) It seems clear that we can and do distinguish human beings from machines all the time and with an amazing success rate. I suggest that this is because the separate categories are very important to us. We actually *care* about whether something is a human or a machine - at least I do, as you can see from my anticipated emotional response during the experiment.

The development of artificial intelligence has a significance far beyond this debate. I have argued (Beardon, 1994) that artificial intelligence is essentially a modernist (albeit an *avant garde* modernist) project that aims for closure and belongs culturally to a bygone era. I also think that it was motivated by the attempt to create a single category for mental labour that spanned both humans and machines which had economic, social and military implications.[3] Certainly, I believe that artificial intelligence is an ill-founded and ultimately incoherent project that represents one of the latest scientific grand narratives.

Two more recent developments have, I believe, provided conceptualisations of digital technologies that are more human-oriented. The first is the concept of 'virtuality' as opposed to the concept of 'artificiality' (Beardon, 1998; Levy, 1998). 'Virtuality' concentrates upon the power of machines to store up future work which can then be made 'actual' on demand. It is like putting money in a bank: it does not solve all problems but, provided that you are careful, it places greater power in your (human) hands.

The second is the network, through which the individual (computer and human being) quickly becomes embroiled in the social. It is no longer tenable to see the computer as a prosthetic extension of the self. The digital becomes a form of communication, a creator of new social spaces, a forum for judgement, values and meaningful action. As Levy says, the important step is not one of computer tools but, rather, 'to recognise the current step in the self making of the human race' (Levy, 1998, p. 9).

Conclusion

Not everyone might be the same as me, but I am far from alone when I say that I would do some things for people that I would never do for a machine. It is not that I could not physically do them for a machine and it is not that I could necessarily produce a good reason why I should not do them for a machine — it is just that, given my priorities in life, I help people and I don't help machines.

You might be different, I accept that: your values might be different from mine. I might think that doing X is a good thing while you think doing X is not good or you may just be indifferent. Just like I said, it is a question of ethics.

Notes

1. A good visualisation can be found in the film *Westworld* (Michael Crichton, 1973).
2. The concepts of *knowledge* and *trust* are, in fact, closely related. In Aristotle, for example, the concept of *entelechy* is used in a sense similar to our concept of *destiny*. A true object is thus one that fulfils the expectations so described. We still use the word *true* in this way today when we say 'He is a true friend'. In the case of our thought experiment, we could say that my friend turned out not to be 'true'.
3. Norbert Weiner warned of the economic consequences of equating human and machine labour. 'Such mechanical labour has most of the economic properties of slave labour . . . However, any labour that accepts the conditions of competition with slave labour accepts the conditions of slave labour, and is essentially slave labour.' (Weiner, 1961, p. 27)

References

Beardon, C. (1994). 'Computers, postmodernism and the culture of the artificial.' *AI & Society*, 8(1) 1-16.
Beardon, C. (1997). 'What Does It Mean to be "Virtual"?' in: J. Berleur & D. Whitehouse (eds.) *An ethical global information society: Culture and democracy revisited*. Chapman & Hall, pp. 143-155.
Dennett, D. (1991). *Consciousness Explained*: Little. Boston, Mass.
Dreyfus, H. (1972). *What Computers Can't Do*:Harper, New York.
Gigliotti, C. (1998). 'What is Consciousness?' *Digital Creativity* 9(1), 33-7.

Haraway, D. (1990). A manifesto for cyborgs: science, technology and socialist feminism in the 1980s. In L. Nicholson (ed.) *Feminism/Postmodernism*. London. Routledge: 190-233.
Levy, P. (1997). 'Welcome to Virtuality.' *Digital Creativity* 8(1), 3-10.
Searle, J. (1980). Minds, Brains and Programs. *Behavioral and Brain Sciences*, 3, 417-24.
Simon, H. & Newell, A. (1958). Heuristic Problem Solving. Operations Research, 6.
Turing, A. (1950). Computing Machinery & Intelligence. *Mind*, LIX, 236.
Weiner, N. (1961). Cybernetics: MIT Press, Cambridge, Mass.

Colin Beardon is Professor of Art and Design at Exeter School of Arts and Design, University of Plymouth, and has higher degrees in Philosophy and Computer Science. He is a founder member of Computers in Art & Design Education (CADE), co-editor of Digital Creativity and Chair of IFIP WG 9.2 (Social Accountability and Computers).
<c.beardon@plym.ac.uk>
http://www.esad.plym.ac.uk/personal/c-beardon

The Metaphoric Environment of Art and Technology

Carol Gigliotti

Virtual environments (VR), artificial life and the World Wide Web (WWW) exist as emergent conceptual systems based on metaphor. The metaphorical systems upon which these conceptual systems are based, and upon which their development and use are based, vary in specifics of origin but are constructed on the experiential basis of all metaphorical understanding. An important consequence of this is the capacity it offers to these conceptual systems to influence other emergent structural metaphors. VR, artificial life and WWW metaphors developed from physical and cultural sources of experience will contribute, are contributing, to new meanings and values of our physical and cultural experience.

Johnson argues that not only are our shared folk theories of morality metaphoric but our refined philosophical moral theories are based on systematic metaphors. Based on findings from linguistics and cognitive psychology[1]. Johnson focuses on conceptual metaphor used generally in human thought and action rather than those of morality. Metaphor plays a large role in the aesthetic experience contributing possibilities for imaginative structure and coherence. Metaphor, then, is one key to how the aesthetic and ethical are linked. The artistic development process often relies on overarching metaphors. In its capacity for providing unity and coherence to a concept while at the same time offering an experiential comprehension of that concept, a metaphor could be defined as an environment in which the aesthetic experience takes place.

Recent work in the cognitive sciences, including psychology, linguistics and anthropology, is helping us understand more about what we call consciousness. These findings are allow us to describe aspects of the products of consciousness, concepts, as both physically based and imaginative in character. A large part of the way our conceptual system is structured is by metaphorical mapping. The artistic development process often relies on overarching metaphors. In its capacity for providing unity and coherence to a concept while at the same time offering an experiential comprehension of that concept, a metaphor could be defined as an environment in which the aesthetic experience takes place. This essay and presentation locates and interprets several metaphors artists working with these technologies are using and how these may contribute to the ongoing processes of emergent conceptual construction of meaning.

Along the way, I would like to highlight how metaphors act as simulators for the future in two ways: offering understanding through the use of empathy and through the use of strategy testing. Imagination, belief and desire are intricately linked in these processes but my interest is not in separating them out for analysis or criticism. In fact, that approach may be counter-productive, since the process involved in constructing meaning in any way, including a metaphorical one, is always bound up with these three linked aspects of consciousness.

Metaphor has developed out of, or has been useful in developing, each of the environments I mentioned earlier: VR, artificial life and the WWW. This process of individual and group experience is evidenced in the complex dialectical use of metaphorical systems at work in the literature, discussion and practice of artificial life, VR and the WWW. Lakoff and Johnson emphasise the way in which physical experience forms the basis of many metaphors. Which aspects of this experience, however, form the core structures of metaphorical systems is affected in large part by culture. Additionally Lakoff and Johnson emphasise the partial quality of all metaphorical structures. As much as is revealed by these structures, so too are other features hidden.

Using Lakoff and Johnson's (1993, 1980) theories of metaphor, we may better understand the overt or latent values assumed in the development of interactive technologies. In this short essay I will concentrate on one of those environments listed previously: virtual reality (VR). This will help to understand what values are implied in particular artists' work with VR and how they serve as simulators for the future . The metaphorical system upon which VR environments are based is: The Virtual Is Real. Many of the entailments or consequences of this metaphorical system emanate from three physically based and interconnected metaphors. They are Immersion, Navigation, and Other Worlds. These metaphors fall into what Johnson calls the Event Structure of metaphorical systems, which includes our concepts of action, purpose and cause. These metaphors may be mapped onto the target domain (i.e. events) from the source domain (i.e. motion in space along a path). This is the Location version of the Event Structure.[2] Examples of this are:

> The participant is *immersed* in VR.
> The participant *navigates* VR.

But there is also the Object version of the Event Structure. In this version, according to

Johnson's theories, a parallel system exists in which 'achieving a purpose through one's action is understood metaphorically as acquiring an object.'[3] An example of this is:

> The VR participant *accesses another world.*

These three metaphors, Immersion, Navigation and Other Worlds, offer us insights into what values have been previously assumed in the development of VR and what values some artists working with VR today are implying in their work.

Huhtamo lists various meanings Immersion conveys:

> 'plunging into water', 'breaking through the screen (or the mirror),'
> 'leaving (or changing) one's body', 'losing oneself in a simulated world',
> 'navigating in cyberspace' [4]

As Huhtamo explains, the transition from immediate physical reality to *somewhere else* implied by these metaphors is complex. Technology itself is a location of *somewhere else*. He argues, 'It is an "*obscure* object of desire" - seductive and repulsive at the same time.'[5] If this is the prevalent version of the metaphor of Immersion, are artists' supporting that version or attempting to change it?

> Navigation as a metaphor of virtual environments entails similar meanings. As Florian Rötzer claims, 'If it is fundamentally possible to enter a completely artificial world, wander through it, interact with people, or even with intelligent beings within this world, then this virtual reality, in which everything is basically manipulatable, triggers the wish to experience this at some time.'[6] What values accompany that wish, and what ethical implications does it have for us as a culture? Of course, the wish to experience other realities and modes of existence has always existed. For what purposes are today's artists using this ancient wish and how do these purposes and goals affect their use of technology to make that work?'

Other Worlds metaphors suggest the possibility of all reality being only a symbolic construct. Slavoj Zizek expresses this idea most succinctly when he says: 'The ultimate lesson of virtual reality is the virtualisation of the very true reality.'[7] Like wishing for totally unencumbered navigation through a virtual world, the need for perceiving the actual world as under our control has been with us a long time. Investigations of the origins and goals of simulation and artificial reality have proven helpful in understanding the ontological and epistemological assumptions used in developing these technologies (Wooley 1992).[8]

Certainly, one of the artist Char Davies's (1997) major impacts on the development of VR environments, is in her stated goal of 'exploring its potential as a site not for escape but return, not towards some transcendent technological sublime but as a context for undoing our habitual perceptions and refreshing our sense of being here now'.[9]

Artist Margaret Dolinsky uses metaphor in her CAVE environments to heighten our awareness of our built-in psychological abilities to 'enlighten our sense of self, time and space.'[10] Julieta Aguilerra's (1997) approach to her wonderfully evocative Quick Time VR environments emphasises the importance of contextualising virtual environments within reality understood both aesthetically and spiritually. Comparing technologically constructed virtual environments to image and environmental meditation environments such as icons and Chinese and Japanese gardens, Aguilerra directs her VR environments to "consider the questions of representation and being."[11] Escape from reality is not her goal. Rebecca Allen's (1997) incorporation of artificial life applications into a virtual environment called *Emergence* is another example of artists working towards reconstruction of the assumed metaphors of virtual reality, and in this case, artificial life as well. Her stated goal in this project is to provide a creative environment that goes beyond the rigid format of commercial video games with their uninspired designs and simplistic focus on violent, "kill or be killed "behaviour".[12] Her use of the West African belief in the 'bush soul' as both a metaphor and a working conceptual system directed towards changing our notions of interdependence with other life forms is a excellent example of how metaphors influence other emergent structural metaphors.

These artists among others, such as Toni Dove and Brenda Laurel, have used their imagination to develop and construct metaphors which correspond to their belief in contrasting values to prevailing notions of the goals and purposes of interactive technologies and to their desire to highlight other aspects of these technologies. In this way we are reframing consciousness.

References

Allen, R. 1997. 'The Bush Soul: Travelling Consciousness in an Unreal World,' in Roy Ascott, ed. *Proceedings of the First Annual CaiiA Research Symposium.* University of Wales, Newport.

Aguilerra J. C. 1997. 'Insights on Art, Virtual Reality and Ultimate Reality', in Roy Ascott, ed. *Proceedings of the First Annual CAiiA Research Sumposium.* University of Wales, Newport.

Davies, C. 1997. 'Changing Space:VR as a Philosophical Arena of Being', in Roy Ascott, ed. *Proceedings of the First Annual CAiiA Research Symposium.* University of Wales, Newport.

Dolinsky, M. 1997. 'Dream Grrrls: A World of Virtual Reality', in Roy Ascott, ed. *Proceedings of the First Annual CAiiA Research Symposium.* University of Wales, Newport.

Johnson, M. 1993. *Moral Imagination: Implications of Cognitive Science for Ethics.* Chicago: University of Chicago, Pp. 36-7 and 39-40. See also Lakoff, G. and Johnson, M. 1980. *Metaphors We Live By.* Chicago:University of Chicago Press, and, more recently, Lakoff G. 1993. 'The Contemporary Theory of Metaphor', in Orton A. ed. *Metaphor and Thought.* Cambrdige: Cambrdige, Massachusettes.

Huhtamo, E. 1995. 'Encapsulated Bodies in Motion: Simulators and the Quest for Total Immersion', in Penny S. ed. *Critical Issues in Electronic Media..* Albany, New York: State University of New York Press, p. 159

Rotzer, F. 1995 'Virtual World:Fascinations and Reactions', in Penny S. ed. *Critical Issues in Electronic Media.* Albany, New York:State University of New York Press. P. 127.

Woolley, B. 1992. *Virtual Worlds.* Oxford:Blackwell Publishers.

Zizek, S. 1996. 'From Virtual Reality to the Virtualization of Reality', in Druckrey, T. ed. *Electronic Culture: Technology and Visual Representations.* New York:Aperture Foundation, Inc. P. 295.

Dr Carol Gigliotti (carol@cgrg.ohio-state.edu) is presently Assistant Professor in the Department of Art Education and the Advanced Computing Center for the Arts and Design at the Ohio State University where she teaches courses on both the theory and practice of interactive technology. She lectures and publishes widely.

– *Values* –

She will be included in The Digital Dialectic: New Essays on New Media from MIT Press due out this year. Her most recent Leonardo essay is 'Bridge To, Bridge From: the Arts, Education, and Technology' Vol. 31, Issue 2, 1998. She has a chapter in New Technologies in Art Education published earlier this year by NAEA. A forthcoming essay, 'Mothering the Future' will be included in S. Diamond and S.Lotringer (eds.) Flesh Eating Technologies (Banff Centre and Semiotexte). She is Project Director of Astrolabe: Ethical Navigations through Virtual Technologies, a WWW site, CD-ROM and on-line journal available at http://www.cgrg.ohio-state.edu/Astrolabe

The Four Seas: Conquest, Colonisation, Consciousness in Cyberspace

Joe Lewis

As a child I pictured myself as a great explorer and coloniser. Sunday mornings, sabre in hand, I'd set off into the wilds of our sparsely furnished three-room-cold-water-flat. Because I was of mixed heritage and lived in a predominately coloured neighbourhood, I was aware, at least as aware as a bright, politically tenacious ten year-old who had read Herbert Apekther's historical posits, could be. Aware of the mechanics of "establishing place", territorial redefinition, ethnic decontexturalisation, and the symbiotic interplay of subjugation, i.e., religious, linguistic and the political assumptions that my soon to be discovered flock would use to develop their new identities. I was not interested in settling, pioneering, planting founding, migrating or opening up anything. No, I was looking for GOLD!

One day, after the hair-raising near-death experience of almost being caught by the native chieftain, while appropriating some ducats from King Dad's pockets, I beat a hasty retreat to the Penny Arcade at Broadway and 52nd Street, on the Great White Way. In those days, it was all about shoot'em up-spender as predator.

But on this day, my life, and in hindsight, the lives of everyone else, were forever changed. My space was invaded by a strange and hypnotic beeping sound. It came from an area where once stood a life-sized Marshall, who upon the deposit of a dime would bellow, 'You cock the gun before you draw. Ready? Draw!' I cautiously moved towards the flattened sounds. That was the first time I laid eyes on PAC Man. I instantaneously realised just how easy it was for 'predator to become prey'.

Nintendo/Sega: Tomb Raider, Shock Troopers, Warriors of Fate, Metamorphic Force, Area 57, Mortal Kombat . . .

Life was a Penny Arcade:

In many ways, the unearthing of cyberspace mimics the so-called West's discovery of the New World. Upon first contact, its lands were mapped, its contents enumerated and all substances with obvious profit immediately extracted. Forthwith, the conquering explorers systematically re-arranged the communities they 'found', forcing the inhabitants and eventually, the physical territory itself, to conform to an alien set of ideals. Then, promptly burying their discoveries under the shifting philosophical paradigms of their day, they 'appealed to divine providence for a justification of their politics'. (Lewontin 1991).

Sardar suggests the West's hunger, particularly for new conquests and its successful cultural erasures of realities both past and present, provides the impetus for the development of new technologies of subjugation, making place for neo-fictions. To look at the inner reality of the West, the darker side it projects onto the other cultural and mental landscapes, we must look at the West's latest conquest, the new domain that it has colonised: cyberspace.

> What the cyberspace 'frontier' is doing as a first step is rewriting history. [It is] An exercise in catharsis to release the guilt of wiping out numerous indigenous cultures from the face of the earth, the colonisation of two-thirds of the world and the continuous degradation of life in the Third World. Why else have these colonial metaphors of discovery been adopted by the champions of cyberspace-particularly, as Mary Fuller and Henry Jenkins note, when these metaphors are undergoing sustained critique in other areas of the culture . . .
>
> (Sardar 1996)

> Columbus was probably the last person to behold so much usable and unclaimed real estate (or unreal-estate) as these cybernauts have discovered'. (Barlow 1990 as cited by Sardar). 'The comparison to cyberspace drains out the materiality of the place Columbus discovered, and the nonvirtual bodies of the pre-Columbian inhabitants who did, in fact, claim it however unsuccessfully . . . The drive behind the rhetoric of virtual reality as a New World or new frontier is the desire to recreate the Renaissance encounter with America without guilt: This time if there are others present, they really won't be human . . .
>
> (Sardar 1996)

Nevertheless, today's exploiters of cyberspace bring their old baggage with them-virtual Pandora's boxes-into the unknown. History does not lack evidence as to exactly what happened once the original colonial receptacles were opened: Viruses, religious beliefs and inflexible value systems erased millions of lives, habitats and community histories faster than any delete key-stroke. Today, plenty of examples exist of this history repeating itself and we do not have to look beyond our own shores to find them.

'In the beginning there was the word and the word was God'

Considering the power of the word, routinely reconstructed by the likes of Said to Rheingold to Haraway *ad infinitum*, the misanthropriation and use of the word Avatar, as an agent for virtual enterprises is particularly repugnant and fraught with subconscious colonial
whatever. . .

Both Eastern and Western spheres of socio-political influence have manufactured and pursued manifest destiny narratives with great determination and vigour. The West's obsession is best characterised by the anthem 'bring the body and the mind will follow'.

In comparison, Eastern acquisition philosophy, while trying to keep its fingers in the honey pot too, was more geographically introspective and transcendental in its approach but as malevolent and acquisitive (there is much cyber-activity devoted to the virtual replication of said transcendental experiences, and for that matter, dumping the body altogether).

Ironically, both East and West are still constructing corresponding xenophobic paradigms attempting to justify their inability to participate in the grand consensual hallucination of future consciousness-each essentialising the other as non-entities by projecting all that they fear about themselves upon the other-self-realisation vs. the de-corporealisation of body, space and time. Moreover, the East's self-proclaimed superiority demonised the West to such an extent as to 'virtually' separate itself from the rest of the world. The Great Wall of China attests to this fact, which worked well before the internationalisation of spaghetti. But fortress philosophy had one fatal flaw, contempt prior to investigation. While the East isolated itself, the West separated God, man and science, literally circumscribing the East, and years later in the form of capitalism, passing dialectic materialism through its porous so-called free will.

How Much of Our Future Will Be Informed by Our Past?

> The process of installing barriers is persistent and ubiquitous. It involves the separation and privileging of the ideational over the material and in such a way that matter is denigrated as the base support of an ascendant entity (mind over matter, male over female, culture over nature, the West over the rest, and so on) . . . it is particularly surprising that, after several decades of sustained intervention within the politics of representation, the new world of cyberspace/VR should so faithfully mimic the old.
>
> (Kirby 1997).

Binary: 1, 10, 11: '1' If by the Great Wall, and '10' If by Sea

> The West is. . . responsible for what Hegel calls the power of the negative, for violence, terror, and permanent aggression directed against life. It has generalised and globalised violence-and forged the global level itself through that violence. Space as locus of production, itself product and production, is both the weapon and the sign of the struggle
>
> (Gregory 1994).

The ascent of the New World serves as an excellent example of Hegel's hypothesis. It came complete with natural resources and an indigenous work force to exploit them. This indigenous work force, however, proved ill-suited and unadaptable to colonial mercantilist agendas. This predicament prompted the mass abduction of millions of African people. Having done this, the conquerors discovered to their chagrin that the New World was a finite system.

Conversely, cyberspace, though lacking inherent communities, is nonetheless a conceptually limitless system. In spite of it having just been discovered, cyberspace is heavily populated and a fixed site for many virtual groups and personas. When contrasted to the exploration of the New World, the probability of discovering exotic fauna and flora or expeditions mysteriously vanishing into T1 line is unlikely. Cyberspace, though, remains a vast and significantly unexplored place.

Which is not to say that some frightening parallels do not exist. A tangible example is Karen Hossfeld article, 'Small, Foreign, and Female'. Silicon Valley is filled with secrets and she reveals one of its dirtiest. Hossfeld focuses her attention on the forgotten work force that turns high-tech dreams into palpable realities. The title was inspired by an interview with a manager of a small contract assembly firm who explained why so many assembly workers are either Asian or Latina women. It seems there is a definite racial/gender pecking order in hiring. There's an assumption that immigrant women, especially Asian women, will be submissive employees.

There are countless stories of sexual harassment, poor working conditions and threats of deportation even to those here legally. But the dirtiest secret Hossfeld uncovered could have deadly consequences.

According to Hossfeld, many of these women are unaware they are being exposed to toxic chemicals on the job as new chemicals are thrown into the manufacturing process without adequate testing. Rumours of toxic chemical use became so widespread that IBM commissioned its own study to quell them. To IBM's dismay, that study, completed in June 1993 found the opposite to be true: Women who worked with a substance called ethylene glycol ether experienced an unusually high rate of miscarriages. IBM announced plans to eliminate the substance by the end of 1994. (Hossfeld 1993).

Yet another example Hegel's 'globalization of violence', within cyberspace this time, is a phenomenon best illustrated by P.J. Huffstutter's *Los Angeles Times* article 'Battling Bullies on the Web'. 'In the world of on-line games, young "player-killers" prey on the less skilled- some of whom hire bounty hunters to strike back. Says one victim, "I wanted them to feel the frustration I felt," ' (Huffstutter 1998).

Having said all this, 'one positive aspect of reliance on Cyberspace for new millennium virtual confrontations might be, as Deborah Tannen suggests in "The Argument Culture: Moving from Debate to Dialogue', is that Western society became less battle-happy with the advent of print. The arguments made on a printed page modified the exaggerations that arose when ideas were conveyed out loud before an illiterate audience'.

(Ann Rauhala 1997).

In closing, will defining, mapping, and enumerating the assets of cyberspace devastate its virtual communities, secluded personas, ecologies and belief systems, as did the exploration and colonisation of the New World? How do current agendas for cyberspace development affect the quality of interactive access, self-determination and behavioural protocols when informed by well-worn strategies of ad-venturism (sic), conquest, and colonisation? If cyberspace as a socio-economic entity were to replace the status quo economic fabric, would it duplicate the malevolent forces of colonial rule? And if so, would it create a proactive majority underclass, organised resistance and civil war?

Without doubt, a significant amount of interactive-cyber-confabulation is informed by these powerful dynamics. Are there alternative strategies? My colleague, Ron Saito recently suggested, we will remember the early days of cyberspace as possibly the only time there was a level playing field.

– Values –

Thoughts become actions. Actions become habits. Habits become character. And character, character becomes one's destiny.

References

Biocca, Frank and Levy, Mark R. 1995. *Communication in the Age of Virtual Reality* New Jersey: Lawrence, Erlbaum Associates, Publishers.
Brook, James and Boal, Iain A. 1995. *Resisting the Virtual Life, TheCulture and Politics of Information.* San Francisco: City Lights.
Druckery, Timothy 1996. *Electronic Culture, Technology and Visual Representation.* New York: Aperture.
Gregory, Derek. 1994. *Geographical Imaginations: Modernity and the Production of Space* Oxford: Blackwell Publishers.
Hossfeld, Karen. 1993. *Small, Foreign, and Female. Wired*, Sept./Oct. p. 24.
Huffstutter, P.J. 1998. 'Battling Bullies on the Web' Los Angeles *Times*, Sunday April 25,1998, pp. 1,18.
Kirby, Vicki. 1997. *Telling Flesh, The Substance of the Corporeal.* London: Routledge.
Laszlo, Ervin.1987. *Evolution: The Grand Synthesis.* London: New Science Library.
Lewontin, R.C. 1991. *Biology as Ideology: The Doctrine of DNA.* Ontario: Anansi Press.
Markley, Robert. 1996. *Virtual Realities and their Discontents.*Baltimore: John Hopkins University Press.
Rauhala, Ann. 1997. Review of Deborah Tannen's 'The Argument Culture: Moving from Debate to Dialogue. (Random House), *Review of Books*, The Globe and Mail, Saturday April 11, 1998, pp. D11.
Rheingold, Howard. 1994. *The Virtual Community: Homesteading on the Electronic Frontier.* New York: Harper Perennial.
Said, Edward W. 1993. *Culture and Imperialism.* New York: Vintage Books.
Sardar, Ziauddin and Ravetz, Jerome R. 1996. *Cyberfutures, Culture and Politics on the information Superhighway* London: Pluto Press.

joe.lewis@sun.edu is an artist, chair of the Department of Art at California State University, Northridge, and a Ph.D. candidate at CAiiA-Star. He is a Board Member of the College Art Association and the 1999 Los Angeles Conference Studio Session Co-chair. Lewis is a regular contributor to Art in America and has written for Artforum, and the LA Weekly. In addition, his artwork has appeared in many solo and group exhibitions throughout the United States.

Future Present: Reaestheticising Life Through A New Technology Of Consciousness

Anna Jean Bonshek and Gurden Leete

While consciousness is understood to be mediated by social and cultural forces, digital artists extend the concept of social determinism to include virtual reality's influence on identity (Gómez-Peña 1996), suggesting that digital media and visual culture can transform notions of the self (Turkle 1997). Whether conceived of as unified (Kuspit 1990), or as deflated (McEvilley 1992), in the social construction model our sense of self is framed in reference to changing circumstances.

The nature of consciousness and its formation is a complex issue. However, the consciousness-based approach to education employed at Maharishi University of Management Art Department operates on a unique principle: that there is a universal field of consciousness - the simplest form of human awareness and the basis of all phenomena - which can be enlivened in individual and collective life, enriching culture and transforming individual and collective consciousness. Physicists refer to this field as the unified field of natural law unifying all the laws of nature (Hagelin 1987). Considering the possible applications of this principle to the area of visual cultural production, it is possible to entertain the idea that life itself can be an aesthetic experience (Bonshek 1996), and that digital media may be the most appropriate technology to enhance that experience.

Unfolding New Technologies of Consciousness

Maharishi Mahesh Yogi, known worldwide as the founder of the Transcendental Meditation technique, has brought to light a new science of consciousness, referred to as Maharishi Vedic Science Maha means great, Rishi means seer or knower and Veda is complete knowledge), which identifies an unbounded, infinite field of consciousness and provides technologies to experience and enliven this level of human awareness. These technologies consist of the Transcendental Meditation® (TM®) technique and the TM-Sidhi® program and Yogic Flying®. Due to the potential for consciousness to structure relative phenomena, through the practice of these technologies spontaneous fulfillment of desires can be realised by mere intention. The practice of Yogic Flying provides a practical demonstration of the ability to project thought from the Unified Field of Natural Law, and develops the ability to act spontaneously in accord with Natural Law for the fulfillment of any desire. The phenomenon of Yogic Flying proves that through the Transcendental Meditation and the TM-Sidhi Programme, anyone can gain the ability to function from the simplest form of their own awareness and can develop mastery over Natural Law. (Maharishi Vedic University 1994). Thus, we have the potential to achieve anything.

Maharishi Vedic Science unfolds experience and knowledge of the universal field of *pure consciousness*. Beyond space, time and relative phenomena, pure consciousness is a field of all possibilities (the infinite diversity of life emerges from this field) and infinite correlation (there are no limitations of time or space - everything is intimately connected). Any action performed from this level is in accord with Natural Law. This field of pure consciousness or Natural Law is responsible for the evolving, ever-expanding universe, the structure of DNA, the human physiology (Nader 1993), all streams of knowledge, different cultures and all possible values of visual culture.

According to Maharishi Vedic Science, individual consciousness, family consciousness, community consciousness and national consciousness are all specific values or expressions of pure consciousness but each has its own value. Any particular value of *collective consciousness*, while encompassing all the various tendencies of which it is comprised, is more than the sum of its parts (Maharishi Mahesh Yogi 1978). While pure consciousness is beyond change, collective consciousness changes over time.

Pure consciousness is a state of pure awareness where consciousness is open only to itself (Maharishi Vedic University 1994) and can be subjectively experienced and verified through the practice of the Transcendental Meditation technique and TM-Sidhi program.

Through the practice of these technologies one can gain a state of consciousness where everything in life is known to be an expression of one's own consciousness, where infinity can be perceived in every point or grain of creation (Maharishi Ved Vigyan Vishwa Vidya Peetham 1995). Such life is reaestheticised through the art of living; every thought and action is promoted from the infinite level of consciousness - that field of all possibilities that generates, supports and nourishes life on all levels. In order to help bring about this condition of life, Maharishi Sthapatya Veda - an aspect of Maharishi Vedic Science - provides knowledge of consciousness in relation to the built environment, the influence of the planets and different Laws of Nature (Maharishi Global Construction 1997). In this context, new approaches to using digital media to simulate the experience of living a life of all possibilities are being developed in Europe and North America.

Promoting Holistic Development of the Subject

With the development of new digital media and technology in this post-colonial era, new ideas about knowledge, learning and identity have emerged. Reality can be influenced by virtual reality. Educators are now calling for the development of the learner rather than specialised knowledge. Despite this, for over 25 years, holistic development of the knower has been the focus of consciousness-based education - increasing intelligence, flexibility and creativity in the individual (Dillbeck and Dillbeck 1987). While artists can enhance their creativity and increase field independence through the practice of the Transcendental Meditation technique and TM-Sidhi program (Fergusson 1992), on a broader scale, world trends improve - reducing conflicts and strengthening cultural integrity (Orme-Johnson and Dillbeck 1987).

Though producers of visual culture equipped with pervasive means of structuring discourse, without enlivening pure consciousness, digital media artists can only transform thinking at a relatively superficial level; only by operating from the source of change and individual and social life can theories of knowledge, aesthetics and ethics begin to represent the structuring values of culture and provide the subject with a vision of all possibilities.

Expanding Identity: A New Definition of the Self

From the perspective of Maharishi Vedic Science, the self can be experienced directly as the infinite, unbounded *Self* or *Transcendental Consciousness* - a state of consciousness beyond the changing states of waking, dreaming and sleeping. In developing higher states of consciousness, the subject can permanently maintain the Self along with these changing states, and perceive finer, more abstract values of the phenomenal world (Maharishi Mahesh Yogi 1969). In the most developed state of *Unity Consciousness*, the infinite value of pure consciousness is perceived at every point in creation, permeating material phenomena. There is no outside of the Self. The Self is wholeness and everything is wholeness moving within itself (Maharishi Ved Vigyan Vishwa Vidya Peetham 1995). The division between self and other is simply an aspect of the play of consciousness within consciousness.

In Unity Consciousness individual awareness has expanded to its full potential. The sense of self is no longer limited by social and historical influences. This does not mean that individuality does not exist but, while previously felt to be only shaped by factors as gender, social and economic forces, individuality has now assumed its infinite status - embracing localised values. With this understanding, aesthetics and culture can be appreciated in a new light.

The Role of Culture in Transforming Collective Consciousness

The underlying purpose of culture is to develop individual and collective consciousness. According to Maharishi Vedic Science, the word culture refers to everything concerning life. Life is made up of so many elements, and the behaviour of all these elements put together constitutes the process of evolution (Maharishi Mahesh Yogi 1978). Pure consciousness gives rise to the structure and dynamic evolution of existence. Evolution is the progressive development of life's full potential.

There is both a universal and specific value to culture. The holistic value of Natural Law handles the holistic value of culture - universal culture - whereas specific values of different Laws of Nature are concerned with the specific culture of an individual country or the specific area of a country; these are the Laws of Nature that give rise to the specific geographic and climatic conditions, accents of speech, languages and the trends of society on all levels of life - spiritual, social, material" (Maharishi Ved Vigyan Vishwa Vidya Peetham 1995).

The mannerisms, behaviour, traditions, language and ways of living of people, the geographic and climatic conditions of a place, are all governed by specific Laws of Nature and have a role in promoting life. These Laws of Nature maintain the life-supporting functioning of a group in a particular environment. Cultural integrity involves the support of Natural Law and specific Laws of Nature; therefore it is vital for every culture to operate from the level of Natural Law or pure consciousness - ensuring the progressive development of life. How can this effectively be achieved in this technological age of cultural hybridity?

A culture is a dynamic display of an evolving collective consciousness; to achieve cultural integrity, with the characteristics of stability, adaptability, integration, purification and growth, individuals can practise technologies of consciousness. Through this practice, the group, while maintaining its diversity and dynamic potential for change, becomes more creative and progressive (Orme-Johnson and Dillbeck 1987). The value of pure consciousness, having the value of infinite correlation, is enlivened everywhere. Governing the principles of nature which structure and support life, pure consciousness begins to be expressed in behaviour, ethics, systems of aesthetic judgement and theories of knowledge. The universal value of culture is strengthened while the diversity of every aspect of collective consciousness is enriched.

The individual and collective sense of identity are defined by the degree to which the individual and group are living this unbounded, infinite awareness. By activating the universal level of pure consciousness through the appropriate technology anything can be changed, transformed, reaestheticised; visual culture can begin to represent the underpinnings of culture.

Digital Media: Representing the Infinite Possibilities of Consciousness

History has shown that new technologies have a relatively limited transformational effect on culture and humanity. Technologies based on partial knowledge of the Laws of Nature are not holistic; shuffling a deck of cards in the hope of dealing a good hand does not change the deck. The 20th century has seen unprecedented technological development and has also seen world wars, environmental disasters and genocide. Digital media can only be transformational to the degree that consciousness has been developed by those engaged in visual cultural production, the generation of meaning and the definition of knowledge and aesthetics. If changing consciousness through a new technology means a shifting of signification with no development of life itself, then it is not profoundly transformational or evolutionary.

Only technologies of consciousness can expand identity to an infinite status, allow the individual to function from the level of infinite correlation and provide cultures with the capacity for dynamic change while maintaining cultural integrity, rather than simply shifting power structures or creating new language games. The digital media artist who practises the Transcendental Meditation technique, the TM-Sidhi Program and Yogic Flying has the potential to draw from an infinite resource.

It could be argued that digital media more adequately enables representation of the capability of the mind to experience and comprehend infinity. Coupled with a technology of consciousness - that expands the mind itself and allows it to access infinite creativity - digital media can begin to indicate the enormous potential of human existence and the future direction for life and culture. New ideas and phenomena, and old concepts that were felt to be incomprehensible, become more prevalent in our increasing appreciation of the potential of human consciousness. With the further development of digital media, the ability to simulate the infinite dimension of human life lived in its full potential will ignite the imagination of all cultures with a vision of all possibilities that can reaestheticise life. The seed of the future is in the present and we have the means to structure limitless possibilities in human life and art.

Acknowledgement
In writing this paper, Anna Bonshek would like to acknowledge the assistance of Gurdon Leete.

Notes
Gómez-Peña, G. 1996. *The New World Border*. San Francisco: City Lights Books.
Turkle, S. 1997. 'Life on the Screen', Keynote Address. Chicago: The Eighth International Symposium on Electronic Art.
Kuspit, D. 1990. 'A Psychoanalytic Reading of Aesthetic Disinterestedness', in *Art Criticism*, 6 (2), pp.72-80.
McEvilley, T. 1992. *Art and Otherness: Crisis in Cultural Identity*. New York: McPherson Press.
Hagelin, J. 1987. 'Is Consciousness the Unified Field?' in Modern Science and Vedic Science, 1 (1), pp 28-87.
Bonshek, A. 1996. Art - 'A Mirror of Consciousness'. *Dissertation Abstracts International*, 57, O8A, p3306.
Maharishi Vedic University. 1994. *Maharishi Vedic University: Introduction*. Holland: Maharishi Vedic University Press. p388.
Nader, T. 1993. *Human Physiology: Expression of Veda and the Vedic Literature*. The Netherlands: Maharishi Vedic University Press. Maharishi Vedic University. 1994. pp53-5.
Maharishi Ved Vigyan Vishwa Vidya Peetham. 1995. *Maharishi University of Management: Wholeness on the Move*. India: Age of Enlightenment Publications. pp285-287
Maharishi Global Construction 1997. *Designs According to Maharishi Sthapatya Veda*. Holland: Maharishi Vedic University Press.

Dillbeck, S. and Dillbeck, M. 1987. 'The Maharishi Technology of the Unified Field in Education: Principles, Practice and Research', in Modern Science and Vedic Science, 1 (4), pp382-468.
Fergusson, L. 1992. 'Field Independence and Art Achievement in Meditating and Non-Meditating College Students' in *Perceptual and Motor Skills*, 75, pp1171-5.
Orme-Johnson, D. and Dillbeck, M. 1987. 'Maharishi's Program to Create World Peace: Theory and Research' in *Modern Science and Vedic Science*, 1 (2), pp 207-59.
Maharishi Mahesh Yogi. 1969. *Maharishi Mahesh Yogi on the Bhagavad-Gita*. New York: Penguin, pp. 314-15.
Maharishi Ved Vigyan Vishwa Vidya Peetham. 1995. pp.13-71.
Maharishi Mahesh Yogi. 1978. *Creating an Ideal Society*. Rheinwheiler, West Germany: Maharishi European University Press, p136.
Maharishi Ved Vigyan Vishwa Vidya Peetham. 1995. p.72.
Orme-Johnson, D. and Dillbeck, M. 1987. pp.207-259.

[*] Transcendental Meditation®, TM,® TM-Sidhi® and Yogic Flying® are registered in the US Patent and Trademark Office as service marks of Maharishi Vedic Education Development Corporation and are used under licence by Maharishi University of Management.

Biography
Visiting Professor of Art at Maharishi University of Management, Anna Bonshek is an artist who has exhibited in England, Australia and America, published in journals and anthologies including New Art Examiner, College Student Journal, Artlink and Visibly Female, received awards from the Royal Society of Arts, the Science Policy Foundation and the National Endowment for the Arts, and is currently writing a book on art and consciousness. She is also collaborating with Gurdon Leete, Director of Digital Media Programs at Maharishi University of Management. <ABonshek@mum.edu>

Conspiracies, Computers and Consensus Reality
Paul O'Brien

Information overload has reached crisis point on the Internet. Depending on how you look at it, there is either an abundance of rubbish (with a few pearls if you have the patience to find them) or the exciting prospect of liberation from the constraints of consensus reality/realities. Consensus realities are the agreed-upon - and sometimes contradictory - interpretations of the world characteristic of different branches of the academic mode of production. This phenomenon of diverse 'realities' is the result of a long process resulting from the divorce of the sciences from the arts and humanities, specialisation/fragmentation, and the industrialisation/professionalisation of knowledge over the last few hundred years. This means that synthesising overviews of a serious nature - for instance from an ecological perspective - are largely ignored. The result, for example, is that economists blithely continue to champion economic 'growth', with some tinkering with the traditional indicators, at the same time as large chunks of Antarctica are falling into the sea owing to the energy-wastage and pollution consequent on such growth.[1]

– Values –

In the bio-medical sciences, a thorough-going materialism holds sway, apart from a few heretics such as Sheldrake (1988).[2] The phenomenon of an entrenched scientific orthodoxy raises serious ethical problems for society in the areas, for example, of cloning and genetic engineering. Neo-Darwinism has exerted a hegemony in biology reminiscent of the most dug-in periods of Catholic Church dominance in Europe. The materialism, reductionism, mechanism and instrumentalism characteristic of the current scientific orthodoxy relegate values to an implicitly relativistic level, leaving nature itself up for radical deformation in terms of the possibilities opened up in genetics, including interference with consumer choice in food, eugenics, and inter-species commingling.[3] The widespread practice of animal experimentation leads to a cultural coarsening which affects large areas of the life sciences, resulting in a bewildered incomprehension on the part of scientists as to why the untutored masses have problems with cloned humans, headless frogs and mice with ears on their backs. (Of course, there have been a few problems with applied science in the past, such as Nazi 'race science', Hiroshima, Chernobyl, DDT, CFCs, thalidomide, Agent Orange and so on, but that was the past. . .).

In the so-called 'hard' sciences, on the other hand, room is sometimes made for a determining, perhaps even fundamental, role for consciousness, in contrast to the reductionism and mechanism characteristic of the life sciences. The physicist Paul Davies (1982) writes:

> Until we make a definite observation of the world it is meaningless to ascribe to it a definite reality (or even various alternatives), for it is a superposition of different worlds. . . .Only when the observation is made does this schizophrenic state collapse onto something that is in any sense real.[4]

While the multi-dimensional interpretation of reality is taken seriously in physics, it would be true to say that it has made few inroads in, for example, sociology (probably for the very good reason that it is not of much use in the social sciences). It is more difficult to understand, though, how the kind of positions approximating to philosophical idealism to which some developments in quantum physics seem to be tending, have made virtually no significant inroads into orthodox biology.

Part of the reason is that an intellectual establishment within the scientific disciplines - dubbed by Richard Milton (1995) the 'paradigm police' - influenced by a pervasive managerialism, creeping commercialisation and the professionalisation of knowledge, holds tenaciously to whatever consensual world-view is to the fore at the moment.[5] This is despite the fact that history discloses the partial, and sometimes fallacious, nature of scientific 'givens' and their susceptibility to successive 'paradigm shifts' in the phrase popularised by Kuhn (1970)[6]. Milton (1995) amusingly describes the official resistance on the part of orthodox science of the time to the possibilities of, for example, electric lighting, heavier-than-air flight and space travel.[7] (For the sake of fairness, one would have to note that this syndrome operates even in so-called 'oppositional' fields of study like gender studies, which have notoriously produced their own orthodoxies, for the most part resistant to findings in areas such as biology which - rightly or wrongly - challenge their premises.) Like the post-modernists and the so-called sceptics, the one thing the intellectual orthodoxies of our time resolutely refuse to do is to turn a critical eye on their own presuppositions.

The search for truth competes with the search for power and wealth through the application of science - leading not just to the commercialisation of science but also to a tendency to its commercial monopolisation. Consensus reality often suffers from 'temporal provincialism' and a neurotic 'fear of the new' (misoneism is Arthur Koestler's term) on the part of those who have carved out careers on what may ultimately prove to be shaky terrain. It is shaped by the unexamined and unconscious motivations of those who help to construct it and embodies deep and often unexamined tensions.[8]

In the media, different levels of consensus operate according to the presumed class or educational level of the readership. While aliens, for example, may be a staple of the tabloids and popular TV programmes, as well as being a central trope within science fiction generally, they are usually treated with a robust scepticism by the broadsheets. Despite the admission by many 'respectable' scientists that extra-terrrestrials may exist and may have the capacity to visit us, anyone who ventures to suggest that they have actually done so automatically shifts themselves beyond the pale of 'serious' discourse. There are some good reasons for such an inconsistent mind-set, of course, including the apparent lack of 'hard' physical evidence and the notorious attraction of this area for the cognitively challenged. But there are also some bad ones, including disinformation, censorship and intellectual conformity.[9]

Of course, there have been academic attempts to vindicate the undermining of 'respectable' discourses, particularly in media theory. Defending the tabloid press, the cultural theorist John Fiske (1989) cites its political importance as lying in its opposition to official regimes of truth.[10] Stevenson (1995) describes how:

> Fiske illustrates this argument by referring to a story concerning aliens landing from outer space, which he claims to be a recurrent one within tabloid journalism. The point about such stories is that they subversively blur the distinction between facts and fiction, thereby disrupting the dominant language game disseminated by the power bloc.[11]

The point that such cultural theorists are making, of course, is not that accounts of 'aliens landing' may be true, but that the falsity of such stories which is assumed - calls into question the ideological hegemony of 'respectable' journalistic discourse: 'For Fiske, stories about aliens landing from outer space subvert the language game of the power bloc.'[12]
Even within 'oppositional' academic critiques, then, there are 'assumed' limits to what may be believed.

The Internet is, on the other hand, gloriously free of such limitations. Alternative/conspiracy sites abound: the New World Order, flying black helicopters and in league with various different races of aliens, many from Zeta Reticuli, are about to take over the world and put all right-thinking Americans in underground concentration camps in the Mohave desert where, deprived of their guns, they will be forced to eat yogurt and watch lesbian dolphins having sex with their grandmothers. This is not just a caricature but also a caricature of a caricature, since conspiracy sites on the Web span a wide spectrum from the sober to the lunatic - and it is sometimes hard to tell which is which. Anyone who has tried to do research on the Internet knows the difficulty of filtering out 'serious' information from the rubbish - sometimes all you have to go on is grammar and spelling. This is the other side of the coin of 'freedom of information' - the prospect of information becoming so free that it ceases to have much value at all.

It is possible to view the information explosion, then, from both a negative and a positive point of view. On the negative side, there is the prospect of the commodification and commercialisation of information, the loss of serious information in a sea of banality and economic monopolisation whereby the media will be neatly divided up between Bill Gates and Rupert Murdoch.[13] On the positive, there is freedom of access and flow of information, the right to say anything you like and as many truths as there are people (perhaps even more). Within the influential Frankfurt School tradition in cultural studies, this maps roughly onto the pessimistic (Theodor Adorno) versus optimistic (Walter Benjamin) view of the role of new media.

Taken to its extreme, however, it is arguable that the breaking-up of consensus reality/realities through the undermining of academic and media hegemonies may lead not simply to a 'freeing-up' of information but to the ultimate post-modern nightmare - the abandonment of any notion of truth and indeed of reality itself. This is of course a goal to which VR may already be tending - the acme of verisimilitude may be the destruction of 'reality' itself.[14]

> In the post-truth world, the people are saturated by a plurality of discourses that are struggling for the consent of the audience, the difference being that the explosion of messages that characterises modernity is no longer stamped with the 'authority' of their authors.[15]

This would, perhaps, be the ultimate triumph of irrationalism, notoriously heralded by Nietzsche in the fragment which is virtually the foundation-stone of post-modernism (if such an inappropriate metaphor may be used). In 1873 Nietzsche wrote:

> What, then, is truth? A mobile army of metaphors, metonyms, and anthropomorphisms - in short, a sum of human relations, which have been enhanced, transposed, and embellished poetically and rhetorically, and which after long use seem firm, canonical, and obligatory to a people: truths are illusions about which one has forgotten that this is what they are; metaphors which are worn out and without sensuous power; coins which have lost their pictures and now matter only as metal, no longer as coins.[16]

Already, the status of truth on the Internet is seriously compromised with the phenomenon of avatars, electronic personae which may or may not accurately represent the original. If I present myself in a virtual sense as a Japanese female teenager - which, as a matter of fact, I am not - this is not simply deception, since it is an accepted rule of the game that people may wear masks in a sophisticated electronic version of Mardi Gras. Thus, the notion of truth - about, e.g. ethnicity, gender, age - becomes not simply compromised but in a sense irrelevant. As is already the case in much post-modern discourse, truth is simply not part of the game.[17] If VR may represent the triumph of Wagnerian maximalism in our culture (i.e. the notion of the *Gesamtkunstwerk* or total art work), the Internet in a sense represents the triumph of Nietzsche (in his anarchist/nihilist rather than fascist/dominator incarnation). If truth is all about power, then the apparent destruction of power through the new media may involve not just the undermining of a traditional cosy consensus but perhaps also the destruction of truth itself. The democratisation of information which the new technology promises may lead not just to the subversion of regimes of truth (including ostensibly oppositional ones) but also perhaps to the blurring of the distinction between reality and illusion to the extent that it is no longer meaningful - if it ever was. Consensus realities may be ultimately arbitrary and power-based but it is a moot point whether their replacement by a general popular phantasmagoria is any better.

Acknowledgement
With thanks to Emer Williams.

References

1. With one or two honourable exceptions: see for example, Douthwaite, R. 1992. *The Growth Illusion*. Dublin: Lilliput.
2. Sheldrake, R. 1988. *A New Science of Life*. London: Paladin.
3. The threat posed by instrumental science has raised deep cultural anxieties for several centuries, from Mary Shelley's Frankenstein to Cronenberg's re-make of *The Fly*.
4. Davies, P. 1982. *Other Worlds: Space, Superspace and the Quantum Universe*. London. Abacus, p. 122.
5. Milton, R. 1994. *Forbidden Science: Exposing the Secrets of Suppressed Research*. London: Fourth Estate.
6. Kuhn, T. 1970. *The Structure of Scientific Revolutions*. Chicago: Univ. of Chicago Press.
7. Milton, pp. 11-23.
8. The other side of the coin of fear of the new is of course desire for, indeed worship of the new, for example the recent fetishism of cyborgs and gene mutation - desire and anxiety being two sides of the same coin.
9. Some 'pro-UFO' interventions by reputable authors have been met by attacks on their methodology, e.g. the case of Mack, J.E. 1994. *Abduction: Human Encounters with Aliens*. London: Simon and Schuster. The other response tends to be to ignore the issues, e.g. those raised by Good (1996) and Corso (1997) in the hope, perhaps, that they might thereby disappear. See Good, T. 1996. *Beyond Top Secret: The Worldwide UFO Security Threat*. London: Macmillan; and Corso, P.J. (with William J. Birnes). 1997. *The Day After Roswell*. New York: Simon and Schuster. Corso is a retired high-ranking American officer who claims, apparently in all seriousness, to have coordinated the covert introduction of alien technology into US research and development. There has been some discussion of this momentous affirmation on the Internet but very little - that I havecome across at any rate - in the print media.
10. Fiske, J. 1989. *Understanding Popular Culture*. London: Unwin Hyman, p. 178.
11. Stevenson, Nick. 1995. *Understanding Media Cultures: Social Theory and Mass Communication*. London: Sage, p. 94.
12. Stevenson, p. 98.
13. Recalling the anti-trust conflicts of the turn of the century - plus ca change. . .
14. See Slouka, M. 1995. *War of the Worlds: The Assault on Reality*. London, Abacus.
15. Stevenson, p. 91.
16. Nietzsche, F. 'On Truth and Lie in an Extra-Moral Sense'. in *The Portable Nietzsche*. 1971. Trans., ed. and introd. by Walter Kaufman. London: Chatto and Windus, pp. 42-7.
17. Of course, the great unanswered question is, 'What is the truth-content of the statement "there is no truth"?' (If it's true it's false.).

Paul O'Brien studied literature at Trinity College, Dublin, and the University of British Columbia, and has a Ph.D. in philosophy from Trinity College, Dublin. He lectures in aesthetics and cultural theory at the National College of Art and Design, Dublin, and also teaches the MA course in interactive multimedia at the Dublin Institute of Technology. He has published numerous articles and reviews on aspects of art and culture.
<pobrien@hadcom.ncad.ie>

The Failure and Success of Multimedia
Sean Cubitt

Convergent media: the phrase trips conveniently off the tongue. Yet there has been little argument to placate the fears of Rudolf Arnheim (1958) that the talking picture was a failure because it interrupted the wholeness of the silent spectacle with its dual focus in sound, especially dialogue, and image, lacking that unifying hierarchy which, he argues, creates the only successful hybrid art forms-opera, song and such. Arnheim's argument is something akin to that musicological strain which insists that only dominance produces melody and without melody, music is not. Thus the emergent democracies of Schönberg's atonal, twelve-tone row compositions, lacking a dominant and therefore without melody, were also heard as anarchic, incomprehensible and jarring because ununified by a hierarchical arrangement. In this paper, I want to address the possibilities for such a democracy of elements in multimedia, a democracy which will, in general, provoke a sense of the unpleasant and chaotic.

It seems that, as a general rule, few interactive arts achieve unity and those that do, achieve it by, for example, focusing on text, with images and sounds reduced to the status of illustration; or by centring on image, with text reduced to the status of explication and commentary; or by concentrating on sound, with image and text marginalised as extras. The major unified counter-examples come from narrative forms, especially in games, with the fragmented narratives of shoot-em-ups or the exploratory narratives of fantasy games like *Myst* and *Riven*. A smaller group of unified interactive arts can be classified by their interest in the algorithmic possibilities of the digital media: sites like *Jodi, Technosphere* and *Tierra*. I would argue that works focused on cross-media effects such as narration and computation are also hierarchically organised. The first part of the paper addresses the implications of hierarchy. The second makes some utopian proposals for dis-integrated media production.

In an important article first published in 1967, Michael Fried mounted an attack on the 'literalism' of minimalist art, contrasting 'the literalist espousal of objecthood-almost, it seems, as an art in its own right-and modernist painting's self-imposed imperative that it defeat or suspend its own objecthood through the medium of shape'. Literalism, he asserts, is damned to theatricality, because 'it is concerned with the actual circumstances in which the beholder encounters literalist work', opposing modernism's aesthetic of the work in itself with minimalism's presentation of an objects in a situation, a situation which, 'virtually by definition, includes the spectator' (Fried 1992: 825) in a process of distancing, especially of the work from its creator, which 'makes the beholder a subject and the piece in question . . . an object' (Fried 1992: 826). 'For', he continues, 'theatre *has* an audience-it *exists for one*-in a way the other arts do not: in fact, this more than anything else is what modernist sensibility finds intolerable in theatre generally' (Fried 1992: 830). Moreover, minimalism and theatre share an effectivity by which 'the experience in question *persists in time*, and the presentment of endlessness that . . . is central to literalist art and theory is essentially a presentment of endless or indefinite *duration*'. By contrast, 'it is by virtue of their presentness and instantaneousness that modernist painting and sculpture defeat theatre' (Fried 1992: 832).

If, from the perspective of digital arts, both the rearguard defence of the traditional media of painting and sculpture and the minimal aesthetic of LeWitt, Judd and Morris seem part of the same modern moment, the internal contradictions of late modernism are well worth exploring, for they have not been resolved but merely repressed. For Fried, theatricality threatens to kill art through misunderstanding the boundaries between art and non-art, to the detriment of the former. That is, there is point at which theatricality, in its emphasis on experience, especially on time-bound experience, undermines art and makes instead a work. Such work, in taking on objecthood, produces a subjectivity, even if one whose content is deliberately left to the individual viewer. In fact, individuality of response is a crucial fact of minimalism in Fried's account.

Our digital aesthetics have a similar tendency. Brenda Laurel's metaphor of theatre in interface design (Laurel 1993) disposes towards a subject-object confrontation, though one couched in a mode of illusionism neither Fried nor the minimalists would have succumbed to. One alternative to self-discovery, or self-construction, in the process of experiencing the minimal artwork, is the process of self loss in the HCI, but a self-loss which is always accompanied by the introjection of alternative selves, 'ego-ideals', in the manner we have learnt from identification with the stars of the cinema and which we can now perform with our own avatars online, or with the heroic ego-ideals of shoot-em-ups. It is also possible to construct multimedia in such a way as to promote introjection of the object as a whole, without further terms of surrender, in consumption, widely familiar on consumer internet sites, where even temporalities of consumption can become the object of consumption (i.e. where websites are, as they mostly are, neither here nor now exclusively). What is missing, and here Fried has I believe caught a problem in the construction of the minimal artwork, is the possibility of a relation which is not object-oriented, relations of the two experiences of making and beholding, between maker and experiencer.

Net objects are perversely static, increasing in mass as they increase in velocity; their solidity and effective endlessness, their perpetuity likewise increasing in direct proportion to speed in communication. Interactivity fails because, even as it aims towards the level of engagement of games, it reintroduces the artistic assumption of critical distance as an artistic code, a crucial element of the resistance to entertainment as thoughtless and durationless consumption reinforced by the acquisition of ego-ideals. Such relations are properties of the interface, not just of GUI, but of the white goods aspect of desktop machines.

Thus we confront the computer as an object but also as a presence, anthropomorphising it but also playing on that hollowness which is the secondary product of velocity (since light alone travels at the speed of light, and light is massless; as objects approach the speed of light, they become entirely surface). Thus the internal objects of the computer become hollow as the machine itself becomes increasingly fixed to the place where it immovably is. One of Arnheim's questions might then be reformulated as: how much should (or, more recently, how much can) be done to prepare and define an experience in advance for an audience?

Virilio is not as radical as he seems when, in 1988, he wrote:

> Malevich, Braque, Duchamp, Magritte . . . Those who continued to take their bodies with them-painters or sculptors-ended up elaborating a vast theoretical tract, in compensation for the loss of their monopoly on the image. This, in the end, makes them the last authentic philosophers, whose shared, obviously relative vision of the universe gave them the jump on physicists in new apprehensions of form, light and time (Virilio 1994: 31).

In 1969 we find Joseph Kossuth seeing the synchronous death of philosophy and birth of art in the linguistic turn of the former and the rise, with Duchamp, of the question of the nature of art in the latter. There is, he argues, 'an "art condition" to art preceding Duchamp, but its other functions or reasons-to-be are so pronounced that its ability to function clearly as art limits its art condition so drastically that it's only minimally art' (Kossuth 1992: 841). The movement from appearance to conception marks the beginning of modern art because such art exists only conceptually. For Kossuth, artworks are analytical propositions, in the sense that they are works on the level of definition-as opposed to synthetic statements, whose validity is determined by experience.

Crucially, what is apparent from these discussions is that digital art, as long as it maintains the centrality of an unchallenged conception and unchallenged definition of art-however contested, unchallenged-will remain outside the kind of art which, 30 years ago, defined art as such. Moreover, as Kossuth and Fried both determine art as that which is beyond experience, it is perhaps time to grasp the nettle of experience in digital works. Art itself is a convergent force in multimedia. Yet convergence is in general purely motivated by the unification, and inherently the hierarchisation, of the object and thus of the subject, the two held apart or subsumed one into the other, again hierarchically (either the work is consumable or I must adjust myself in order to be worthy of the work).

Since the first unassisted readymade, and in the midst of Einstein's, Poincaré's and Planck's revolutions, 'artists' are those communicators who are particularly conscious of the media in which they speak, and by the seriousness with which they speak ontologically (not about ontology, but speaking in ontological statements, in *realia* over which they relinquish interpretative, epistemological or even physical control as they abandon the question representation in favour of a querying of and experimentation in presence or, to use another vocabulary, presencing). It is this issue of the object as presencing that re-emerges in the cinema of digital effects and questions whether cinema itself, the projection of films, is an assemblage of *realia*, a mode of presencing and an exploration of the non-existent: i.e. an ontological quest.

(Narrative, of course, is an attempt to reacquire the novum for the epistemological form of knowledge, especially of inner speech, of explication, naming and placing. It is an attempt to maintain transparency in media that, increasingly, *because* of their speed, become concrete, metastable and massy, even as velocity renders matter immaterial and art conceptual). What matters is then not narration, nor even interaction as some mythical core of digital media, but communication and the questioning of communication as it exists presently, as the diffidently separated transport of messages between (inter) entities presumed to exist. It is that presumption that must be challenged and questioned, for in it lies the secret of hierarchy in communication generally and critically in convergent and interactive media.

Convergence depends on the totality of shared, organic meaning and form, in the older Romantic aesthetic, or on the epistemologically grounded superiority of either maker or observer either over one another or over the object through which they communicate. As such, it is both totalitarian (because it cannot and will not recognise dialectic, and/or because it is incapable of democracy, and/or it seeks a massification of one and only one of its terms) and hypostatising (of authorship, of medium, of message and of receiver). The objectification of communicative artefacts, and the presumption of polar opposition between sender and receiver, produce contemporary concepts of interactivity as gameplay, strictly encircled by the parameters of rule-governed structures, systemic rules which in no way challenge the free running of social systems of injustice and exploitation. As Peter Sloterdijk writes, 'advanced industrial civilisation produces the embittered loner as a mass phenomenon' (Sloterdijk 1984: 191-2), generating a cynicism (dramatically similar to the 'philistinism' addressed by Beech and Roberts, 1996 and 1998) captured in the ironic detachment of young British art's conceptual play with popular cultures.

The defeat of cynical reason, of ADILKNO's 'innocence' (ADILKNO Geert Lovink 1996), of Baudrillard's Code (Baudrillard 1975) cannot be achieved by resistance, which such systems exist by encompassing. The necessity is for a synthetic mode of mediation, a concretisation of the communicative as conscious artifice, and therefore an advanced formalism, drawing on the strengths and challenges of minimal and conceptual art to go beyond the now redundant, because systemic, concept of art itself. The task of new media is not convergence but divergence, a radical democratisation of the media of communication.

The possibility of a democratic relation between the elements of multimedia will depend on their atonal lack of hierarchy, the absence of domination by narrative, symmetry, perspective, tonality and other systemic technologies of containment; and further on the lack of closure, completion and autonomy of each element, such that each depends upon but also supports the others, making of each an incomplete virus requiring a symbiotic host, but in a relation in which sound, image and text are mutually virus and host, and in which there is no difference in prestige according to status, a convergence therefore which is purely formal, and whose motivation and construction would be determined at first by a formal working through of motifs, narratemes, emblems, and so a work in which the mathematical rigour of formal synthesis makes use of the creative participation of its audience as a randomising engine.

Yet this formalism will remain susceptible to containment as long as the audience's work is restricted in this manner. The first new medium will not only be radically divergent, but will be invisible and inconceivable until interaction begins to bring it into being, piecemeal, and forever without conclusion. There, there is no artwork, but only work, be it productive, unproductive, laborious or consumptive, or some further synthetic form engaging its own new dialectic. For the foreseeable future, such work cannot be representative, as representative media and democracy have consistently failed to bring about even the simplest of communications, the distribution of adequate food and water, without which there is no other communication. I have been wrong to target hope as the last virtue: without charity, the generosity of communication, there is not even hope.

References

ADILKNO Geert Lovink. 1996. 'Contemporary Nihilism: On Innocence Organised', in Timothy Druckrey (ed.), *Electronic Culture: Technology and Visual Representation*. New York: Aperture, 385-9.
Arnheim, Rudolf. 1958. 'A New Laocoön: Artistic Composites and the Talking Film', in *Film as Art*. London: Faber and Faber, 164-89.
Baudrillard, Jean. 1975. *The Mirror of Production*. St Louis, Mo: Telos Press.
Beech, Dave and Roberts, John. 1996. 'Spectres of the Aesthetic', *New Left Review*, n.218, July-Aug, 102-27.
Beech, Dave and Roberts, John. 1998. 'Tolerating Impurities: An Ontology, Genealogy and Defense of Philistinism', *New Left Review*, n.227, Jan-Feb, 45-71.
Fried, Michael. 1992. 'Art and Objecthood' in Harrison and Wood (eds), 822-34; first published in *Artforum*, Spring 1967.
Harrison, Charles and Wood, Paul. (eds) 1992. *Art in Theory 1900-1990*, Oxford: Blackwells.
Kossuth, Joseph. 1992. 'Art after Philosophy' in Harrison and Wood (eds), 840-50; first published in *Studio International* v.178, nn 115-17, Oct-Nov-Dec 1969, 160-1, 212-13.
Laurel, Brenda. 1993. *Computers as Theatre*, Reading, MA: Addison-Wesley.
Sloterdijk, Peter. 1984. 'Cynicism-The Twilight of False Consciousness' in *New German Critique*. n. 33, fall, 190-206.
Virilio, Paul. 1994a. *The Vision Machine*. trans Julie Rose, London: BFI.

Sean Cubitt is Reader in Video and Media Studies and Route Leader in Screen Studies at Liverpool John Moores University, England. He has published widely on contemporary media, culture and arts. His most recent book is Digital Aesthetics (Sage, 1998).

Abstract Virtual Realism
Nik Williams

> Man is constantly outside himself, in projecting himself, in losing himself outside of himself, he makes for man's existing; and on the other hand, it is by pursuing transcendent goals that he is able to exist; man, being in this state of passing-beyond, and seizing upon things only as they bear upon this passing beyond.
>
> Nietzsche

Frederick said a mouthful and not only because of his fondness for semicolons. What strikes me about this quotation is its unwittingly trenchant insight into issues regarding some of the conditions affecting artistic creation in what I call abstract virtual realism.

For the purposes of this essay, I will assign parts of the quotation to areas of inquiry that have occupied my thoughts and work for many years. To begin then,

> 'Man is outside himself, in projecting himself . . .'

Technologies developed for research employing haptic devices and total immersion techniques have provided modern artists with the opportunity to extend our senses beyond the physical body and to create the prototypical cyborg guest, complete with mind/world altering prosthetics and telepresent virtual environments. Epistemological concerns within virtual realism are encapsulated and converted into a Klein Bottle universe where points of reference fold into themselves and turn inside out simultaneously.

The exoskeletal prosthetics, existing in reality, generate an immersive, sensorially convincing world. To accommodate this wearable world, our consciousness projects and wraps itself around these partially overlapping realities and interpolates a third or separate reality, something akin to Castenade's shamanistic world of divination and hyper-perception. The ontological act of being in the world is accomplished in part through a projection of consciousness into and onto a synthetic sensorium which is further distilled into a narrowly articulated yet palpable experiential field. Artists now have the capacity to envelop and control the viewer to an unprecedented degree. I use the term abstract virtual realism to define the fusion of the potentiating technologies with the subsequent genesis of a new artistic plane of personal expression.

> '. . . in losing himself outside himself, he makes for man's existing.'

For those of us who have enjoyed or endured a total immersion experience, what we encounter most often is a visually dominant environment which imparts a distinct sense of presence with a corresponding lack of place. I will not discuss the problems of motion sickness and the collateral effects on the oculomotor system since these problems are well known and have been targeted by engineers around the world.

In the real world, our minds and bodies are fully engaged. We navigate through a complex and highly nuanced combination of hard - and soft-focus mental states, transitioning seemlessly between attention-dominated to emotion-dominated modalities. We drift between controlled, analytical thought streams characterised by a distance between the thinker and the thought and an emotional, memory dependent level where sinking focus serves to open thought processes to connections and recollections. The objective of

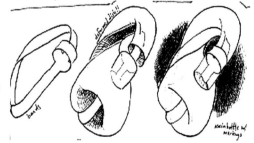

Klein Bottle sketch courtesy of the Geometry Center

the abstract virtual realist is to author an experience which invokes the emotional stream of the viewer, freeing him or her from the sense of inhabiting an illusion. In a related sense, what we require in our . . . .of abstract virtual realism is the capacity to exercise what T.S. Eliot called the 'objective correlative' or 'a set of objects, a situation, a chain of events' that represents and summons forth a particular emotion. This calls for nothing less than the objectified representation of an irrational world, whether a low resolution VRML rendition of a shopping mall or highly articulated experiment in synaesthetics, which makes a clear distinction between representation and the idea as a thing in itself. Merleau-Ponty refers to the language of self-mirroring worlds within poetry when he writes:

> If a poem is to pure, the poet's voice must be stilled and the initiative taken by the words themselves, which will set in motion as the meet unequally in collision. . . "

What he is saying is to create meaning is to pass beyond practical language and its reliance on pre-established signs to direct, authentic language where the idea itself takes hold. In abstract virtual realism, we have the same issue with representations of reality and the suspension of the apprehension of illusion. We require a collision between concrete and emotional content. Without it, we cannot obliterate the line between the physical and the immaterial.

> '. . . and on the other hand, it is by pursuing transcendental goals that he is able to exist.'

This is a key concept in understanding the relationship of art to the evolution of personal expression and human communication. It is also a stumbling block for those who have a problem with the spiritual connotations of 'transcendental'. Spirituality, as I understand the term, is an attribute common to most serious artists and is reflected in their works. However, one need not be intent on imparting a spiritual 'message' to have a transcendent quality to their work. In a general sense, rising above gives way to moving beyond. When we cross a psychological threshold, whether through personal adversity or the application of effort towards an achieved goal, we experience a relative transition from a lower or preoccupied state to a higher or clarified state of awareness. The impact of this movement varies from individual to individual but all cultures of which I am aware have an idea of transcendence embedded deep within their ethos, regardless of the mechanism employed or status accorded to it. The quest for meaning, the linking of mental states to forms of language and expression, these are constants within the sibling disciplines of art and science. The notion of transcendence within the art context is crystallised as an experience beyond the bounds of human cognition and thought, something extraordinary and beyond the immanent.

In general, evocative aesthetic experience in abstract virtual realism cannot be derived from real world experience, since we are not able to invoke the objective correlative (a notable exception to this is Char Davies 'Osmos'). Rather, it is based on *a priori* elements of experience which are most often de-contextualized and situational. Research has shown that we are indeed hard-wired to 'get' the referential simulacra presented by the immersive experience but only through the linking of memory to representations of reality can our low-focus flow of emotional streams connect to a sense of the extraordinary. Once again poetry opens a door, this time to reveal Shelley standing at the foot of Mont Blanc:

> The everlasting universe of things
> Flows through the mind, and rolls its rapid waves,
> Now dark, now glittering.

Artists working with advanced technologies are challenged to create the kind of psychic framework alluded to by Shelley and the quest for this humanisation of virtual reality is a way of striving for existence.

> '. . . man, being in this state of passing-beyond, and seizing upon things only as they bear upon this passing beyond.'

The idea of the otrepasser, the passing-beyond, has its roots in the works of Becket, Malraux, Camus, Merleau-Ponty and Sartre. Whatever it means to individual sensibilities, it has a general meaning which implies that all is in process at all times and a part of such process is to transcend the immediate givens of a situation and to understand that there are no facts or realities in and of themselves. If we begin from a state of concrete unreality, as we do in abstract virtual realism, linking the process of passing-beyond to the psychic framework of the virtual world is crucial if we are to achieve an 'authentic' emotional response.

In the real world, it is not possible to do actions that are not possible. In virtually realistic environments, the impossible is the norm. There is a flourishing Gibsonian sub-culture within which reality is constructed and consciousness expanded through philosophical and technological links across a deeply embedded internal neurology and external network topology. This is a vibrant, global community of socially and technologically sophisticated people who too often see their world delimited by the extent to which technology has spread. Without it, they (we) see no way to effectively interact with the world. This community of web idealists has been energised by attacks from politicians and reactionary 'culture thugs' who mask their neophobic proclivities as conscientious, socially responsible interventions designed for the greater good. The Byzantine twists of political opportunism will not be miraculously transformed by the subtle interweaving of global telecommunications. However, it is becoming increasingly clear that the egalitarian ideals formulated in the 1960s regarding man as an enlightened citizen of planet earth have been gradually stripped of their utopian tone and now serve as practical guidelines for solving problems of intercultural relations, world hunger relief and environmental degradation. This socio-political transition has been bracketed by the creation of the tools of the telepresence trade. But technology in pursuit of an aesthetic is a reversed ideal. There is significant evidence over the past 30 years of the fruitful collaborations between artists and engineers. Yet technologists continue to dominate the debate over the legitimate artistic use of technology. There is a sense that art in its purest form has taken a back seat to the engines of commerce as a significant partner in the development of tools for 'artistic' use. Without commerce there is less motivation for technical research or funds for product development. One can argue that brushes and pigments, welding torches and steel are elegant and adequate requisites for the act of creation (leaving aside the toil of the artist and all that implies). Widespread acceptance of abstract virtual realism is not simply a matter of reducing prohibitive costs or lack of cultural validation. We need to balance our drive to create new modes of expression with the understanding that much of what we are involved with is the actualisation of conjectures and probings by artists of another time and place.

To put it another way, how different is it for a painter to make an important picture than it is for an abstract virtual realist to create a world with an psychic framework? Technical considerations aside, my sense is this is a question of almost total uninterest for most electronic artists. There is sometimes an arrogance of mind that comes from working with languages and physical systems incomprehensible to the lay public. My own experience has shown that the best artists are acutely aware of the evolution of art, not just within their own milieu but as it affects their lives and their perceptions of the world. We should occasionally do the obvious thing and look for clues outside our own investigations. In the spirit of this self-administered admonition, I will close with the mention of an artist whom I met only briefly several decades ago but whose work continues to exert a strong influence on me.

Mark Rothko was profoundly motivated by Nietzsche, since it was he who had successfully identified the desire for transcendence as the Dionysian impulse, an urge to lose the sense of self and achieve a primal state of oneness with the world. This observation presupposes a world wherein the self and the world are separate and distinct entities such that the self, contained within the physical body, is also contained by the world. The self was not a part of the world that housed it. Rothko was unique among the Abstract Expressionists in his desire to dematerialise the art object by suppressing nearly all of the painter's touch. His ultimate goal was to convey a state of exultation in which the viewer is subsumed in the totality of the colour field. Rothko made no apologies for the spiritual content of his work; in fact his was a singular focus on the redemptive qualities of free expression coupled with a highly ethical philosophy of social justice. All of this is exuded by the intense capacity of his work to make the viewer disappear, in effect, to transcend the normal for the extraordinary. Rothko understood that once that state was achieved, the mind was free to contemplate ideas and feel connections that are normally suffocated by daily activities and distractions. This was not escapism. This was fine tuning. A visit to the Rothko Room at the Tate Gallery might convince you that one need not resort to advanced technology to create a virtual world.

So the obvious question is, what does a painter have to do with immersive worlds and advanced technology? Nothing, much in the same way that Maya Deran has nothing to do with Spike Lee or John Cage with D.J. Spooky. Perhaps I am begging the question here, but are we are really so different from our predecessors simply because of the times we live in or the tools we choose to work with? Each artist is confronted with the problem of how to destroy without negating, how to explore without a lifeline, how to find a way through the tangle of style and hype. What I am advocating is that if we want to see our way clear, objects in the rear-view mirror are closer than they appear to be.

Nik Williams - Project Manager: Beecher Center for the Electronic Arts. BFA/MFA California Insitute of the Arts. 30 years in electronic heaven and hell
<nik@panix.com>
http://www.panix.com/~nik

'Beyond Film and Television-Whose Consciousness is it Anyway?'

Clive Myer

Just when we thought it was safe to switch off the television...

The spread of new technologies throughout the world is not working to advance human freedom. Instead, it has resulted in the emancipation of market forces from social and political control. By allowing that freedom to world markets we ensure that the age of globalisation will be remembered as another turn in the history of servitude.[1]

What, then, of this alleged technological freedom of information, currently celebrated as individual, artistic communication, extolling the virtues of non-linearity as if by messianic apparition and heralding a great nirvana?

What, also, of this second coming of interactivity, which will free the viewer/participator from her shackles of passivity and miraculously enable multilinguisity and international communication?

> In the 1990s, a new technology will be approaching its maturity, one which treats human activity not so much as a stimulus to be recorded for later interpretation but as a dynamic agent which provokes and incites reaction, participation and communication. In short, interaction.[2]

New media would seem to wish to reinvent the wheel, supercharged with acceleration from 0 to 5 gigabytes in only 12 months. Apart from the speeding up of the process of built-in obsolescence, new models of hardware and indeed, software carry with them the means of production and the means of access and distribution. In essence, each new computer and/or software system is a microcosm of the global system of capitalism, spawning its way across the far reaches of the planet and replanting its seed in ever-expanding territories.

Globalisation means lifting social activities out of local knowledge and placing them in networks in which they are conditioned by, and condition, worldwide events.

> ...The global economy de-skills people and organisations. It does so by making the environments in which they live and work unrecognisable to them. It thereby renders their stock of local and tacit knowledge less and less serviceable to them.[3]

But no proto-linguistic system is free of the constraints of it own and shared histories of explicative methodology. Within the box of images and sounds that necessarily encode the multifarious functions of the new technological media and offer up to us a re-reading of our very own individuality comes free, at the bottom of the packet, a set of ideological transfers that when allowed to soak in for a while, rub gently on the mind. They evoke memories of poor media past and call forth visions of new media to come. Television is fast becoming a poor medium, unvitalised since the popular advent of colour in the late sixties. Television is the main purveyor of degrees of collective consciousness, whether reflecting popular opinions or formulating them. Television is on the threshold of a major transformation, or rather, convergence. However, it is destined to carry with it the traces of the history of its own making or, more pertinently for the purposes of this paper, the traces of the formulations of the codes of representation that flourish within the make-up of what might be called social collective consciousness.

Hop, Skip and Jump

There can be no doubt that there exist chains of signification between theatre, photography, film, radio and television. The chains can be linked in several historical and temporal directions to include painting, printmaking, sculpture and writing. The primary link is representational, the secondary link is caught in a split fusion between entertainment and information. Within this oxymoron imaginarily exists the real world and its Other. Esher-like in its cyclical re-existence and constant re-emergence, we are left with the residue of its voracious appetite-(some form of) knowledge.

There can be only doubt that there exists a chain of knowledge that is free of the representational values that created it. Within each segment of that knowledge is contained the trace of each aspect of construction from which it was formulated. Between *tabla rasa*, the moment of birth and *rasa tabla*, the moment of death, is another space-that between conform and conflict. The binary pattern of the shifting dialectic allows and encourages knowledge of ourselves as individuals only inasmuch as we are part of a mass. From within this broad concept the question can be raised as to the status of the individual in relation to the social mass organisation (society) which on the one hand attempts to justify itself-law, order, regulation, education-and on the other hand attempts to deny itself-freedom, chaos, chance, opportunity.

In order to understand this fluctuating phenomenon we must seize its moments and, by default, participate in our own transmogrification. By bringing to the forefront an awareness of specific cultural moments and codifications through the examination of certain cinematic and televisual homogenisations, it is possible, though transitory, to examine elements of the continuous restructuring of certain aspects of collective consciousness, through which the individual plays his or her role, negatively or positively, as a part of that collective practice.

In 1937 a group of young intellectuals[4] realised the sociological implications of transforming what until then was the province of primary myth, oral history, to that of its secondary level, written and visual language. Whether they were able to consider the ideological implications for the proceeding 60 years is doubtful and understandable, given the incredible rate of the development of technology within this time.

With hindsight, one might see the philosophy of the Mass Observation movement as being misplaced or misguided-a group of socialist roaders whose good intention was to document and preserve the thoughts and actions of the working class. This idealism was to recognise who would make use of such information but was not to recognise the in-built contradiction inevitable in any process of dialectical materialism. Humphrey Jennings and Charles Madge expressed its potential usage in the forward of their 1937 Mass Observation publication on the coronation of George VI:[5]

> . . . of interest to the social worker, the field anthropologist, the politician, the historian, **the advertising agent** (my bold), the realistic novelist and indeed any person who is concerned to know what people really want and think.

Did Mass Observation inadvertently encourage audience categorisation and identification through a false sense of what was 'seen to be' working class? What could be seen as the continuation of the idealisation of the subject, derived from the procreation of the Renaissance dream, quickly became, through the creative juxtaposition of filmic and textual editing, the manipulation of the viewer/subject relationship rendering appropriate Marx's assumption of a false consciousness. Jennings' film work in particular testifies to this position and his output during the Second World War can easily be seen as state propaganda. These films invoke the notion of nationhood more strongly than would be possible through lived memory. In 30 years time, when representations are all that are left of this 20th century milestone, let us hope that film education has rendered this 'evidence' as faction, not fact.

Film and television give the place of the spectator as shifting struggle, somewhere between an imaginary participator and a passive consumer. New technology gives the spectator the 'chance' to be an active consumer.

Is the notion of chance similar to that of a national lottery? Does the notion of an active consumer, within these terms, promote a higher value judgement than that given to a passive consumer, or is it, more discreetly, a variable or arbitrary judgement? With all eyes and wallets on consumption, false consciousness may renew itself as consumption controller in a non-binary (East/West) cultural and economic dilemma. Global capitalism looks towards constant renewal of old and new messages. New colonialism means new global markets. Supply and demand requires a global marketing consciousness. As Gray points out:

> A truly global economy is being created by the worldwide spread of new technologies, not by the spread of free markets. Every economy is being transformed as technologies are imitated, absorbed and adapted. No country can insulate itself from this wave of creative destruction. And the result is not a universal free market but an anarchy of sovereign states, rival capitalisms and stateless zones.[6]

The notion of individual independence fragments its parts to re-unite as world consciousness. Foot soldiers identify specific audiences, locating special needs. New technology speeds towards the consumer to reach the parts older technologies couldn't reach.

— *Values* —

From Mass Observation to Mass Participation - the Funeral Procession of Diana, Princess of Wales

Some might say that the large number of people participating and sharing in a collective grief was a positive step for such a nation, historically imbued with the notion of a stiff upper lip. Some might also say that this display of solidarity represented a final rejection of upper class values, flying in the face of royal reason.[7] But what was really happening on that eventful day of 6th September 1997? If Mass Observation (rekindled in the 1980s) were to have recorded the views of some of the thousands of ordinary people gathered in the Mall, Kensington and Westminster, would they indeed know what people really want and think?[8]

Employed by Chinese television to shoot news footage on the day, I was able to witness the proceedings both in person and 'through the lens'. In one sense, seeing the representation as it happened (often the cameraman/woman has his/her right eye to the viewfinder-which is itself a miniature television screen, and his/her left eye open on the live event. He/she could be described as being in two worlds at once, interacting with two different levels of myth simultaneously). For the viewer at home, the process would be different. Whereas, on the one hand, through the operation of a broadcast video camera, the operator is watching both the theatricality of the staged event and its first level of representation (unedited) through the viewfinder, on the other, the viewer will see the event only as an edited (whether cut between cameras in real time or broadcast several minutes after filming in the case of taped recordings) second level myth. For the viewer, the whole process could be described as a television lived event as opposed to 'live' television. There exists a strange, fascinating and even exciting feeling knowing the contradiction and self-participation involved in assisting this process of giving the viewer what the producer thinks they want. In this instance my brief was to show the crowd reactions to the cortège. I knew instantly what this meant-weeping faces and big close-ups of running eyes and rolling tears.

The sensibility involved in the knowing participation of an enacted and extended moment of collective consciousness of this magnitude has an aura of sado-masochism around it. There exists both a sense of guilt and a sense of thrill. It is at once both an honour and dishonour, an amazement and a disgust. Why and how could so many people believe they really knew this icon? I spoke to and interviewed people who had journeyed from Scotland and from Wales, from Newcastle and from Cornwall. I felt I was in a Humphrey Jennings film. Each attested to the friendships they had struck with people they had never before met. There was, truly, an atmosphere of nationhood perhaps not invoked to this extent since the enforced national pride of 50 years ago. Suddenly, silence fell. The cortège appeared from around the corner of St.James's Palace and for four or five minutes the crowd was hushed. All eyes were on the princes. Cutting backwards and forwards between coffin and young bowed faces, the crowd enabled it's own framing and editing. There was no need for representation. The eye as lens, true to Vertov,[9] the brain cortex as editing table, memory, nostalgia, family as context. Slowly and still very quietly, the sobbing began. Like a Mexican wave, emotions swelled, tears rolled down the cheeks of the old and the young. Mothers and daughters hugged and wept, boyfriends and girlfriends comforted each other, young men's arms across each other's shoulders. Nobody spoke. As if on a film set of highly rehearsed actors, the camera traversed the scene, lens doubler engaged (20 x 1 magnification) as tears dripped off noses-it had to be recorded; it had to be exploited. Minutes later the cortège reached Westminster Abbey. Without warning, the princess's brother spoke. The crowd, only minutes earlier stood in unison, now sat in unison-against walls, against trees, their heads propped in hands, and the second stage of mourning had begun.

In a recent edition of Screen[10] a number of approaches were taken towards an analysis of the event. Rosalind Brunt discussed Diana as icon. In articulating the historical representational processes that iconicism traces, she concludes that in this instance we are virtually taken back to its original religious meaning through our still popular need to deify and sanctify. Jenny Kitzinger discusses her image in another way, that of visual symbolism and the constant barrage of her placement within different contexts, wearing different clothes. Perhaps the most memorable of this particular duality were flack jacket/landmines or white dressed blonde/sick black baby. This problematic, of the displacement of social constructs through the famous star system, was reflected upon in Godard's film in 1972, 'Letter to Jane', where he criticised the well-meaning of Jane Fonda's visit to war-torn Vietnam, by allowing herself to be photographed by fashionable magazines, recuperating the politics/act of sympathy within coffee table journalism. In the same set of papers, Roger Silverstone reflects on the dilemma created between individuality and the ability to take control of one's own life in relation to explanations of behaviour that consider mass hysteria, religious fervour and media manipulation.

The notion of collective consciousness is, of course, complex and cross-disciplined. We can, I believe, begin to gain some insight into its definition through an analysis of certain specific moments such as the death of Diana. However, there is no doubt in my mind that in order to achieve this (ongoing and without closure) analysis, it is just as relevant to understand the ordinary, everyday actions of our social exchanges. What is called for is a re-reading of movements such as Mass Observation in the 1930s or the work of the Glasgow Media Group[11] in the 1970s. Even more so, the numerous archiving currently taking place in different institutions should be regarded from the perspective of what, when and why are we recording and preserving from within the assumption of a neutrality of context, supposedly non-mediated reflections of our social exchanges. This debate, it is hoped, will continue in the broadest of fields.

As to whether the framing of filmic consciousness has penetrated beyond its own codification to emerge as the reframed intraconnected, newly-developing, but apparently 'free' consciousness, one has to ask if that media codification has or will emerge through to the notion of the net as individual sites of multitudinous variation. Jenny Kitzinger[12] sites (with some reservation) Bryan Appleyard (*Sunday Times*, 7 September 1997), noting that in the final episode of the soap opera of Diana's life, 'Diana dies everywhere and instantly-on the internet, CNN and every television screen in the world, on the radio, in every newspaper, she was the first icon fully to live and die in the global village'. When opening up the Diana condolency website this week I was the 65,132nd person to have presumably wished to communicate with the Princes William and Harry. I wonder what I should have said?

Notes

1. Gray, John. 1998. *False Dawn, The Delusions of Global Capitalism*. Granta: London.
2. Rickett, F. 1993. 'Multimedia', in *Future Visions, New Technologies of the Screen*. British Film Institute: London.
3. Gray, John. op. cit.
4. The Mass Observation Project developed by Charles Madge, Tom Harrison, Humphrey and Stephen Spender, Humphrey Jennings, Tom Driberg, and Kathleen Raine.
5. Jennings, H. and Madge, C. 1937. *May the Twelfth, Mass Observation Day-Surveys 1937*.Faber and Faber: London. It was not with preconceived intention that the group should study this particular event. The volunteers across the country had agreed to supply reports on the twelfth day of each month since February 1937. The coronation was on 12th May 1937.
6. Gray, John. op. cit.
7. I refer here to the actual event rather than the televised event.
8. Jennings, H. and Madge, C. op. cit.
9. *The Man With a Movie Camera*, USSR, 1927.
10. Vol. 39, no.1, Spring 1998.
11. The study of right wing bias in the news through publications such as *'Bad News', More Bad News'*, etc.
12. *Screen*. op. cit.

c.myer@newport.ac.uk
myer@videoin4.demon.co.uk

Chimera for the 21st Century: Mis-Construction as Feminist Strategy

Terry Gips

Since our first depictions of the human form, we have intentionally distorted or taken liberties with its representation - perhaps as often as we have sought to produce perfect likenesses. In many cases, renderings have abstracted or altered the human figure for formal and symbolic purposes without suggesting that the body is anything but physically normal. However, artists have also conceived of bodies that deviated from biological fact and took on imaginary shapes and identities. During the past two decades, most disciplines in the humanities and social sciences were engrossed with what came to be known as the 'construction of identity.' Such identity - building certainly included the work of artists. Even though artists do not necessarily begin with theoretical constructs, their images of the body resonate with the broader, more encompassing discourses on identity, including, now, the post-biological.

Thinking about identity has invariably engaged the nature/culture debate - biological essentialism vs. social and environmental determination. Theory originating from several perspectives - and probably none as much as feminist theory - has oscillated among various positions on the nature/culture spectrum. Interestingly, it is the cyborg or cybernetic body, part biological and part electronic, which, according to Anne Balsamo (1996), 'disrupt[s] persistent dualisms that set the natural body in opposition to the technologically crafted body'[1] and facilitates a reconceptualisation of the body as a boundary figure belonging simultaneously to natural and cultural systems. Or, as articulated by Steve Tomasula (1998), 'the body has served as a firewall against complete subjectivity. . .body and self were inextricably linked. But body technologies have pried even this relationship loose.'[2] It might be argued that women working on issues of identity seem more ready than men to use the cyber-era lens and embrace destabilised bodies and multiple selves. At the same time that we have decried others' objectification of our bodies and contested cultural constructions of our 'selves,' we have also chosen to practise our own identity construction. Some of our practices reinforce various stereotypes and conceptions of the 'normal' female with a fixed identity; others intentionally distort, destabilise, deform, disfigure, pervert, corrupt, debase, warp and vulgarise the female body - making it strange, *etranger*, foreign. This paper investigates these 'mis-constructions' as strategies developed for the purpose of decoupling the body from the self and navigating the post-biological universe.

Even though various technologies have been employed throughout history to mask, modify and extend the boundaries of the self, we sometimes behave as though the present cybernetic technology is delivering unprecedented possibilities - and demands - to transcend the flesh-and-blood body. We may also may be tempted to think that history, particularly ancient history, has few clues to today's enigmas of consciousness. These views are partially justified by the extraordinary shift in consciousness of the late nineties; however, certain parallels between past and present variants of the body can illuminate the critical issues of human 'being' in the imminent millenium.

I will survey works by artists who invent figures to which I give the name 'chimera.' They are hybrids of two or more elements: human, animal, mechanical and electronic. One might ask, why not choose the term 'cyborg,' since it is widely used to describe human-machine hybrids as well as human organisms embedded in a cybernetic information system? I have opted for chimera, first of all because this word jolts anyone so addictively 'plugged in' to today's future-focused technology that the past has been dismissed from his or her field of information. Derived from the Greek word meaning 'she-goat,' the chimera prevalent in classical mythology was a tri-part animal: a fire-breathing monster with a goat's body, a lion's head and a serpent's tail. Second, chimera is a biological term referring to an organism combining genetically distinct tissues, thus prompting us to consider the potential of human hybrids and clones. Third, chimera are usually thought to be grotesque, monstrous, powerful *and female*. Finally, chimera evoke the rituals of countless cultures and the seemingly inherent human inclination to invent 'unnatural' forms in order to do battle with new terrain, whether it be geographical, psychological or spiritual. Post-biological consciousness is essentially *terra incognita*. Artists, like everyone else, are searching for strategies to ease exploration and transition.

Such exploration is central to Marge Piercy's 1991 novel, *He, She, and It*.[3] In this and other cyber-fictions, shaman-like guides abound, sometimes in the form of electronic devices and disembodied intelligences. Authors (and readers), though, seem reluctant to let go of the crutch of physical beings, so guides are often border figures who inhabit two worlds. Piercy's guide and protector is Yod, a robotic figure not of human origin but one who is so like a human that he feels emotions, engages in bodily sex and dies an anguished death. As a parallel to Yod's narrative, Piercy intersperses chapters which retell the old Jewish folk story of the Golem of Prague, thus underscoring the quandaries of origin, biology, sexuality, spirituality and mortality as well as the similarities among the Golem formed from an inert mass of clay; Frankenstein's corpse reincarnated through an electrical charge; robots of moving metal parts and electronic control units; and cyborgs of melded flesh, plastic, wires, batteries and various prosthetic elements. The novel also implies an urgent need to examine the human fascination with creating life from inanimate material.

Perhaps we are heeding Piercy's advice in our almost obsessive return to animal and anima, to carnal and corporeal, to flesh and blood, even in this time of increasingly disembodied day-to-day life. At the same moment that millions of people around the globe are engaging in virtual visits with friends and family, travel agents, museums, libraries, and shopping malls via the Internet, we see ample evidence of body-focused engagement. We can peruse the immense database of the Visible Man;[4] read about and partake in endless amounts of exercise, eating, health and sex; order embryos and sperm from catalogues to match our reproductive desires; study diagrams of the genetic cloning of Dolly; and become acquainted with 'Techno Sapiens' in *Time* magazine's supplement.[5] To celebrate the new millennium in Great Britain, we will tour the inside of a two-hundred foot tall replica of the human body. What explains these indulgences? Do we believe that extensive knowledge, close attention and tangible monuments will slow the body's slippage?

In the art world, too, the human body is rampant. Portraits and traditional figurative work are popular, and more importantly, art preoccupied with the pulsing and performing body. There is art that reveals a determination to penetrate and know the body on both sides of the skin; art that projects the body into the social, physical and psychic arena of the cybernetic age to see how well it will fare; and art that both imagines restructuring the body and literally does so. Such tactics, including the extreme cases, are not without precedent, however. In fact, it is useful to revisit mythological chimera as well as some singular works produced earlier this century to better understand body images of the nineties.

For example, it seems impossible to look at certain contemporary depictions of cyborgs and not recall the robot Maria of Fritz Lang's 1926 film *Metropolis*. Maria was not the typical robot, a purely mechanical structure programmed to perform human tasks; rather, she attained being when life was transferred from a real/biological woman, thus making her a true boundary figure, inhabiting the murky area between conscious human and cultural object. According to Claudia Springer (1996)[6] and Andreas Huyssen (1881-2),[7] Maria's overt sexuality was aggressive and threatening and represented patriarchal fear of female sexuality which, in turn, was conflated with machines and technology running out of control.

In the troubled time between the two world wars, many contemporaries of Lang wrestled with the nature of machines and technology, sometimes celebrating modernism and the future, sometimes warning of their evils. Surrealists and Dadaists probed the ramifications of technology by combining human and mechanical elements in their work: George Gross depicted his fellow artist Heartfield with a mechanical heart. Raoul

Hausman's 1919 *The Spirit of Our Time - Mechanical Head* is a wooden mannequin adorned with tape measure, ruler, watch and other items. It parallels his cynical questioning of the need for a spirit or soul in a world which operates mechanically
(Foster 1985)[8].

Hannah Hoch also warned of technology's consequences through collages such as *Beautiful Girl*, a visual conglomerate of BMW emblems, a wheel, part of a crankshaft, a hand holding a pocket watch, and, centrally located, a woman in a bathing suit. The figure is faceless and anonymous, however, her head having been replaced with a light bulb. Less well known but perhaps intuiting a more distant future - our current time at the end of the nineties - is Hoch's collage that depicts a sexual encounter between two figures with chimeric identities. Both are female but one has a winged insect as upper torso while the other shows an African child's head pasted on a slender white torso which is, in turn, attached to voluptuous legs. Hans Bellmer's famously bizarre puppet dolls of the 1930s may also be interpreted as efforts to plumb the psychological issues of consciousness and question fixed identity. Some of his works, such as *Machine Gun in a State of Grace*, might be read as cyborgs, organic figures with mechanical prostheses. An unusual drawing from 1920 by Surrealist Max Ernst is part biological/electrical diagram, part fantasy creature and part philosophical gesture. *Sitzender Buddha (demandez votre medicin)*, as it is titled, appears to be an organism evolving under simultaneous biological and technological forces.

Nineteen years later, Frida Kahlo painted *The Two Fridas*, a pair of clone-like figures seated side by side whose breasts have been opened to expose blood-red hearts interconnected by tubes. Overflowing with symbolism and multiple meanings, the picture clearly alludes to dualities of being and the transferral of life through an exterior/artificial conduit. In retrospect, it is hard not to see Kahlo's work as harbinger for the feminist body art of the seventies and work being made today that investigates modes of consciousness. Many women have used their bodies as site and substance for their work. Rather than changing identity through masking or costume, these artists punctured the boundaries of self. In the seventies, Ana Mendieta made a series of drawings called *Body Tracks*, rubbing her own blood on large sheets of paper. Carolee Schneeman performed *Interior Scroll* in 1975, pulling a long scroll of paper from her vagina and reading the text as it was unfurled. Artists Niki de St. Phalle and Pat Olesko both fabricated giant female sculptures whose vaginas served as doorways through which visitors and performers gained entrance to the body's interior. Olesko also remade her body and emulated the ancient goddess Diana of Ephesus by attaching dozens of huge plastic breasts, as did The Waitresses who performed at the Los Angeles Women's Building.

Alongside such works which reiterated female biology, other feminists meddled with essentialist definitions of gender: Lynda Benglis presented herself with a large dildo in a full page *Artforum* ad in 1975. Jana Sterbak turned her arm into a hybridised phallus/bestial horn. Louise Bourgeois, who had been blending natural and cultural definitions of woman in her work for decades, produced some eerie forms in the 1980s which satisfy the definition of chimera and recall ancient goddess figures that were part human, part animal. Kirsten Justesan and Nicola Hicks also mixed genetic material. Their human/animal hybrids have correlates in mermaids and minotaurs. Annette Message 's *Chimeras*, her title for this body of work from the 1980s, are more ephemeral: these winged creatures and insects float on walls and envelop space like a clinging vapour. Faith Wilding's *Recombinants* are also mythical, playful and fleeting, sometimes crossing into a world of hermaphroditic and trans-species chimera.

In contrast, much of Kiki Smith's work is about the body we know first hand, its tangibility, flesh, fluids, viscera, excrement and its fundamental animalness. On the surface, there is nothing post-biological about this work. However, she repeatedly invokes sexual ambiguity and defies us to see either the body contained by skin or consciousness constrained by body. Other works, including her glass *Breast Jar*, are suffused with the aura of science and artificiality. Likewise, both Mona Hatoum's and Helen Chadwick's work have distinctly medical associations, a strategy that can be horrific and frightening to those reluctant to know their own biologies and acknowledge that technology may be eclipsing the gods in matters of life and death.

Other women who have employed the power of the abject to question biology as the essential site of identity include Cindy Sherman. Her grotesque females challenge us to view natural bodies and artificially assembled ones under the same light and speculate about which are more real. Using models and props photographed in the studio, she strategically places the monstrous alongside the venerated and transgresses conventions of the sacred. Katie Toivanen, on the other hand, works her alchemy by manipulating photographs on a computer to produces images of flesh that are seductively lush and repulsively gory. Perhaps she is indicating that lipstick and liposuction are not that different, or that the body can be sculpted at will along the lines of Orlan's ongoing project to reconstruct her body through cosmetic surgery. As Orlan states, 'my body is my software,' implying that her consciousness is elsewhere and that she can continually recreate herself if she chooses. *Roberta*,[9] the virtual person created two decades earlier by Lynn Hershman through a series of events, public communications and visual documents rather than body surgery, was, in the artist's words, both real and artificial. Is not Orlan likewise both real and artificial?

Lynda Dement's interactive CD-ROM *CyberfleshGirlmonster* includes scanned body parts and voices from about 30 participating women assembled and animated into a series of conglomerate 'monsters.' The user encounters the monsters relatively blindly, without aid of a menu system or controllable interface. With similar perverseness, albeit in traditional painting and sculpture, Nicola Tyson and Peggy Preheim confront us with bizarre futuristic hybrids that refuse categorisation. These are not just any hybrids, but like Ginevra Bompiani's (1989) characterisation of ancient chimera, these chimera for the 21st century are supreme hybrids, supreme examples of freedom, sensual, light and fiery, elusive, indefinite and unbelievable.[10]

Notes

1. Balsamo, A. 1996. *Technologies of the Gendered Body*. Durham and London: Duke University Press.
2. Tomasula, S. 1998. 'Art in the Age of the Individual's Mechanical Reproduction,' in *New Art Examiner*. April, pp 18-23.
3. Piercy, M. 1991. *He, She, and It*. New York: Alfred A. Knopf, Inc.
4. http://www.nlm.nih.gov
5. Time digital. April 27, 1998. New York.

6. Springer, C. 1996. *Electronic Eros: Bodies and Desire in the Postindustrial Age.* Austin: University of Texas Press
7. Huyssen, A. 1981-2. The Vamp and the Machine: Technology and Sexuality in Fritz Lang's "Metropolis" in *New German Critique.*
8. Foster, S. 1985. *Dada/Dimensions.* Ann Arbor: UMI, pp 138-41
9. http://www.arakis.ucdavis.edu/hershman/roberta/roberta.html
10. Bompiani, G. 1989. The Chimera Herself in *Fragments for a History of the Human Body*, Part One, M. Feher, ed. New York: Zone, pp. 365-409.

Terry Gips is Director of The Art Gallery and Associate Professor of Art at the University of Maryland, College Park, Maryland. She curated The Digital Village exhibition in 1995 and edited a special issue of Art Journal: Computers and Art/Issues of Content in 1990. She is a practising artist working with photography, digital technologies and installation.